CHESAPEAKE
REQUIEM

CHESAPEAKE REQUIEM

A YEAR WITH THE WATERMEN OF VANISHING TANGIER ISLAND

EARL SWIFT

DEY ST.
An Imprint of WILLIAM MORROW

FIRST EDITION

Designed by Suet Chong
Map illustration by Paul Pugliese
Photograph on page iv courtesy of the author

Library of Congress Cataloging-in-Publication Data

Names: Swift, Earl, 1958- author.
Title: Chesapeake requiem : a year with the watermen of vanishing Tangier Island / Earl Swift.
Description: First edition. | New York, NY : Dey Street, [2018] | Includes bibliographical references and index.
Identifiers: LCCN 2018024659| ISBN 9780062661395 (hardcover : alk. paper) | ISBN 9780062661401 (trade paperback : alk. paper)
Subjects: LCSH: Crabbing—Virginia—Tangier Island—Pictorial works. | Blue crab—Virginia—Tangier Island—Pictorial works. | Tangier Island (Va.)
Classification: LCC SH400.5.C7 S95 2018 | DDC 639/.560975516—dc23
LC record available at https://lccn.loc.gov/2018024659

ISBN 978-0-06-266139-5

18 19 20 21 22   RS/LSC   10 9 8 7 6 5 4 3 2 1

For Mark Mobley

CONTENTS

―――

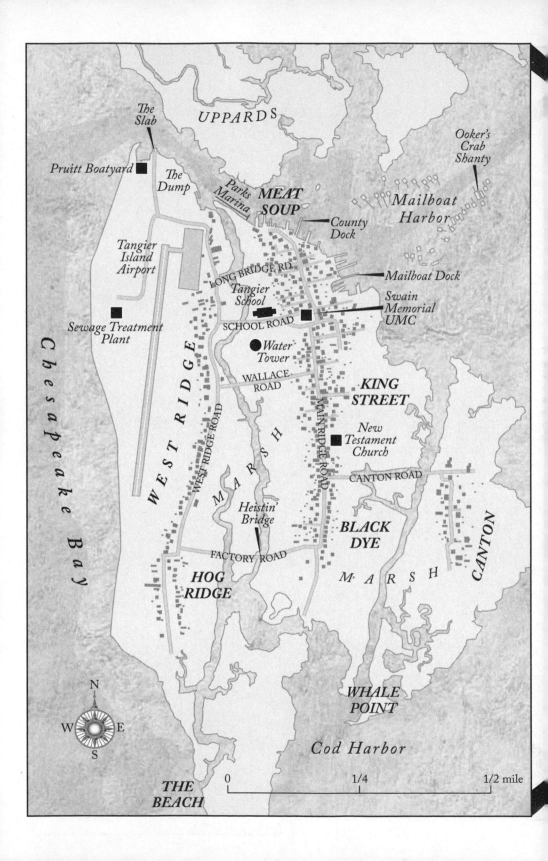

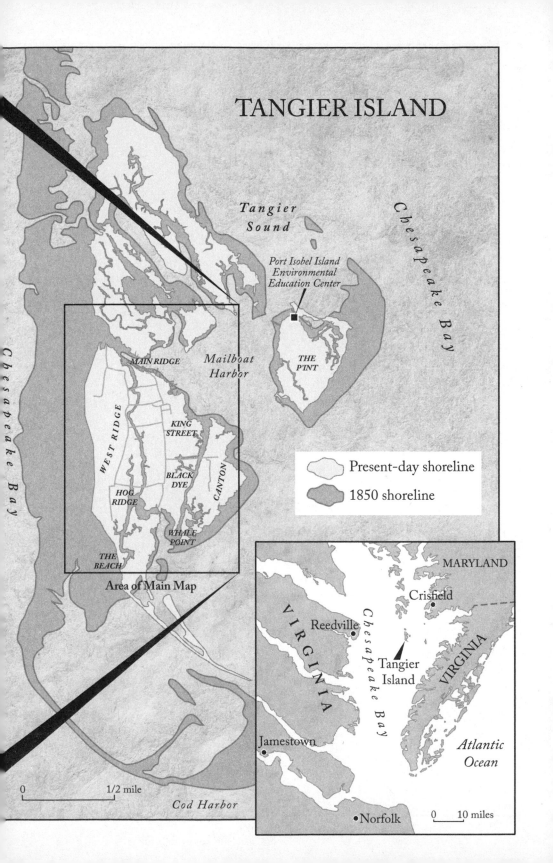

TANGIER ISLAND

Tangier Sound

Chesapeake Bay

Port Isobel Island
Environmental
Education Center

THE P'INT

Mailboat Harbor

MAIN RIDGE

WEST RIDGE

KING STREET

BLACK DYE

CANTON

HOG RIDGE

WHALE POINT

THE BEACH

Chesapeake Bay

Area of Main Map

| Present-day shoreline |
| 1850 shoreline |

0   1/2 mile

Cod Harbor

MARYLAND

Crisfield

VIRGINIA

Reedville

Chesapeake Bay

Tangier Island

VIRGINIA

Jamestown

Atlantic Ocean

•Norfolk

0   10 miles

CHESAPEAKE
REQUIEM

Carol Pruitt Moore undertakes her regular pilgrimage to Uppards.
(EARL SWIFT)

INTRODUCTION

———

A DAY AFTER THE STORM PASSED, CAROL PRUITT MOORE climbed into her skiff and set off for the ruins of Canaan.

It was her habit to strike west from the harbor, into the channel that split her island in two. She'd throw open the throttle and crouch forward as the boat's nose came up and its stern dug deep, her six-foot frame rocking against swells that thudded the fiberglass hull. One hand on the tiller, ball cap pulled low to cheat the wind, she would speed past the county dock and twenty-odd workboats parked off to port and, to starboard, a ragged line of crab shanties—weather-scoured sheds perched on stilts a few feet over the drink, surrounding decks stacked high with wire-mesh crab pots. A watery little business district, home to the greatest fishery of its kind in the world.

Out past the last of the shanties, the channel opened into the Chesapeake Bay, and most every day she threw the boat into a banking turn to the north. This time of year, she was running afoul of the prevailing winds, and if they were blowing against the tide, the journey would not be smooth. But like most of her neighbors, Carol Moore had been born into a seafaring family: She was the eighth generation of her kin born on Tangier Island, sixteen open-water miles from the nearest mainland town in Virginia. She could handle a boat before she could read.

On a typical afternoon she'd hug the edge of a treeless, uninhab-
ited marsh that islanders call Uppards, until a mile along she reached
Tangier's northern tip, where a century past a community thrived.
Canaan had long since washed away, but she found sanctuary on the
deserted beach there, an escape from the close quarters and relative
bustle of town. And more: She came upon tiny bottles once filled
with patent medicines, offered by visiting drummers to island women
a hundred years dead—Hamlin's Wizard Oil, a nineteenth-century
tonic for cancer, quinsy, and "bites of dog," and Greene's Infallible
Liniment, a cure-all "For Man or Beast," and Turlington's Balsam of
Life, said in its day to cure kidney stones and "inward weakness." She
found eighteenth-century clay pipes and headless porcelain dolls. She
collected bits of crockery, edges tumbled smooth by the surf.

Each relic she plucked from the sand was a tangible link to her
forebears, for her father's people had lived at Canaan for generations,
back when it had its own school and a general store and thirty-odd
houses, and had stayed until the last holdouts quit the place. By the
time she was born, in 1962, that abandonment was decades passed,
and none of Canaan's buildings remained. But on childhood visits
she had seen chickens and goats still roaming loose there, along with
thickets of wild rose and rhubarb and so much anise that the whole
place smelled like licorice. She remembered, too, walking deep into
the marsh with her father, weaving among water bushes to a small
cemetery far from the water. Its simple marble headstones testified
to the lives of good Christians, loyal wives, hardworking watermen.

Here was where her great-grandparents had passed their days,
and their parents, and theirs—people who persisted in her blood, her
looks, her habits and manners. Here lay reminders of what she would
become, and of the impermanence of all things. Some days she spent
hours there, walking the water's edge. Others she stayed for just minutes. Almost always, she left for home feeling more sturdily, steadily
grounded.

Only this afternoon promised to be different. For the previous
three days, Tangier had been raked by Hurricane Sandy, its winds

blowing a steady fifty knots or better, gusting past seventy. The gale had conspired with the tides to swamp the place. White-capped waves had rolled in from the east, breaking over streets and into homes. Water had surged through the channel, sweeping away anything not anchored firm. At high tide, only a few knobs of high ground and a single humpbacked bridge had escaped the flood.

Now, with the air scrubbed clean, whatever caught Moore's eye revealed itself in preternatural detail. A few crab shanties had been flattened, and even from a distance their staggered plywood showed off its grain, and the wounds in their smashed planking gleamed a sunny, surreal yellow. Crab pots were strewn in the shallows. Pieces of roof and pier lay twisted among the reeds at the water's edge. Skiffs were sunk to their gunwales.

She sped into the bay. It was a sunny afternoon, cloudless, the sky a deep and flawless blue, the water flat—"slick calm," as the watermen say, which leaves Tangier lips as *slick cam*. Back in town trees were down, yards lay piled with trash that the tide had lifted from the marsh, and families were dragging ruined furniture and carpeting outside; to experience such peace out here, just hours after the storm's mayhem, seemed unreal.

Sure enough, when she rounded the island's northern tip, she saw that the gale had not spared Canaan. The shoreline had been broken into pieces, cleaved by three new channels. Much of the sand that she'd walked for years had been ripped away, and minus this protective buffer, the Chesapeake now lapped against the dark sod in which the island's marsh grass was rooted. The soil crumbled with each pulse.

She motored along the water's edge until she found a place to land the boat, and as she scrambled out she noticed something a foot or two offshore: a stained brown sphere half submerged in maybe six inches of water, rocking gently with the push and pull of the waves. As she stepped nearer she saw there were two of them. After a moment's hesitation she bent down for one, and came up with a human skull.

It was well preserved, its hollows rinsed clean by the agitation of the surf. That of an adult, it appeared—though all of its upper teeth were missing and its lower jaw, too, so it was hard to reckon much beyond that. She rescued its mate, finding it in similar condition, and rested both on the ground well above the tide line.

She was struggling to digest her discovery when her eyes fell on a succession of familiar shapes at her feet—a rib cage first, then clavicles, then a pelvis and all the other components of a human skeleton, down to the tiniest bones of the toes. The remains lay faceup in bas-relief, topmost inch exposed, the rest still encased in clay. Around them lay the traces of a coffin, and around that, a sharp-edged rectangle cut into the earth: a grave.

And lo, a few feet away was a second grave, and in it a second complete skeleton, likewise only partially exposed, with hair combs held fast by the clay on either side of its skull. And beyond that, a third grave, containing a tiny casket inside a bigger coffin, and nested in the middle the skeleton of a small child, perhaps no more than a toddler. Its shroud had long since succumbed to rot, but two bright white buttons that had held the garment closed were aligned on the child's breastbone.

Carol Moore looked around her and saw that she stood among a half dozen additional graves arranged in a haphazard cluster at the water's edge. A fourth complete skeleton occupied one. Others were a jumble, their contents stirred by the surf. A spine lay topside. Shin bones. A mandible. And a few yards away lay white marble headstones, those she remembered seeing as a youngster. In the forty years since, the bay had chipped away at the shoreline, had little by little stolen the ground under Canaan, until a quarter-mile of island was gone.

Skip ahead to a clear, breezy afternoon four years later. I'm at Uppards combing the clay shoreline and struggling to picture the Canaan of years gone by—a stretch, for it "got drownded," as Tangiermen say, and is now on the bay's bottom a fair piece offshore. Crabs scuttle over the vegetable gardens that nourished the place, the foundations of its homes, whatever remains of the school that doubled as its Methodist meetinghouse.

Actually, it was underwater years before Carol Moore's macabre discovery on Halloween 2012, just after Sandy passed on its way up the Atlantic coast; the graveyard, after all, was hundreds of feet farther inland than Canaan itself. Now the graves she stumbled upon are well out to sea, too.

But what the bay has stolen, it returns in pieces: Stranded by the last high tide are square handmade nails and rusted bits of machinery, shards of china rimmed in blues and pinks, and the roots of trees that once shaded Canaan's homes. I collect a nail, the neck of an ancient ice-blue bottle, and a wave-worn knot of tree limb, stepping around headstones that every few weeks Carol has dragged clear of the advancing water. In loving memory of Polly J. Parks, who died in December 1913, age thirty-seven. Nellie A., beloved daughter of S. E. and Eva I. Pruitt, buried before her second birthday. Margaret A. Pruitt, born in 1836, gone in 1901.

I consider the stones for a long moment, wondering whether the people bearing those names ever imagined that the bay would one day claim all but these scant traces of their existence—that it would plunder their homes, their entire village, then come for what remained of them. Doubtful, I decide. The Chesapeake had been stealthy here.

The Pruitts and Parkses might scarcely have noticed the waves nipping at the land's edges, carrying off a few grains here, another few there. It wasn't until these three were dead and buried that the cumulative effects of those tiny incursions became plain: that half an inch a day added up to more than a foot a month. That in a year, a merciful year, the water drew fifteen feet closer. That eventually came a dawn when it pushed at the door.

So it went all around Tangier Island; so it continues. From where I stand, I can see Tangier proper across the marshy expanse of Uppards—the steep roofs of weather-boarded homes bunched along lanes not much wider than sidewalks; the spire of Swain Memorial United Methodist Church, for generations the island's cultural and spiritual centerpiece; and, hovering above, a sky-blue water tower decorated with a giant orange crab on one side, a cross on the other.

It is a community unlike any in America. Here live people so iso-lated for so long that they have their own style of speech, a singsong brogue of old words and phrases, twisted vowels, odd rhythms. Its virtually amphibious men follow a calendar set by the Chesapeake Bay blue crab, and they catch more of the prized delicacy than any-one else. It is a near-theocracy of old-school Christians who brook no trade in alcohol, and kept a major movie from filming in their midst over scenes of sex and beer. And not least, this is one big extended family: All but a few islanders can trace their lineage to a single man.

For 240 years they've occupied a speck of mud and marsh that nowhere reaches more than five feet above the tide, seldom tops three, and most often fails to clear one. Now it is washing ever faster into a bay on the rise. What's more, the island is slumping, actually subsiding into the earth's crust. Full moons pull water not only over its edges, but straight up through the ground, turning yards into ponds. In fact, the lower Chesapeake's relative sea-level rise—the one-two punch of water coming up and land going down—is among the highest on Earth, and of all the towns and cities situated on the estuary, none is as vulnerable, none as captive to the effects of climate change, as Tangier.

Near the water's edge I find a silvery metal bracket perhaps two inches long and shaped into a scroll, like the head of a violin. I have no clue what it might be. Carol Moore is scanning the shore about fifty yards away, and I carry my find to her. She identifies it with the briefest of glances: "That's from a casket."

A few minutes later, examining a tidal pool in the sod, I find what appears to be an interesting piece of driftwood. It's pale gray, four inches long, and resembles a tree stump in miniature. It feels feather-weight, leached of substance, and I see that it's laced with tiny holes.

With a jolt, I realize it's bone.

I gently return it to the ground.

PART ONE

AND
EVERY ISLAND
FLED AWAY

Houses on the Main Ridge hover over the marsh and the Big Gut.
(EARL SWIFT)

ONE

T USED TO BE THAT PEOPLE LIVED ON DOZENS OF THE LOW-LYING islands in the Chesapeake, raising crops, grazing livestock, and culling the bay's vast bounty of sea life and waterfowl. They endured much, these islanders: mosquitoes rising in clouds from the marshes, summer squalls filled with lightning and waterspouts, howling winter nor'westers, and loneliness most of all. Still, they built farms and villages that thrived for generations.

Then, starting about 1900 and culminating a few years either side of World War I, most of them fled for the mainland. They left behind James Island, at the mouth of Maryland's Little Choptank River, which had once boasted twenty houses, a church, a school, and a boatyard. What spurred the exodus is plain on modern charts of the middle bay: Water has whittled James to three tiny islets. It is fast becoming a shoal.

Settlers quit Sharps Island, not far to the south, which had stretched across 449 acres before the Civil War and was still big enough to merit a three-story hotel and a steamship pier in the 1890s. Sixty years later it was the size of a small bedroom. By 1963, it had vanished completely.

They abandoned Holland Island, which in 1900 was home to 253 people and big, fine houses shaded by hardwoods. It had its own two-room school, church, and post office, along with several stores and a large fishing fleet—and a first-rate baseball diamond and a team said to be among the best on the bay. Today, watermen pull crab pots from water a fathom deep over the infield. The bay toppled Holland's sole surviving house in 2010, and is now poised to erase the final vestige of the island's long human presence: a cluster of head-stones rising from a shrinking patch of marsh grass.

Worrisome ghosts, those places, for Ooker Eskridge, speeding across the water toward his crab pots in the chill of a daybreak in May 2016. Three miles off to the east, silhouetted against first light, lurks a closer reminder of nature's restlessness, a backdrop to most of his workdays. Fishermen and farmers lived on Watts Island for more than one hundred years, grazing cattle and sheep. At its height, it boasted a village of twenty-odd houses and a small church. Forty summers ago, when Ooker began following the water full-time, Watts was a long, fat crescent of high ground, much of it wooded, and overrun with goats descended from its farming days. Right before his eyes, Watts has since dissolved away, and now a fragment of the island and its swooning straggle of pines seem to shrink by the day.

Just three hundred yards to starboard is more poignant evidence. Only two of the bay's offshore islands remain peopled today, and here is one of them: Tangier, Ooker's birthplace, the only home he's known, fast disappearing itself. His people have lived here since 1778. The island is dotted with the graves of his ancestors, some of whom he shares with Carol Moore, some with his wife, some with most every Tangierman he encounters in the course of a typical day. In his youth he assumed he'd live out his years here, working the water into old age as his father did, his father's father, his great-grandfathers, until he earned a final rest beside them in the soupy island ground. That's no longer a safe assumption.

He steers his twenty-foot Privateer toward a bright blue Styro-foam buoy bobbing in the water, cuts the outboard to idle, and leans over the side to snag the marker with a boat hook. The float is con-nected by twenty-four feet of nylon line to a wire-mesh cube resting on the bottom, eight feet down. He pulls in the line hand over hand until the crab pot breaks the surface, then lifts it aboard. Why it's called a pot is anyone's guess—it's more a cage—but by any name, it works: A half-dozen crabs crouch inside. He springs open the pot's top, shakes the crabs out onto a wooden tray, refastens the top, and tosses the pot overboard. With a tap on the throttle we ease toward another buoy, thirty yards away.

Well into the twentieth century, the island's relentless wear seemed to provoke little concern on Tangier Island. Not so today: Its 481 inhabitants now fret that the end times are near, and Ooker Eskridge is more preoccupied with the dilemma than most. For eight years he's served as their mayor. He's read the reports about what's happening to his island, has discussed what it means with government officials and scientists. All affirm what he has seen for himself, the anecdotal evidence he gathers every day of crabbing season. "I had a pot hung up on a tree stump yesterday," he tells me. He's armored against the cold with a hoodie and jeans stuffed into oilskins, or waterproof overalls. "I pulled and pulled and finally got it loose, and when it came up it had a piece of tree root stuck to it."

In the fifteen seconds before the boat reaches the next float, he culls the catch, tossing an undersized crab overboard. One of the others is a mature male, or jimmy. Its shell, or carapace, is six inches across. He tosses it into a wooden bushel basket reserved for "number ones"—the biggest jimmies, laden with meat and the standard at picnics and crab boils. The remaining four crabs are mature females with bulbous orange egg masses ballooning from their bellies. Any mature female is called a sook; in this condition they're busted sooks, sponge crabs, or lemons. Ooker tosses the four expectant mothers into a second basket.

"It was hard to picture land, or even marshland, growing around it," he says of the stump, which was 150 yards into the bay on the island's far side. "But that was high ground once, to have trees growing out of it." The next pot comes up with light gray mud smeared across its bottom—clay, a signal that we're right now over some of that recent high ground. Inside the mesh wait another two lemons and Ooker's real quarry—a peeler, or a crab about to molt, soon to be a soft-shell crab and a summertime delicacy at restaurants up and down the Eastern Seaboard. He identifies it with a glance and a squeeze, and drops it into a third basket.

We putter from float to float, down a row that runs straight for a quarter mile along Tangier's southeast flank, the deck rocking gently beneath our feet. A few more peelers come out of the pots, along with number ones, some number twos—a little smaller, a bit less meaty— and an abundance of lemons. Osprey fly out from the shore to circle the boat. The sun bursts from a broken ceiling of dark clouds but fails to cut the chill. Ooker swings the Privateer around to start a second row. Thirty pots down. Another 180 to go.

"Out crabbing you see these tree stumps well offshore, and a light goes on," he tells me as he hauls up the next pot. "I think about it sometimes—that I'm crabbing on ground that used to be shaded by large trees." While he dumps crabs into the tray, I gaze over to the Tangier shore, now a quarter mile away, and try to imagine land in place of all that water.

To look at it on a map, Tangier seems the least likely of the old island burgs to have lasted so long. There never was much to the place, and it's way out in the middle of eighteen trillion gallons of water.

At about two hundred miles long, north to south, the Chesapeake is America's largest estuary, a mixing bowl for the tidal waters of the Atlantic and fresh water flowing from the Mid-Atlantic's big rivers—the Susquehanna and Potomac, the Rappahannock and York, the James. It stretches from a northern "peak" at the Susquehanna's mouth near Havre de Grace, Maryland, just a few miles south of the Pennsylvania border, to a twelve-mile-wide union with the ocean off Virginia Beach.

The bay's upper half is narrow—in one spot, less than three miles across, and never so wide that one shore isn't plainly visible from the other. But about a hundred miles south of its peak it promptly doubles in width; Maryland's Eastern Shore, which separates bay from ocean and forms the Chesapeake's right edge, is deeply notched there. Thirty miles farther south, give or take—and right at the line where Maryland and Virginia meet—the Eastern Shore again abruptly narrows, and the bay widens to roughly thirty miles across. It's there, dead center at the Chesapeake's broadest point, at the mercy of nature's wildest whims, that you'll find Tangier.

Little surprise that since its settlement it has felt more outpost than town, a place and people removed from the rest of their country. Come a hard January, the surrounding waters can freeze up so thick and tight that the military has to fly in food and heating oil. Rough weather any time of the year can maroon Tangier as well. Great expanses of open bay stretch to its south, southwest, and northwest— long "fetch," in maritime parlance, which means there's plenty of room for winds from those directions to build waves. Those compass points happen to correspond to storms that all too often sweep the bay, churning its relatively shallow water (it averages just twenty-one feet deep) into a heaving froth.

Even on calm days, reaching the place poses challenges that few other towns in America can equal. You can fly in aboard a small plane, if you have the wherewithal; an asphalt airstrip runs down the island's western edge. Otherwise, the most convenient Virginia port, Onancock, is sixteen miles to the east-southeast, nestled at the head of a meandering creek. A summer-only passenger ferry takes seventy-five minutes to make the trip. The nearest town to the west, Reedville, is about the same distance away; that crossing by summertime tour boat takes ninety minutes. For the shortest transit, and the only year-round, reliable passage home, Tangiermen have to travel outside their own state, to Crisfield, Maryland, which lies just above that second notch in the Eastern Shore. This self-anointed "Crab Capital of the World" lies twelve miles to the island's northeast. That voyage takes forty-five minutes in fair seas.

A typical run out of Crisfield sees the *Courtney Thomas*, Tangier's mailboat, back away from the town dock at 12:30 P.M., its afterdeck piled with luggage, UPS boxes, and grocery bags, its enclosed cabin busy with chatting passengers, its wheelhouse the province of lounging island men and, at the helm, Captain Brett Thomas—distant kin to both Carol Moore and Ooker Eskridge and the fifth generation of his family to skipper the island's chief link to the rest of America. At sixty-four feet, the *Courtney Thomas* is the largest boat homeported in Tangier, and its twin diesels hurry it along at about seventeen knots. In ten minutes it's left the protection of the Maryland shore and ventured into an often tempestuous arm of the bay called Tangier Sound.

The marshes of Smith Island, six miles out, form a low skim of greens and browns off to starboard at about the time that Tangier's water tower first appears, faint and spindly, on the horizon ahead. Over the next few minutes Swain Memorial's church steeple rises slowly beside it, along with a clump of woods to the east. Rooftops emerge next, becoming a snaggled silhouette of houses, most of them smallish, two-story, and wood-framed.

Eleven miles out of Crisfield, Tangier remains a board-flat green wafer just above the water, but by now it's becoming apparent that this is not a single island but a tight knot of three. The first you reach, home to the woods, is small and roughly circular. It's owned by the nonprofit Chesapeake Bay Foundation, which uses it as an environmental education camp. It's called Port Isobel these days, but to many a native it is still known by its original name, East Point, or the Point, which the locals (who turn any *oy* sound to *eye*) pronounce *the P'int*. It slides by to port.

Ahead lies the navigation channel into Tangier's harbor, lined on both sides with an entire village on stilts—crab shanties, workboats tied up alongside, their sterns stenciled with the names of watermen's wives and children. Most of the shanties are attached via decking to long rows of wooden shedding tanks, where watermen like Ooker Eskridge dump their peelers, then pluck them back out when they shed their exoskeletons to become the soft-shell crabs savored by gourmands.

If Crisfield is indeed the Crab Capital of the World, it's largely because of this odd, watery little knot of industry, because Tangier's shedding tanks produce more soft-shell crab than any other single source on the Chesapeake. Ooker's shanty is halfway down the line and easy to pick out: Leaning against the hut is a sheet of plywood painted with an enormous ichthus, or Jesus fish, and the words WE BELIEVE.

The mailboat's engines slow here, and its bow settles into the water for a calm six-knot chug into town. The channel makes a southwestern beeline for the town's water tower, then swings to the right and heads west through a man-made cut to the island's far side. Uppards stretches away to the north, the cut's right side. To the south looms Tangier proper.

The *Courtney Thomas* doesn't make the turn to follow the channel. The mailboat dock juts from the shore dead ahead, and Brett Thomas slides the boat alongside the long pier at ramming speed, only to brake to a stop with smooth precision in exactly the same spot he did yesterday, the day before, and hundreds of days before that. The mailboat's arrival is among the island's principal daily occasions, and it rarely occurs without an audience. Six afternoons a week, a gridlock of golf carts and curious old-timers is waiting.

I FIRST STEPPED off the boat in the summer of 1999, as a reporter for the *Virginian-Pilot* of Norfolk. My assignment was to explore how the place was faring eleven years after opting out of the Virginia Lottery, Tangier having been one of the state's very few towns to ban the sale of tickets on moral grounds. It was an iffy premise for a story, but I was excited by the chance to experience a "quaint" place supposedly "lost in time"—phrases I'd found oft repeated in past stories about the island. I called on the preacher at Swain Memorial, talked with town leaders and crabbers, dodged the bicycles and golf carts that plied the narrow streets—I could count the cars and trucks on two hands—and over the course of my stay slapped a thousand times at vicious, green-eyed deer flies that left deep, bloody craters in my arms, legs, and scalp and seemed to crave the taste of bug spray.

I left impressed by the town's friendliness and its snug dimensions. Tangier proper is only a little higher and drier than Uppards, tidal wetlands accounting for something like 70 percent of its area. Solid ground is limited to three slender, parallel ribs of sandy loam that rise above the marsh—"ridges," in island speak, though so low in elevation that if it weren't for the buildings and the odd tree rising from their surface, you'd be hard-pressed to pick them from the surrounding wet. They amount to islands within an island. To put their relative size in landlubbing terms: All of Tangier, including Uppards, the P'int, and a few outlying, marshy islets, is a bit smaller than New York's Central Park. The ridges, lumped together, would fit inside the reservoir at the park's center, with plenty of room to spare.

A few months later, as the media fretted over what might happen to the world's computers on January 1, 2000, the *Virginian-Pilot*'s editors dispatched me again to Tangier—this time because they figured that if Y2K brought global economic chaos and social collapse, the island and its old-fashioned, analog ways would plod on, unfazed. After all, it got by without cars. Its dial-up internet was too glitchy and glacial for much besides email. Its population didn't get home phones until 1966 and lacked reliable electricity until 1977— and most every household still had a VHF two-way radio and kerosene lamps on hand, just in case. My editors reckoned the town might be the safest place in America that New Year's Eve.

While experiencing what had to rank among the quietest millennial celebrations anywhere, I spoke with a lot of islanders. A consistent thread ran through our conversations, one that was far more elemental than a breakdown in technology. The very water that had provided Tangier's sustenance for more than two centuries, they told me, now posed a real threat to its future. So in March 2000 I returned with a photographer, and over the course of six weeks we gathered evidence of the erosion that was steadily claiming the place.

That's what everyone called it then—erosion. Global warming had been discussed by scientists for more than a century, and the term itself dated at least to 1975, yet in 2000 neither the phenomenon nor the handle had wormed far into the public consciousness, especially on Tangier; islanders had not yet been schooled in the notion that the bay was rising or that the ground underfoot was sinking. But they sure enough saw that their home was getting drownded, and the anointed representative for their views was James Wyatt Eskridge, also known as—*always* known as—Ooker.

This was eight years before he became mayor. His résumé, had he had one, would have painted him an unlikely spokesman. Ooker spent every summer day crabbing for peelers. He passed the late autumn catching eels for the Italian Christmas market and the winter dredging for crabs and oysters—which is to say, he spent a lot of time in boats, mostly alone. Even so, he was emerging as the community's public face. Then, as now, he was a tall, lean, ruggedly handsome fellow, with ice-blue eyes, a limber vocabulary, and a fondness for conversation. The camera loved him, and unlike many Tangiermen, he didn't seem to mind the camera.

Ooker carried me in his skiff around the P'int, where the bay had snuck beneath a forest of loblolly pine; the remaining trees were falling, one by one, into the surf in that spring of 2000. We rode to the sandy spit guarding Tangier's southern end, where he landed the boat and we strode among cordgrass and quivering blackberry to the beach's low crown. From there, he pointed out places where he'd played as a teenager that were now a hundred yards or more offshore.

Ooker was forty-one at the time. This had happened fast.

The spit ran south for nearly a mile before hooking eastward, then doubling back on itself, forming a tail that gave the island the unmistakable shape of a seahorse. But that sandy coil had withered since his youth, Ooker told me, and it had shifted eastward. Down at the hook there'd been a big house in the seventies and, before that, at least three fish meal plants—factories, unlikely as that seemed, where oily little fish called menhaden were processed into fertilizer. The house was survived only by its water well, a rusty pipe jutting from the waves. The last vestige of the plants was a broken concrete slab overwashed by the surf.

I TOOK A sea kayak to the island late in that 2000 project, and while paddling around Uppards one morning I beached the boat at Canaan. A copse of tall pines stood at the old townsite, which was still a fair piece inland and well above sea level. You could see the trees plainly from a mile south, down in Tangier proper. A single-wide trailer, left over from a hunting lodge that Carol Moore's uncle had operated in the seventies, sat there as well, perhaps 150 feet from the water.

Sixteen years later, what's left of the trailer's rusted chassis is in the bay, and the trees are long gone. The land at the north end has morphed profoundly: When I met Carol Moore shortly after my return, she showed me pictures of Canaan she took just four months before, depicting fingers of peaty land jutting into the water alongside pockets of sandy beach. "Everything in this picture is gone," she told me. "I can see it at Uppards from one week to the next. In the past few years, I know I've seen—I know it sounds crazy—but I know I've seen maybe a hundred feet gone in maybe three years.

"You used to, if you stepped off that north shore, you were on sand for quite a ways out. Now you step off and you're in the *bay*," she said. "It's all gone."

Ooker Eskridge thus presides over a smaller island than existed just months ago, let alone on my previous visits. Everyone agrees, in the Tangier way of putting things, that the place "is going away from here in a hurry." The island is far from convinced that scientists are right about what's causing the trouble, however. Erosion had been stealing land in the Chesapeake decades before anyone talked about climate change, they'll tell you. Why, look at Holland Island. Rising seas didn't kill it. Waves simply tore at the shore where most of its households were concentrated, until people there had to move. Like Tangier, Holland had little high ground, or "uplands," meaning the refugees couldn't move from one part of the island to another— another wasn't available. So they were forced to the mainland, which slashed Holland's population, undermined its economy, and sparked a cascading failure of the entire enterprise. By 1916, the population had dropped to 169. Four years later, not a single person was left.

Count the mayor among the doubters. "Our main concern is the erosion," Ooker tells me while we're crabbing. "Sea-level rise, that might be occurring, but it's small-scale next to the erosion.

"I'm not so sure it's man that's the cause of it," he says. "That's my thing, is how much of it is man's cause. I don't think man can do a lot to affect it." Tangier's dilemma seems to him more a product of nature's regular cycles, part of a divine plan far beyond the influence of mere humans. And, perhaps, a harbinger of the coming Judgment. "We've been taught from early on to look, as the Bible says, for signs of the latter-day times."

He's standing at his console as he says this, steering the boat and casting a steely gaze off into the middle distance. "One of the signs to look for, it says, is knowledge and travel greatly increased—and just look at how much knowledge and travel have increased." He switches his focus to me. "The other thing," he says, "is that things that used to be bad will be seen as good, and things that were good will be seen as bad." He raises his eyebrows.

Carol Moore shares his skepticism. "When glaciers melt the sea probably rises, but that's not what is going to take Tangier away," she told me as we sat at her coffee table, which doubles as a display case for bottles, clay pipes, and arrowheads she's found up at Canaan. "Tangier's demise is going to be erosion."

But hers is a political distinction, more than biblical. "If the government officials insist that it's sea-level rise, what can you do about sea-level rise?" she asked. "Nothing. Not a thing. And if that's what they see this being, then they won't want to spend any money to try to stop it.

"We're one strong storm away from being a Holland Island," she said. "If we don't get help, we're going to be history. The end."

Perhaps just as daunting, the mayor's constituency has withered even faster than his island. Since 2000 the population has dropped from 604 to 481, or by a fifth in sixteen years. It has aged, too: Young islanders now tend to strike for the mainland after high school, leaving few couples of childbearing age, at least by the standards of Tangier's fertile past. The one schoolhouse serving all the island's children, kindergarten through twelfth grade—the only combined school left in Virginia—had one hundred students in 2000; enrollment has dropped by a third since and is projected to dwindle to fifty-three in 2020. Incidentally, the class of 2020 already has its valedictorian and prom king: Matthew Parks is its sole member.

TRUTH BE TOLD, all but a minuscule minority of scientists are certain that the forces threatening Tangier and other beleaguered sections of America's eighty-eight-thousand-mile shoreline are indeed the products of humanity's hand on the environment. We're years past the point when a human role in climate change was only a theory.

In fact, a February 2016 report in the *Proceedings of the National Academy of Sciences* found that the jump in sea levels during the twentieth century "was extremely likely faster than during any of the twenty-seven previous centuries," which is to say, since at least 800 B.C. The authors, scientists from Singapore, Germany, Britain, and the United States, cited research suggesting that without man-made global warming, sea levels might actually have fallen around the world during the twentieth century. Instead, they rose by nearly half a foot.

As climate change emerged as a recognized crisis in the years after my 2000 Tangier stay, I found myself wondering from time to time how the islanders were faring. That was especially the case during the six years I owned a house near the water in Norfolk, sixty-five miles south of Tangier, and saw each successive northeaster bring higher flooding into my neighborhood. If things were this bad in the city, I couldn't imagine but that the island's situation was growing dire. Finally, in the late fall of 2015, I resolved to revisit the place and see for myself.

Before I made that trip, the journal *Scientific Reports* published a study titled "Climate Change and the Evolution and Fate of the Tangier Islands of Chesapeake Bay, USA." Written by three researchers associated with the U.S. Army Corps of Engineers, it compared maps dating back to 1850 to chart the island's land loss. Using analytical software, it then extrapolated what Tangier might look like twenty-five, fifty, and one hundred years in the future.

The view wasn't encouraging. Modern Tangier had shrunk by two-thirds since 1850, the article said, from 2,163 acres to 789. That amounted to 8.43 acres per year, and this loss did not include uplands that had slumped into lower-lying marsh; this was island, wet or dry, that had turned into *bay*. Tangier seemed to have little of itself left to give.

But give it would, the article promised. Uppards, which accounts for about a third of the remaining acreage, would vanish, leaving the town's northern flank unprotected from winter storms. The marshes south of town would retreat and, with them, any shield against damaging summer winds. At the same time the village itself would subside until its houses were split among three islets. Not long after, the so-called ridges would turn to marsh.

Based on their projections, the authors figured Tangier would likely be uninhabitable in fifty years. And that was an optimistic view, the paper acknowledged, because the authors had used three models of sea-level rise for their projections, reflecting best-case, middle-case, and worst-case scenarios. Their predictions were based on the middle model, which was beginning to look overly conservative: "A recent study has indicated that a higher SLR [sea-level rise] scenario is becoming more likely," they wrote, "as humanity fails to take effective action to reduce carbon emissions." Under the worst-case scenario, all of Tangier, including its three ridges, would be underwater by 2060. If that was accurate, the town didn't have fifty years left but significantly less time. How much, exactly? Twenty-five years? Twenty? Fifteen? The article didn't say.

The authors suggested that the loss could be slowed by constructing stone breakwaters and sand dunes on Uppards's east and west shores, and by blowing sand dredged from the bay's bottom onto former uplands and planting the new ground with pine trees. They reckoned that such a project would cost roughly $20 million to $30 million—or somewhere between $41,000 and $62,000 for every man, woman, and child on the island. The article ended with a bleak forecast: "The Tangier Islands and the Town are running out of time," it read, "and if no action is taken, the citizens of Tangier may become among the first climate change refugees in the continental USA."

I read the *Scientific Reports* piece shortly after it was posted online in December 2015. My first reaction was that Tangier didn't have anywhere close to fifty years and that even twenty-five seemed fanciful. The bay had always done efficient work there, and layered on that physical plight were the town's demographic issues. Even if Tangier somehow avoided drowning for, say, twenty years, there might not be anyone left to celebrate.

I arranged with an islander to rent the second floor of her house on the westernmost of the three ridges, and dispensed of mainland obligations so that I might spend the six-month peeler-crabbing season, and months beyond, on Tangier—joining its watermen on their boats, absorbing its odd and long-standing customs to discern what we'd lose with its demise, and plumbing its collective anxiety over what the future holds.

The three-hundred-mile drive from my home in Virginia's Blue Ridge Mountains took me across the state's rolling Piedmont, over the fall line at Richmond, and down the broad, flattening coastal plain to the metropolis clustered around Norfolk and Virginia Beach. From there I crossed the Chesapeake Bay Bridge-Tunnel, nearly eighteen miles long, over and under the bay's wide mouth, the Atlantic boundless to starboard, the bay wind-roiled and heaving and only slightly less intimidating to port. Northbound up the Eastern Shore, I passed platoons of trees killed by trespassing salt water, their trunks stripped and bleached to a ghostly silver, and real estate billboards hawking bayfront property with HIGH BANKS— NO MARSH—SANDY BEACH. I drove past roadside stands selling fireworks and fresh crab, and restaurants touting crab cakes and steamed crab, and gift shops selling souvenir crab figurines. A few miles into Maryland I turned onto a doglegging two-laner for the final twenty miles into Crisfield, traversing vast fields of soybean and long, low-roofed chicken houses, and after parking the car walked two blocks to the mailboat.

Which is what brings me, freshly unpacked, to the south end of the island, to the beach Ooker showed me in 2000—the sandy spiral that served as a landmark to mariners since the seventeenth century, the miles-long tail that gave Tangier its distinctive shape for as long as man had mapped the Chesapeake.

About a mile into my hike, the sand simply stops. I walk past the point where the beach slips underwater, wade knee-deep into the shallows beyond, confused: Where is the hook? The spit should cut to the left, double back on itself. I should be able to look east across a cove of shallow, protected water, to see the end of the island's tail.

Except that I can't. The beach ends here. Tangier's tail has washed away.

A westerly blows stiff and steady as breakers pound the stump.

Ooker Eskridge aboard the *Sreedevi*. (EARL SWIFT)

TWO

—

WELL PAST THE MIDDLE OF THE LAST CENTURY, TANGIER had more in common with the hollers of far Appalachia than with its neighbors on the Virginia mainland. Families were large, incomes meager, and the conditions rustic: Islanders cooked and washed with water fetched by the pail from a handful of community wells. Some used privies and slop jars into the late sixties. Food on the table was, likely as not, caught or shot by a member of the household. Visits to the Eastern Shore were special events, and as radio and television were slow to make landing, trends sweeping the country often passed unnoticed. The motorcar remade America's cities decades before Tangier got its first.

These days, though modernized in countless ways, the island continues to get by without many seeming essentials of contemporary life. Cell phone signals die halfway through the boat ride over. The tap water is unfit to drink. No doctor or dentist is in residence, and the nearest emergency room is thirty minutes away by helicopter—assuming, that is, the weather's fair enough to fly. Eight months of the year, a visitor can't buy much with the cash the lone ATM spits out: a meal at the one year-round restaurant, perhaps, or a room at the one year-round inn, or chips and a soda from the single small grocery.

Cars and trucks remain so few in number that they're easy to list:

an ambulance, fire engine, and a subcompact police car, a few pickups used for hauling freight, three or four careworn sedans, a couple of thrashed station wagons, a Jeep, a Mini Cooper, and a Smart car. Most everyone gets around by bicycle, scooter, ATV, and golf cart, meaning there's no demand for a traffic light—and not much attention paid to the few stop signs. Rare is a ticket from the lone cop.

Trips to the shore are common enough but require forethought: The mailboat to Crisfield leaves at eight each morning, a smaller afternoon boat at four, and as one-way passage on either costs twenty dollars, seeing a movie or shopping at a mall is complicated and costly before it starts. All of which explains why first impressions of Tangier, and much that's been written about it, tend to fasten on what the place lacks, rather than its one great asset—the geographic advantage defining its very existence. For all of its deprivations, Tangier could not be better situated to harvest *Callinectes sapidus*.

Even if they've never heard of the island, millions have savored its catch. No other food is so instantly identified with Maryland and the Virginia tidewater as the blue crab. Bought by the dozen, steamed and sprinkled with Old Bay, it has been central to backyard crab pickings for generations. Its sweet, tender meat spawned the region's famed crab cakes, which have inspired imitations around the world. And in its most adventurous form—as a newly molted softshell, eaten legs and all—it is perhaps at its best. Deep-fried or sautéed, as entrée or sandwich (always on white bread, and minus any garnish but a little cocktail or tartar sauce), the softshell's juicy, flavorful meat and pleasing texture—firm, with a pleasant snap to the skin, not unlike a good hot dog—belie its admittedly challenging appearance. In the big cities of the East, a significant share of the softshells you find on your plate have been supplied by the small but single-minded brotherhood of watermen at Tangier.

Witness Ooker Eskridge, again off to his pots on a late May morning. For peeler potters, the season traditionally starts on the month's full moon, which this year fell on the twenty-first. A cold front lingers over the bay three days later, however, and the water remains cool for this late in the spring; the peelers are dawdling in warmer deep water, beyond the reach of crab pots. But they'll show up soon enough. Every waterman out here knows they pretty well must: The blue crab is a far-ranging creature, but its travels are choreographed by forces as old as time and as predictable as clockwork, and they're bound to bring peelers to Tangier.

A couple of lemons crouch in a pot Ooker pulls up. When they hit the culling tray, one assumes a fighting stance, holding her red-tipped claws wide and open, as if daring him to approach. Right she is to be agitated, for she's being kept from an important task: riding the outbound tides to the saltier water at the bay's mouth, to produce a new generation of her kind. She'd have reached the spawning grounds at about the time the orange sponge attached to her belly turned black, signaling that her millions of eggs had matured. Her babies would have hatched in an inky billow, each of them bug-eyed and shrimplike and not quite one hundredth of an inch long. These larvae would have drifted into the Atlantic to become part of the stew of planktons near the ocean's surface and, in most cases, to become a meal for some larger species. The survivors would have fed on even smaller fare, however, and grown—and that means they'd have molted, because like insects, crabs get bigger only by shedding their hard exoskeletons. After a few molts they'd have caught flood tides back past the Virginia Capes into the bay and started toward Tangier. Only after eight or nine molts would they assume the blue crab's familiar shape.

From the boat I can see that form painted in silhouette on Tangier's water tower: carapace, or top shell, shaped vaguely like a football, its front edge serrated with short spikes, its ends drawn into long, sharp points; five pairs of legs—the middle three for walking, the rear pair flattened into paddle-like swim fins, the much beefier front pair ending in powerful pincers. For some reason the Tangier water tower, like another that looms over Crisfield, depicts a cooked crab. In their natural state, they're not the least bit orange. They're a dark olive on their top sides and creamy white or pale gray on their

bellies, with coloring on their claws that varies by sex—those of jimmies are a vibrant sapphire, while sooks' are tipped in bright red.

Those claws can deliver a nasty pinch. "They've got a good reach
on 'em," Ooker tells me. "And a crab's very accurate, in or out of the
water. They got good eyesight." He moves in on the posturing lemon,
feinting with one gloved hand to draw her eye as he snatches her from
behind with the other, then tosses her in the basket. Lemons can be
especially aggressive. "Very feisty women," he says. "They might have
fingernail polish on, but they're still a rough bunch."

Another pot comes aboard. Several large crabs crouch inside,
along with a sand mullet that's been clawed to bits. Ooker grunts and
shows off the mangled fish. "He went in the wrong pot."

By the time a blue crab looks like a crab, it has grown to ten
times its original larval size, yet remains a tiny creature, just a tenth of
an inch wide. It grows rapidly as it travels, however, molting another
eighteen to twenty times before reaching adulthood, emerging a third
larger with each molt. The process is perilous: A crab grows a complete
new shell under the old before the latter splits at the rear and the creature backs out of it, but that new shell is as soft and pliable as skin for
several hours and takes up to four days to fully harden. Until it does,
the young crab is defenseless, and an easy meal for sea turtles and a
wide variety of fish—striped bass, croakers, eels, rays, cobia, and red
drum among them. "I've cleaned a rockfish before," Ooker tells me,
using the regional nickname for striped bass, "and found twenty-six or
twenty-seven small crabs inside it." It's also fair game for other crabs.
The evidence suggests that these merciless cannibals eat a *lot* of crab.

We motor on. As Ooker hauls in the next pot, I'm surprised to
find that I can see it coming up from the bottom. The wind's calm
today, the water glassy and exceptionally clear, and I now notice dark
splotches on the bay's floor, six feet down: marine eelgrass, which
grows in large underwater meadows off Tangier's east side. Such subaquatic vegetation is vital to crabs—for hunting, mating, and, most

especially, hiding, because it's in these grasses that they hunker down until their shells harden and it's safe to move on.

Their travels take them up the bay until the late fall, when they burrow in the mud and sand on the bottom to wait out the coming cold. They neither molt nor eat during this hibernation; they spend it in a senseless torpor until the following spring, when they're on the move again, migrating into the lower-salinity waters of the middle and upper Chesapeake. There, they reach adulthood and pair off to mate. Afterward the jimmies, which prefer a less salty environment, stay put and continue to molt. The sooks hang around to build their strength before heading back down the bay to spawn—a journey that sees them again burrow into the bottom to wait out another winter. When they emerge from the mud a few months later, they use the sperm they've carried since mating to fertilize their eggs and acquire a bright orange sponge.

Tangier watermen thus have a shot at them both coming and going: They catch peelers and newly matured crabs as they move up the bay and lemons on their way down. Jimmies and "clean" sooks— mature females that haven't yet produced sponges—are in residence all summer long.

Ooker and his kin catch just a fraction of the hundreds of millions of crabs that happen by the island each year, but even so, their haul is impressive. A hard crabber with a top-level license can put out 425 pots and can legally catch forty-seven bushels of crabs each day, six days a week. Each bushel contains roughly five to six dozen number ones or six to seven dozen number twos. Assuming an average of six dozen per bushel, that's 3,384 crabs per crabber per day, so that over a typical season, from mid-March to November, a single waterman catching his limit could theoretically haul aboard more than 660,000 crabs.

Peeler potters catch far fewer—their season is shorter, they're limited to a maximum of 210 pots, and only some of those pots will yield crabs approaching a molt; a bushel or two of peelers marks a decent day on the water. But they land a "by-catch" of hard crabs, too, and over the summer those by-catches add up. Today, Ooker has amassed four bushels of sponges, two bushels of number ones, and a half bushel each of clean sooks and number twos. Each crab has defied one-in-a-million odds to find itself aboard his boat. Of the two million eggs in the average sponge, only one or two will reach maturity. The murky world beneath the bay's surface is a stew of tiny life, and all those critters down there are trying to eat one another.

Amid all this drama, it can be easy to overlook the important link that underwater grasses represent in the food chain. When the grass beds off Tangier are big and lush, they offer the island unparalleled advantages in chasing the blue crab. But lose the grass beds, and you lose not only crabs but many of the smaller creatures they feed on. The island itself is an essential element in the health of those under-water meadows, which thrive in the calm of leeward shores, shielded from the winds and waves that create turbulence on the bottom.

Ooker is keenly aware of this interdependence. "I'm not a scientist or highly educated, but it doesn't take a rocket scientist to figure some of this stuff out," he's told me. "When we lose this land to erosion, we're not only losing land, we're losing the wetlands and the seagrasses in the shallows just offshore. Crabs need those grasses. We lose ducks, shorebirds, blue heron, egrets. When you lose the habitat, you lose all of them.

"I throw that out there because some people who aren't interested in saving people are interested in saving wildlife," he said, winding up to a pitch that many Tangiermen hurl in discussions about the island's physical dilemma. "The government has the money. That's one of the frustrating things. They have the money, and they waste so much. A lot of it is spent overseas on people who don't really give a crap about the United States."

The mayor knows that by any conventional measure, his island is hardly the most important place in America imperiled by rising seas. It's statistically insignificant next to big coastal cities where millions of lives and properties are at stake—such cities as New York, much of which risks being underwater by the century's end, and New Orleans, which seems doomed to a watery grave without vastly expensive intervention, and the metroplex around Miami, where storm sewers have been known to run backward, spewing water *into* the streets during astronomical high tides. Boston, Norfolk, Jacksonville, San Diego, Los Angeles—all stand to undergo profound and painful change, along with scores of other cities and towns.

But little Tangier is important in one respect. As the *Scientific Reports* article concluded, it's likely to be the first to go. That experience—and the uncomfortable questions it forces the country to confront—will inform what the rest of us on and near the coasts can expect in the decades to come. What makes a community worth saving? Will its size alone prompt the nation to fight for its survival—or are other, less tangible factors just as important? Which such factors count the most? And if size is the chief consideration, what's the cutoff, the minimum population, that's worth rescue? What, in short, is important to us?

And there's the matter of the Chesapeake Bay blue crab: Without Tangier, big-city restaurants will be serving a lot less softshell, and many more will have to substitute imported crabmeat for the genuine article in their crab cakes. "Here we are, *in* the United States and just a few miles from Washington, D.C.," Ooker says, "and we need protection."

OFF THE P'INT an osprey flies out to the boat and hovers a few feet overhead. Ooker pulls half of a menhaden from a bait box beside the console and flings it into the air. The bird watches it splash down in the water, then folds its wings to become a missile and jets in after it. Ooker postpones pulling the next pot to watch it surface and lift off for home, shrugging water from its back, the fish clenched in its talons. "That one's Klinefelter," he tells me. He's named all the osprey nesting on Tangier, along with several gulls, and can identify each from fifty yards off. They know him and his boat, too, and most mornings fly out for a free meal. First comes Klinefelter, nested on the P'int's

east side and named for a former owner of that islet. He's followed by
Transmitter—so named because he wears a tiny backpack that tracks
his movements, which range to Brazil over the winter—then Fishy,
nested in marsh on the island's southeast shore, and Old-Timer, from
what's left of the spit. They'll circle for a minute before Ooker coaxes
them closer with a whistled approximation of their call. Gulls form a
tight swirl around them, hoping to beat the raptors to the food. Now,
with Klinefelter a diminishing speck against the sky, gulls continue
to noisily circle the stern. Ooker eases the boat forward.

His behavior is not typical of Tangier watermen, who tend to pay
the wildlife around them little mind unless they're pulling it aboard
their boats. An example of the norm is working one hundred yards
off to our west, closer to shore: Leon McMann, Ooker's father-in-law
and a full-time waterman at age eighty-five. Leon leaves the dock
each morning only to catch crabs; he spares no time to admire the
beauty of the bay, the quality of the light, the grace of a gliding fish
hawk, and he'll tell you so. In rough seas or calm, whether beset by
wind or fog or furnace heat, he judges a day by the number of peelers
he snares and not much else. "He's tough. Very tough," Ooker says as
we gaze at the older man, who's stooped over his culling board at the
stern of his boat, wearing the Tangier crabber's uniform of ball cap,
oilskins, high rubber boots. "On the water you don't have retirement,
so you pray for good health and keep working until you play out."

It may be from Leon that Ooker's wife, Irene, inherited her attitude toward Ooker's communion with the birds, which he considers
"one of the great things about being out here," but about which she evidently harbors some doubt. "My wife says I kill too much time feeding them," he tells me. "She'll say, if I get home late, 'You been feeding
the birds again?'" And it may be from Leon and Irene that Ooker's
thirty-eight-year-old son, James—a.k.a. Woodpecker—inherited his
own approach to crabbing, which sees him fish up so many pots each
day that he pays two island men to crew for him. James roars up now
in the *Rebecca Jean II*, a big, high-sided craft built for ocean work,
which he takes up in the winter. The deck of Ooker's boat is a full two

feet closer to the water than his son's. The redheaded Woodpecker looms over us as he pulls alongside and stops. "How's it going?" Ooker asks him.

Woodpecker shakes his head. "Ain't nothin' but lemons." He leans over the side to place two peelers in Ooker's hands—as a hard crabber, he has no use for them—and, with a wave, guns the *Rebecca Jean II* away. Ooker watches him go. "I'm not sure I ever had the ambition he's got," he murmurs. "I had more than I do now, but he's really working at it. He's *good* at it."

We chug to the next pot. Ooker's interest in birds goes way back. He earned his nickname as a toddler, trying to imitate a pet rooster his family kept. He is the youngest of eight children born to Will Eskridge, a seventh-generation Tangierman, and Mildred "Mish" Pruitt, herself six generations removed from the island's first settler. Will was twenty when Ooker's oldest sibling, brother Ira, was born in 1931, and forty-seven when Ooker came along in July 1958. He shrugs. "I was a surprise."

Two of his older brothers, William and Warren, were twins; Warren died in combat in Vietnam when Ooker was ten. The island was accustomed to wartime loss—Tangier sent more of its young men into World War II, per capita, than any other town in Virginia, and eight didn't come home. But Warren's death, the town's sole casualty in Southeast Asia, devastated the population, and it remains an ample source of grief for the Eskridges. To this day, William keeps a bronze plaque eulogizing his brother on his front lawn.

Otherwise, Ooker enjoyed a fairly typical Tangier childhood. He mudlarked in the marshes. He played with other island kids on an abandoned menhaden fishing boat beached down on the spit, climbing its rigging and swinging into the water from its mast. He worked summers for his father, shaving ice blocks so that Will could cold-pack his soft crabs for shipment and, later, crewing on Will's boat, as Will had on his own father's. He sang in a mediocre island rock band called Hot Ice; Carol Moore's brother-in-law, Tracy, played the drums. One atypical element of his youth: Ooker stayed in school.

Most Tangier boys in the seventies quit at sixteen to follow the water, which promised an income at least equal to blue-collar work on the mainland. When he was in fourth grade, there were ten boys in his class; when he graduated in 1976, he was one of two.

In the next pot, a few crabs are joined by a northern puffer, a small, pop-eyed fish with spines covering its entire body and tiny fins that look to be afterthoughts. Ooker picks it up, flips it onto its back, and rubs its belly with the tip of a finger. The fish instantly swells into a sphere. "Ever seen a puffer fish bounce?" he asks. He flings it to the deck and it rebounds into his hand like a tennis ball. "I did that with a group of visitors one day," he says, "and this lady says to me, 'Should you be doing that to that fish? Aren't you being hard on it?'

"I said, 'Well, normally I'd be cutting its head off and pulling its meat out of its skin. Comparatively, it don't mind this at all.'" He looks at me and chuckles. "I guess I was a little harsh."

After Ooker has fished up the last of his pots, we head for the harbor and pull alongside a big, unpainted aluminum workboat belonging to Lindy's Seafood. A wholesale outfit, Lindy's sends the vessel sixty miles down from Woolford, Maryland, to buy up Tangier's hard crabs every day but Sunday. The number ones Ooker's caught will bring in $100 a bushel—this early in the season, they're hard to come by and thus fetch the highest price of the summer. Clean sooks and number twos are going for half as much. Lemons earn just $20. "Good meat and lots of it," Ooker says of the pregnant females, "but there are fewer crabs in a bushel because the sponge takes up so much room."

In all, he's made $280 on his by-catch and has a half bushel of peelers that he'll sell by the dozen once they become softshells. Before setting out this morning, he sold two boxes, or eighteen dozen soft crabs, to New York's New Fulton Fish Market for $460. He put them on the mailboat, and by now they're on a truck rolling up the coast. "All in all," he says, "it's been a pretty good day."

It's far from over, however. After sweeping mud, slime, seaweed, and bits of eelgrass and crab from the boat, Ooker ties up at his shanty and dumps his peelers into the shedding tanks arranged in long rows out back. Each tank holds crabs sorted according to the time remaining before they molt. Greens or snots, a day or two away, crouch in a couple of tanks under six inches of water pumped from the harbor below. Ranks, just hours from shedding, are in a third tank. Busters, actually in the process of leaving their old shells, are in another.

Ooker must check the tanks every four hours from now until bedtime, because in that time ranks can become busters and are open to attack from their neighbors. More important, busters become soft crabs, and left in the water, their new shells will firm up. Pulled out, however, they won't harden, and they'll be ready to pack off to Crisfield or New York.

It's WELL INTO THE AFTERNOON, most days, when Ooker motors from the shanty to dock space he rents on the Tangier waterfront and climbs from the boat onto the Main Ridge, the middle and most important of the three low strips of upland on which the town is built. It's here, on either side of a road a dozen feet wide—just big enough to permit golf carts to pass—that the island's commerce, culture, and most of its population are concentrated.

The Main Ridge begins at the water's edge at the far northeast corner of Tangier proper, or clear across the island from the spit, at the weedy and hardscrabble Parks Marina, the point of entry for visiting boaters: They reach the road on foot via a narrow, zigzagging path that takes them from the slips past a succession of hand-painted signs, some rustic to the point of folk art, touting the town's restaurants.

At the path's end, the asphalt begins. Main Ridge Road runs south past a couple of tumbledown crab shanties, a few houses, and a fine, rambling Methodist parsonage into what passes for the island's downtown: two restaurants (one open just for the four-month tourist season) and the Daley & Son Grocery; a couple of summer-only gift shops; a combination outdoor diner and cart-rental joint, likewise open from late May to early October; the power plant and post office; and, beneath a simple but stately bell tower peaked with a four-sided steeple, Swain Memorial, its gabled south, east, and west sides lit by enormous arched windows of stained glass. Here and there along the way, narrow lanes branch off to the docks for fuel, bait, the mailboat, and summertime tour boats, and to houses clustered behind those facing the main road. No path travels far before dead-ending at water or marsh.

For longer than anyone can remember, and for reasons no one can say, this northernmost four-hundred-yard stretch of the Main Ridge is called Meat Soup. Just past the church the road jogs through an S-curve, and for another 830 yards it's lined mostly with houses. About halfway down, give or take, stands the New Testament Church, austere home to a flock that broke away from the Methodist congregation just after World War II, and which now rivals Swain in its head count. The stretch of the Main Ridge from Swain to New Testament is called King Street, an old name for the road. Again, skinny lanes sprout east and west to houses tucked back at the ridge's verge into wetland.

The southernmost third of the Main Ridge, from New Testament down to the road's unceremonious end—where the houses peter out and the asphalt gives way to a rutted dirt track into the marsh guarding the town's southern flank—is called Black Dye, another name no one can explain. The border between the lower two neighborhoods is indistinct, and Tangiermen will disagree as to where King Street ends and Black Dye begins. But if the line isn't at the New Testament Church, it's close.

So runs the stem around which Tangier is built—about two-thirds of a mile long, at no point much more than two hundred yards wide, and in places less than half that. Most afternoons will find Ooker pedaling his beach cruiser to a low, nondescript building in Meat Soup, across from Swain Memorial: the former Gladstone Health Center, named for a longtime island doctor, and supplanted in 2010 by a big, modern clinic farther south on Main Ridge Road. The building's back door is unlocked. Just inside, a dark, twisty corridor leads to the center's former birthing room. Every weekday from

2:30 to 4:00 P.M., it hosts a fraternity of island men who convene to discuss the weather, crabs, erosion, and the shortcomings of government regulators and marine scientists. About half are retired; a few others probably should be. Ooker, among the youngest regulars, calls it the Situation Room.

A typical session might begin with Leon McMann, the presiding elder, ranting about the state's crabbing rules, as he did one afternoon in late May. "They're trying to make laws that'll make more crabs, but they don't know what they're doing. They don't have no idea," he said, pointing at his fellow members with a finger curled and swollen with arthritis. He leaned back in his chair. "Well, they ain't watermen," he added with a sigh. "None of the people making the laws are watermen. They don't know nothin' about it. They know the water's wet, is all."

The group might mull the wonders of the strange animals they hunt, as they did when Richard Pruitt, a sixty-year veteran of the water, still chasing peelers, walked in with a freshly molted soft crab and the empty shell, or "shed," that it left behind. He displayed them both in the palm of one hand. The newly emerged crab seemed impossibly big next to the translucent husk it had occupied until minutes before, and inside the shed was a maze of cavities, each an obstacle to molting, a potential snag to a soft crab striving to pull free. "That's amazing, isn't it?" Richard asked quietly. From the eyebrows down, his face was already tanned the color of beef jerky. Under his ball cap—which, like all Tangier watermen, he wore indoors and out, except in church—he was fish-belly white.

"It is that," said Bruce Gordy, a retired schoolteacher, the island's unofficial historian, and Richard's lifelong friend, leaning in close to study the creature.

"The people who make our laws, they don't know that right there," Leon said about the marvel before our eyes. "They don't know nothin' about that."

They might discuss the discomforts of growing old: "I got up this morning, and I took two Alka-Seltzer Plus," Leon told the group. In old photos, he's a big, barrel-chested fellow with bushy sideburns and a thick, dark head of hair. He has lost his brawn and several inches of height since, and his scalp is visible through a thinning tousle of white. "When I got on my boat I took three ibuprofen. When I got home I took two Aleve. And I imagine I'll take two or three more pills before the evening's through. I do that every day."

All of this is delivered in Tangier's odd tongue, a tuneful confluence of accent and dialect that stretches one-syllable words and knots them into two, warps vowels, and, to an untrained ear, can be as indecipherable as Tagalog or Navajo. "Hard" comes out as *howard*, "island" as *oyalind*. Visiting journalists have long pegged it a stubborn vestige of Elizabethan English preserved by the island's isolation. That's not the case, but it does evoke a time long passed—and it's rendered all the more foreign by the islanders' habit of saying exactly the opposite of what they mean.

Talking "over the left," it's called, or simply talking backward. It's signaled by subtle inflection, often by slight stress on the subject of the sentence. Thus an islander might react to a beautiful woman by saying, "*She* ain't nothin' to look at." I witnessed an example before I even arrived on Tangier, when a Mustang took a corner too fast in Crisfield and spun out in view of a crowd on the dock. "That car—it didn't go sideways," one islander commented. Replied another: "I'll bet his face ain't hot."

Now Leon peered through his glasses around the Situation Room. "I know some people who never take a pill," he said. "I think they're the only thing keeping me *alive*." He shook his head. "Pills, I ain't took none of them."

On some days, too, the group might take up the subject that preys quietly on every island adult, as happened when Leon, glaring sourly from under his ball cap, announced: "Now they're saying that we're sinking."

Bobby Crockett, a former waterman who spends weeks at a time off-island on a tugboat, waved the suggestion away. "I'll tell you one thing," he said. "I hope I live until this island sinks. I'd live to a ripe old age."

"That's what they say," Leon countered.

"They do," Bruce Gordy agreed.

"I'll tell you one thing. I hope that's what brings the end," the redheaded Bobby declared. "I hope it is, because if that's the case, this island will be here for a good long time. We'll all be dead and buried before then, and our children, too." Well, now, Leon pointed out, nobody much fretted that the island was washing away until an islander started taking measurements on the western shore fifty-plus years ago, but that didn't mean it wasn't happening. Bobby allowed that erosion was another story. He remembered the western shore stretching a "right long way" beyond where it does now.

He added that up at Tylerton, one of the three villages on Maryland's Smith Island, the state had spent a pile of money on protecting the shoreline. Why wasn't Virginia doing that for Tangier? "I'll tell you one thing," Bobby said. "In Tylerton they got the prettiest bulkhead, all along the town and even up in the marsh. And I heard they spent $10 million on a jetty all the way around the wildlife refuge [north of town]."

Leon fixed him with a scowl. "Save the birds," he muttered. "Kill the people."

Tourists walk Main Ridge Road in King Street. (EARL SWIFT)

THREE

——

MORNINGS COME EARLY ON TANGIER. HOURS BEFORE daybreak—in many houses, closer to midnight than dawn—bedroom, bathroom, and kitchen lights flick on, coffee is chugged, lunches are packed, and watermen straggle in the silent dark across the marsh and up the ridges to Meat Soup. In the harbor, outboards whine and diesels burble to life. The scent of four-cycle exhaust hangs over the docks. Under long strings of bare bulbs, peeler potters exchange shouted hellos as they buster up, or sort, their molting crabs. Then, while much of the town sleeps on, the island's crabbers take to their boats and head west to the bay, or east to the sound, and their pots.

By four on a springtime morning, the waters off Tangier are dotted with moving light—the sallow glow of boat cabins, the powerful beams that captains use to find their buoys, and blue-white LEDs illuminating open decks. They glide among the fixed green and red flashes of navigation beacons—a sight that evokes both lonesomeness, for their being surrounded by so much blackness, and an odd reassurance, for their constancy in all but the most fearsome weather. If the wind is right, the sound of boat motors will warble to shore from more than a mile out.

The schedule is but one reason I'd make a lousy waterman. I'm rarely hungry for breakfast at three. At what seems a reasonable hour to rise, the workday is half spent—the part on the water, at least—and I'm eating lunch when I'd otherwise be finishing my second cup of coffee. By noon, or soon after, we're headed back to port. I spend the subsequent afternoons in a state akin to jet lag—physically drained, slow thinking, and disoriented by so many remaining hours of bright sunshine. My internal clock stays scrambled for days.

All the year round, Tangier's circadian rhythm is disrupted by this maritime take on a factory's midnight shift. It's dictated by necessity: An early start puts crabbers on the water at the coolest part of the day—not for their own comfort but for that of their catch, because crabs crammed into bushel baskets can't long tolerate high temperatures. The other players in the bay's seafood industry—buyers, picking houses, marine police—synchronize their workdays to the watermen's schedule, and Tangier's business hours skew early as well. One winter-only restaurant opens at three in the morning, to serve oystermen before they head out into the cold. Bait's available around the clock. The grocery closes at five, and the biggest summer-only restaurant, Fisherman's Corner, at seven—a late hour, as island suppers go. For generations the island's school started and ended classes earlier than its mainland counterparts, the better to send children home at about the time their fathers returned to shore.

The unnatural schedule poses a struggle even for those who know no other workday. One afternoon in the Situation Room, Ooker announced that he rose at two that morning, dozed off on his porch while tying his shoes—and remained sleeping, bent over double, until he was stirred by a passing scooter on West Ridge Road. "A whole hour, I was sitting there," he said. "If that scooter hadn't passed the house, I'd probably be there still."

Leon McMann nodded sympathetically. "When that clock goes off I'd like to take a bat and beat her to death sometimes."

It doesn't take me long to discover that I can manage, at most, a day or two on the water per week. So most of my mornings begin hours later, when the sun breaks from behind the Eastern Shore and the rooster in the yard two doors down greets the light with fifteen minutes of crowing. The first time I woke to the rooster it came as a surprise, a discordant barnyard note in a marshy seagoing town. And the creature was *loud*: I slept with the windows open and with a set of French doors thrown open as well, and when the rooster let loose I was up in an instant.

My quarters are the entire second floor of a Cape Cod situated at roughly the midpoint of the West Ridge, the island's second-most populated upland. The house is owned by Cindy Parks, niece of a past *Virginian-Pilot* colleague. I have my own entrance via an outdoor staircase in back, which leads to a deck with expansive views of the island's airstrip and the bay beyond. From there I can step through the French doors into my bedroom, where Cindy has converted a walk-in closet into a rudimentary kitchen equipped with a dorm fridge and microwave. A hallway off the room leads to a full bath.

When taking up on an island you expect that water will play a dominant role in your sense of the place, and so it does on Tangier, insinuating itself into every vista, figuring in every aspect of daily life. But lacking trees, discernible elevation, or tall buildings to interrupt the view, this is a landscape defined by sky as well. The heavens are an immense, unbroken bowl. Thunderheads form towering piles visible forty miles off, and approaching fronts split the sky as cleanly as lines on a weather map. The island's dim streetlights leave the night sky unsullied; even a half-moon casts shadows, and stars gleam with uncommon strength.

From the deck I can quickly surmise what the coming day holds and how I might prepare for it. Most of the bay's routine summer weather—meaning the stuff not part of a big storm system—comes from the west, and I can see it an hour or two before it arrives. No need to consult a forecast: I'll know to pack a rain jacket. Is the wind blowing? This is a consideration on those days I'm headed off-island, especially if the wind is running against the tide to produce a hard, white-capped chop. But it also informs an important aspect of terrestrial life. Is it blowing hard enough—at ten miles an hour or better, say—to keep the flies away, especially the bloodthirsty deer flies, better known throughout the Chesapeake as greenheads? If not, I'd better baste myself in bug spray, especially if I plan to cover much ground on foot. Is the wind shifting to the east, which seems to bring out the smaller black flies, or ankle-biters, that can swarm a person by the score and seem to prefer the tender flesh of mainlanders? Again, it's time to reach for the DEET. If it's blowing hard from the east the flies will stay grounded, but the wind may push the tides up over the roads and into the yard. If it's hard from the west, the tides will swing low, and the island's tidal creeks will show their muddy bottoms.

No matter the weather, I'll pull on a ball cap before leaving the house, not only because the mostly shadeless island broils at midday, but because to go hatless—or to opt for any hat but a ball cap—would identify me as a tourist. Said cap, preferably advertising a college or pro team, tugboat company, or Tangier itself, should be sun-bleached and salt-encrusted and blackened with sweat. Born-here authenticity further requires that its brim be misshapen and frayed and stained with slime from a thousand crabs' backs.

My first days on Tangier, and most after, start with a bike ride over the marsh to the Main Ridge, a journey termed "going over" or "crossing over" by the natives, and notable for the wildlife one encounters en route: blue herons, snowy egrets, osprey, several species of gull, and the occasional water snake or muskrat—and, especially, a superabundance of feral cats. They roam the ridges in all weather, huddling under parked golf carts, colonizing yards and porches, and lounging in gangs on the roadsides, where they display little fear of approaching traffic. Cats often outnumber people on the Main Ridge, for Tangier's central spine can seem curiously deserted on a weekday morning. Minutes might pass between golf carts trundling by, on their way to the grocery or post office. Few souls walk the streets before the tour boats arrive at midday. Quiet prevails, save for bird-song and rustling marsh grass. The growl of a lawn mower engine is distinct and intrusive from a half mile away.

Indeed, much of Tangier is missing. Seventy-two islanders, all men, are out in their workboats. Another forty-one men work on tug-boats and live half of their days (or more) off the island, working two weeks on and two off, in most cases, muscling barges up and down the East Coast. Four men are boat captains, meaning they hold master's licenses to carry passengers; they include the mailboat's skipper, Brett Thomas, and the islanders who run the afternoon boat to Crisfield and the warm-weather ferry to Onancock. And two island men are employed by the state as marine police officers, spending long hours on the bay enforcing Virginia's crabbing and oystering laws—a duty that can put them at odds with watermen, which often as not means family.

In sum, about half the adult male population is elsewhere. So as I pedal my bicycle up the Main Ridge one late May morning, the few people I encounter are women—white women, because Tangier is that rare southern community without a single racial minority. Among them is Carol Moore, who's standing inside the low fence ringing the house she shares with her husband, Alonza J. Moore III, a.k.a. Lonnie, who like his fellow crabbers is out on the water. I brake to ask her what's going on, assuming she'd know: The Moores live on Main Ridge Road in Meat Soup, across from the Daley & Son Grocery, and their place borders a smaller east–west lane that ventures across four hundred yards of marsh and a fat snake of dark water to the West Ridge. She has a commanding view of the busiest crossroads in town.

Carol replies that a funeral is scheduled for the coming weekend. Henrietta Wheatley, at ninety-one the oldest woman on the island, died two days ago. Her passing came little more than a week after that of a ninety-year-old man, one of Tangier's two remaining veterans of World War II. It's been a bad month, she says, and there are bound to be worse to come. On this day in May 2016 the island is home to 481, a number that includes young people in college and the military who keep Tangier as their legal address, but don't actually live here and aren't likely to return. Cross them off and the head count drops to 470. Of that resident population, 108—nearly one in four—are past retirement age, and a fair share of those are decades past. Carol's uncle, Jack Thorne, is now the sole surviving World War II vet. He's ninety-two and the oldest of seven siblings, three of whom are close behind him in age; one, Ginny Marshall, just turned eighty-nine and, according to Carol, "can remember when God said, 'Let there be light.'" At eighty-three, Carol's mother, Grace, is among the younger Thorne children. "When one goes, they'll all go," she tells me. "Like flies. It'll be fast."

"Well, there's a happy thought," I say.

"That's how it happens around here," she says. "Which will mean more houses will sit empty, which we don't need."

"How many empty houses can there be?" I ask. When I put out word that I was looking for a place to stay on the island, I received nary a response for weeks before Cindy Parks stepped up. And besides, the island doesn't seem to have enough of anything, let alone houses, to qualify as a surplus.

"A lot of them," she says. "You have time to take a tour?"

We climb into her golf cart and strike west on Long Bridge Road, the northernmost of four lanes linking the Main Ridge with the West. Eight feet across at its widest, it runs from Carol's house past eight compact and closely spaced cottages, most well maintained, two of them unoccupied. We stop in front of one. Its abandonment is not obvious: Weeds sprout around its foundation and the siding needs a scrubbing, but those blemishes don't set it apart from many island homes. "These people from the mainland bought it," Carol tells me. "They came and gutted it—they borrowed some of our tools." She purses her lips. "I don't know what happened, but they haven't been back in years."

Out past the houses we glide across the marsh and thump over the wood-decked Long Bridge. Beneath it curves the Big Gut, a tidal creek that loops and curls between the Main and West ridges and technically makes separate islands of them. The bridge earns its name—it's twice the length of the next biggest span on Tangier, as the Big Gut here is at one of its widest points, a good one hundred feet across. From the bridge one gets a sense of the island's snug dimensions and its fragile relationship with the surrounding Chesapeake. The marsh extends an unvaryingly flat, treeless mile to the south, its willowy grasses turning from their wintertime bronze to pale greens and hissing in a light breeze from the southwest. It is equal parts water and earth—flooding to the road's edge at high tide, its exposed black mud emitting a gassy stink at low, and laced with rivulets, small ponds, and straight ditches dug in generations past. Ibises and egrets high-step in its shallows. It's even quieter than the Main Ridge. The whir of the cart's electric motor, the crunch of its small tires on the road, and the music of the marsh grass are the only sounds.

A few yards beyond the bridge we come to the West Ridge, which is longer than the Main, at just over a mile from end to end, but so slender that for most of its course, houses line only its west side. To turn right here, heading north, would take us past a few houses and single-wide trailers, as well as the airport's parking apron and a prefabricated steel recreation center, its three-story sides painted a dingy beige.

Carol instead turns left on West Ridge Road, and we roll past single- and double-wides and houses ranging from small ramblers to large, century-old Victorians and colonials. Most are set behind deep front yards that aren't much to look at; susceptible to flooding and slow to drain, the lawns are sodden for most of the year, and salt water has killed all but a few of the ridge's trees and large shrubs. As the houses slide by to our right, we come, on the left, to three other roads connecting the Main and West ridges like rungs on a ladder. The first sidles up close to the island's largest building, Tangier Combined School, its elementary classrooms in one wing, high school in the other, the whole of its great wood-framed mass suspended five feet over the ground on piers of heavy timber.

Roughly two hundred yards farther on comes Wallace Road, so skinny that passing carts must veer off the blacktop to squeeze by. It meets the Main Ridge roughly halfway between Swain Memorial and the New Testament Church, and is the crossing I'll most often use to go over. We pass Ooker's house, a mobile home to which he's made additions over the years, and three doors farther along, my place.

Down at the end of West Ridge Road we come to the southernmost crossing, which spans the Big Gut on a bridge that rises at its turtle-backed crest to more than five feet above the tide. The Hoistin' Bridge, it's called—a reference to an earlier span fitted with a deck that was slotted to make way for masted boats. Tangiermen pronounce it *Heistin'* and often drop its surname, as in "I was down to the Heistin'."

Just past the intersection, West Ridge Road does a short jog to the right, and narrows as it continues southward for a couple hundred yards, lined on both sides by tightly bunched houses and mobile homes. This is Hog Ridge, and though its old-timers don't care for the name, at least it has an explanation: Farmers once raised pigs there. At its southern end, the asphalt gives way to a rutted sand path into the marsh. It crosses a small gut on a rickety wooden bridge, then bursts from a thicket of tall reeds onto the beach. The sand is fine and white, and refreshing breezes encourage walks to the spit's end, but not many Tangiermen take those walks. The beach is usually empty.

We're now at the far southwestern tip of the island's road system and halfway through a counterclockwise circumnavigation of the town. Carol has pointed out unoccupied West Ridge dwellings as we've rolled along, and as we turn around I tally them: five houses and six trailers, plus two buildings that are uninhabitable—a large inn on Hog Ridge, closed for years and ruined since by vandals, the elements, and spraying cats; and a house near my place that burned three or four years back. It stands blackened and gaping, its interior a soggy heap of broken furniture, ripped paneling, and fiberglass insulation—an eyesore, but undemolished because it's covered in asbestos shingle, and no one's figured out how to dismantle the wreck without breaking the law or the bank.

We roll back through Hog Ridge and turn east across the marsh, rattle over the Heistin' Bridge, and swing north on Main Ridge Road. The firehouse, low and modern, waits a short distance up on the left, an ambulance and small pumper visible behind glass-paned garage doors. The all-volunteer fire department was thinking strategically when it built here: Of the bridges to the West Ridge, only the Heistin' is stout enough to accommodate the emergency rigs, so it made sense to situate the station near the crossing; the most distant home on the West Ridge, up by the rec center, is thus only a minute or two farther away than the northernmost home in Meat Soup.

The station is also just yards from the main drag's intersection with Canton Road, which drills eastward across another four hundred yards of marsh and two guts to the smallest of Tangier's three inhabited ridges. Twenty-seven homes are clustered there, the largest with their backs to Canton's rapidly eroding eastern shoreline. For most of Tangier's history, Carol tells me, erosion targeted its western shore, but over the past few years, it seems to have shifted its aim. Ooker's brother William, who has the monument to his lost twin in his front yard, lives in one of the waterfront homes, and out back of his place the bay is getting uncomfortably close.

We return to the Main Ridge, having counted six empty houses on Canton, and resume our northward journey through Black Dye and into King Street. The road here is lined with low chain-link and white picket fences, guarding tiny front yards shaded by Tangier's largest and densest cluster of trees—a mix of hardwoods and cedars, a few tall pines. Years ago, Carol tells me, island boys trapped a couple of squirrels on the mainland and released them in Tangier's meager woods; one soon vanished, but the other survived for years. "So," Carol says, "when people here told you, 'I saw the squirrel today,' they meant they saw *the* squirrel. People were obsessed with that squirrel." When at last the squirrel sightings stopped, Carol's husband, Lonnie, brought another pair to the island. The rodents had babies in 2015, which the cats dispatched, but the adults endure. Maybe. Carol says she hasn't seen them for a while.

Also rising from a few front yards, and discussed with wearying predictability by generations of travel writers, are headstones. So much has been made of the graves that Tangiermen have grown thin-skinned about them, and react poorly to tourists training cameras at the markers. The graves date to a time when the population was larger and the ridges more crowded, and open ground suited to burials was in even shorter supply than it is today.

Here, in a small cottage in King Street, is where Carol was born in 1962. Here she lived until she married at eighteen, and for much of that childhood she did not stray far. Until recently, Tangier youngsters stuck to their own neighborhoods and viewed the other ridges, and even other parts of their own ridge, as foreign turf. From King Street, Carol came to know Meat Soup as "up the road," Black Dye as "down the road," the West Ridge as "over the road," and Canton as too remote and seldom visited to think about. "When we were children we didn't go to West Ridge, and I didn't go down the road," she tells me. "Lord, going down the road was taking a long trip, and going to Canton was like going to an amusement park. We didn't go out and rove. I've heard West Ridgers say they didn't go over the road until they were nine or ten."

As luck would have it, Lonnie Moore, though born in Meat Soup, spent most of his youth as a King Streeter; he lived the equivalent of a city block away, behind a famed inn and restaurant, Hilda Crockett's Chesapeake House, that was started by his grandmother in 1939. He was eight years older than Carol—"And trust me," she's said, "that man had sowed his wild oats. He'd not been passing out Bible tracts."

Even so, they started dating the summer after she graduated from high school, in 1980. He was more distantly related to her than many island men—they shared fifth great-grandparents—and "he was fun to be with, to hang around with." He proposed that August. They married the following April.

As we approach Swain from the south we pass Hilda Crockett's, consisting of two big, deep-porched houses that face each other across the road, and several other King Street landmarks: a couple more gift shops, a summer-only ice cream and pizza parlor, and a fascinating museum of island history manned by volunteers four months of the year. The northernmost structure in the neighborhood is a surprisingly big and opulent health center, staffed by a physician's assistant on the days a doctor hasn't flown in from the mainland. It replaced the cluttered old building in which the Situation Room convenes. Its Wi-Fi is powerful enough that on many evenings, Tangiermen who lack wireless access of their own can be found sitting on its porch steps, reading their cell phones.

At the bend where King Street gives way to Meat Soup, the main road widens to accommodate cart parking outside the church and post office, and a narrow lane branches off to a small cluster of houses tucked behind those facing the road. Built on dredge spoils in the sixties, the enclave is known as Ponderosa, after the ranch in the old *Bonanza* TV series—to which it bears zero resemblance.

The tour boats have disgorged their passengers during our exploration of the island, and we weave among them as they wander Meat Soup. Customers are turning up at Lorraine's, the island's year-round eatery, and at Fisherman's Corner, just across the street, and at Four Brothers, the outdoor café and cart-rental place, which is festooned with TRUMP FOR PRESIDENT placards. Golf carts are bunched at odd angles outside the grocery. A few tour buggies—golf carts stretched to accommodate several rows of seats—are loading up with passengers, who spend five dollars each for a brisk tour of the island's roads, with running commentary from the native at the wheel. Meat Soup, all but deserted in the morning, is glutted with humanity. The throngs will vanish as quickly as they materialized, because the vast majority of the visitors are day-trippers whose boats leave Crisfield, Reedville, or Onancock at ten, arrive at Tangier in time for lunch, and head back to the mainland by midafternoon.

The hordes might, on a good day, consist of two hundred people—not an overwhelming number on its face, but at nearly half the resident population, a test of the narrow roads, the shops, and the restaurants. Many islanders take their arrival as a signal to retreat out of sight, as they feel some visitors study them a bit too closely; many have stories of being photographed performing such mundane chores as taking out the trash. Carol is among them. Years ago, she managed a now-shuttered restaurant here in Meat Soup, and found that while strangers were drawn to her quick mind, sly humor, and striking looks—she is fashion-model tall, with high cheekbones and full lips—she didn't much enjoy their company. "I don't really like people, and I had to pretend I liked people," she's told me. She drove a tour buggy for two years, too, which was even worse, "the absolute worst job I've ever had," again because it forced her to make nice. Her near-daily summertime trips to wind-scoured, barren Uppards often coincide with the presence of tourists outside her house.

We negotiate the clotted road, having counted on our tour fifty-two houses and twelve trailers that stand empty, most belonging to islanders who've died or can no longer care for themselves, and whose families have been unable or unwilling to part with them. That's about 20 percent of the total housing stock. And as Carol worries, the number is bound to rise. Of Tangier's 210 occupied homes, sixty-six shelter just one person, and the vast majority of those islanders are elderly.

BUSY AS MEAT SOUP becomes on a summer day, one might assume that it's always been the island's port, as well as its cultural and commercial center. But time was that the focus of island life was at Canton, for it was there, on what's now the smallest and most remote of the ridges, that Tangier's first intrepid pioneers built their homes.

The truth about those days, and much of the island's early history, has been clouded by legend and rather shaky oral tradition. What's certain is that long before Europeans turned up in the Chesapeake, the island was well known to the Native Americans who inhabited the Eastern Shore and lands to the west. Most historians figure tribes used the place as a hunting and fishing ground, rather than residence, but they evidently spent a fair amount of time here. Carol Moore and other "proggers"—island speak for beachcombers—have found hundreds of arrowheads stranded by the tides at Uppards.

The first Englishmen to lay eyes on Tangier did so in early June 1608. Captain John Smith, fabled hero of the Jamestown colony, was sailing up the Eastern Shore on an exploration of the bay when he and his fourteen men sighted islands off to port. When they turned that way, they encountered an afternoon squall packing "such an extreame gust of wind, raine, thunder, and lightning" that it was only "with great daunger, we escaped the unmercifull raging of that ocean-like water." Forced to the mainland, the party again made for the "Iles" the next day and searched the spongy ground for fresh water. Finding none, they sailed on to what's now Maryland.

So it is that John Smith is usually credited with having "discovered" Tangier. In fact, there's no evidence he actually set foot there; he offered too little description in his memoirs to clarify which "Iles" the expedition visited. Tradition holds that he named the island, too. He did, but he lumped Tangier and its neighbors together as "Russels Isles," named for a doctor in his party. "Tangier" didn't come into use until decades later, for reasons unexplained.

In 1666, it's said, an Eastern Shoreman traded two overcoats to get the island from the Natives—more dubious legend—and twenty years later sold a big piece of it to one John Crockett, "a gentleman of

English descent" who moved to Canton with his wife and child and who went on to have seven more children while farming the uplands. Again, not the way it happened.

The earliest published account of these traditions is an 1891 book called *Facts and Fun: The Historical Outlines of Tangier Island*, by Thomas "Sugar Tom" Crockett, an oysterman and eventually the principal of Tangier's school. "The reader may want to know who told me these things," Sugar Tom wrote in the book's opening passage. "I answer by saying my grandmother told me so, and she lived to be one hundred and five years old, and had good recollection to the day of her death; and what she was not an eyewitness to, her mother told her." That was evidently good enough for the islanders, along with a long procession of authors and journalists, who repeated Sugar Tom's version of Tangier's early days in newspaper stories, magazine features, and books into the late 1990s. Even the state historical marker erected outside Swain Memorial testifies that Captain Smith named the island and that John Crockett and family settled it in the seventeenth century.

But with all due respect to Sugar Tom, his granny, and the Commonwealth of Virginia, the first documented white settlers didn't arrive until 1778, when Joseph Crockett (not John), the father of ten children (rather than eight), bought 450 acres of what was then a much larger island and built a house on a patch of high ground in Canton. A native of Maryland's Somerset County and a longtime resident of Smith Island, Crockett was in his midfifties when he made the move, likely with an eye to farming the uplands and grazing livestock in the marshes.

It must have been a lonely and unforgiving existence. The shoals offshore prevented all but small boats from visiting. As John Smith had discovered, fresh water was elusive—far into the twentieth century, Tangiermen used cisterns to capture rainwater from their roofs—and summer brought fierce afternoon storms. Ravenous insects rose from the wetlands like mist. Winters were windswept and frigid, and the surrounding waters often froze solid.

In fact, the island had been viewed as a bit of a hellhole since early in England's Virginia experiment. When colonial leaders sought a place to exile participants in a 1644 Indian uprising, Tangier struck them as ideal. The marooned Indians likely didn't last long. It may have been their remains that Sugar Tom would report seeing more than two hundred years later. "I have had many of their relics in my hand," he wrote, "and they had grave-yards there, for I have seen the bones and had their teeth in my hand."

MEAN THOUGH ISLAND LIVING WAS, over time Joseph Crockett was joined by new settlers who married into the family. Among these come-heres was a stammering, barely literate waterman named Joshua Thomas, who roundabout 1799 bought seventy acres of upland on the West Ridge. His wife of two years, Rachel Evans, was Joseph Crockett's granddaughter.

Thomas had been born on Maryland's Eastern Shore in August 1776. He was a toddler when his father died of complications from a dog bite, and his mother married a Smith Islander named George Pruitt. The American Revolution was raging at the time, and the Chesapeake's islands were frequently used as hideouts by Loyalist picaroons, or pirates, and when Joshua was five or six a band of these marauders burned down the family's house. The loss apparently broke his stepfather's spirits: George Pruitt spent the next several years in a drunken haze before falling from his boat and drowning when Joshua was in his teens.

The young Thomas apprenticed himself to a Smith Island waterman to learn the trade, and by the time he moved to Tangier he was an expert fisherman and boat handler. He found the island less a community than a scattering of small farms, its few dozen inhabitants separated by wide expanses of roadless marsh. At about a mile from Canton, the Thomas homestead must have seemed at the edge of the world. "Our entire stock of provisions consisted of three bushels of meal, and two little pigs," he'd recall. "Of furniture we had barely enough to make out with."

One summer's day in 1807, Thomas was fishing offshore when he was approached by three boats crowded with passengers. They were pilgrims bound for an outdoor Methodist camp meeting on the Eastern Shore, and they hired him to pilot them there. The next morning he and John Crockett, Joseph's youngest son, skippered the party through a maze of boats parked in Pungoteague Creek and stepped ashore into a gathering of thousands. Thomas was intrigued by the Methodists' noisy style of worship, though he didn't fully understand it. While a bearded, wild-haired traveling evangelist named Lorenzo Dow "was preachin' very powerful," he said years later, "a woman in the audience begun to shout. Dow stopped and cried out: 'The Lord is here! The Lord is here!' Immediately I jumped to my feet, and stretched my neck every way to try to see the Lord, but I could not see him."

John Crockett, Thomas said, "appeared to be in such distress and alarm about the shouting, and singing, and people falling all around us, that I consented to go." But Thomas left with much on his mind. Founded as a lay ministry to reinvigorate the Church of England, Methodism stressed a personal relationship with God over stuffy liturgy, and its plainspoken, unrestrained spirit appealed especially to the rural and unschooled—which is to say, people like Joshua Thomas.

Later in the summer, Thomas was drawn to a second camp meeting near present-day Crisfield, and while listening to a sermon "felt something drawing me right to the feet of Jesus," and "went to the altar, and kneeled down and began to lift up my voice in earnest prayer." He returned to Tangier eager to organize a prayer meeting there, and not long after the island saw its first Methodist service. It lasted six hours.

From then on, singing, praying, and shouting became a Sunday staple, the services rotating among island homes, neighbors joining the flock one by one. The following summer a pair of traveling lay preachers visited the island and set up a tent. Preaching under the canvas brought more islanders to Jesus, so the year after that—1809—Tangier hosted its first homegrown camp meeting. The gatherings became annual events down on the spit, attracting more mainlanders from one year to the next, and Joshua Thomas, whose stammer vanished at the pulpit, became an exhorter of growing renown.

So it was that Methodism took root on Tangier and captured the hearts and minds of its people. And as they grew in number, they found themselves ever more dependent on Providence, for over time they not only farmed but turned to the water for sustenance, and much of what was essential for survival lay beyond their control. They prayed for fish in their nets, and protection from the Chesapeake's fast-shifting and inscrutable moods, and courage in the face of freezes and floods. They prayed for fair winds to bring them home and for nourishing rains to feed their corn, potatoes, and greens. They prayed most of all for salvation, and an afterlife free of their island's hardships.

The Methodism here was, in those early years, much like that practiced elsewhere in the Mid-Atlantic. But as the mainland church evolved over the subsequent decades, Tangier's isolation preserved its style of worship, so that today, elements of the island's Methodism seem trapped in amber, throwbacks to a Victorian version of the faith. It is not an uncommon occurrence, for example, for Pastor John Flood to seek a heavenly cure for one of Swain Memorial's flock. At an evening service early in my stay, I watched as he called the congregation to the altar and urged its members to lay hands on an ailing woman. "Most gracious Heavenly Father, we come before your throne this night," Flood said, his soft country drawl amplified throughout the high-ceilinged sanctuary as he anointed the woman's forehead with oil. "We ask your blessings on this oil, Lord, that it would become an oil of blessing and an oil of healing, and as [she] receives this on her forehead, Lord, that you touch her and you touch the body, that you work through this body."

Eyes closed, heads bowed, the congregation formed a tight circle around the sick woman, those unable to reach her from its edge instead grasping the shoulders of those closer in. "We pray now that as you touch [her], that she is receiving that touch from the great physician, the great healer," the bearlike, crew-cutted Flood intoned. "She can feel the power of the Holy Spirit going through her body, from the top of her head to the bottoms of her feet, touching every part, and as it's touching her she is being healed, Lord. And when [she] goes to have those tests this week they won't find anything, Lord, because she is healed here and *now*."

Likewise, it's not unusual for a service down at the New Testament Church, a more literalist splinter of Tangier's already conservative faith, to evoke the exhortations led by Joshua Thomas two centuries past, as was the case when I heard schoolteacher and church elder Duane Crockett preach one Sunday morning in late May. "An unsaved person may not forgive, but if it's Christians who do not forgive, that is an entirely different matter altogether," the balding, fair-haired Duane told us. "Oh, sometimes we can be slick. Sometimes we say, 'I love them, I just don't like 'em.' I wish one time somebody would open the Scripture and tell me where that's said in the Bible.

"It's not in there," he said. "Or: 'I forgive them, but let me tell you what they did to me.' *No.* If you're going to tell fifty people what someone's done to you, you haven't forgiven them at all." Thirty-eight years old and bespectacled, he has an accent that ranks among the island's strongest. "Twelve" is *tway-elve*. "Me" comes out *may*.

"We're supposed to make things right with our brothers and our sisters," he said. "We get afraid that people will get mad at us, and on Tangier it's very easy to fear that because we know everybody. It is a lot different being on Tangier than it is being anywhere else. And I want to say that it is a normal thing, to want people to like us. If you don't care that people like you, or if I don't care that people like me, then there's something wrong there.

"But that doesn't mean you turn your head to everything because you're afraid it's going to offend somebody. You have to address them." He peered at the congregation. "We have an obligation to the world. We have an obligation to God. If there are things that are not right, you shouldn't make them right; you *have* to make them right."

Tangier proper, as seen from the northwest. At upper left is Canton, below it the Main Ridge, and nearest the camera, the West Ridge. The building at far right is the sewage treatment plant. (EARL SWIFT)

FOUR

———

N JUNE 1812, WAR AGAIN BROKE OUT BETWEEN THE UNITED
States and Great Britain. Two springs later, a British fleet advanced into the Chesapeake. Finding Tangier strategically located
and easy to defend, the invaders sent troops ashore to the spit—to the
very grove of tall pines that had shaded the Methodist campers—
and built a fort, a command post for a planned campaign against the
region's big targets. The islanders became prisoners: Redcoats requi-
sitioned their livestock and crops, drew water from their wells, and
expected their help navigating the shoals offshore.

As such things go, the occupation was civil—thanks in large part
to Joshua Thomas, who served as the island's chief negotiator with
Rear Admiral George Cockburn and his officers. Thomas walked out
to meet the Redcoats on their first march up the island, and persuaded
them to detour around his neighbors' crops instead of flattening them.
When he learned the soldiers were felling trees on the Methodist
campground, he told the admiral that those were the Lord's trees,
that a great many souls had been saved in their shade and that others
would be in the years to come. Cockburn spared the trees. Thomas and
the admiral got along so well that when Rachel Thomas fell ill, the
British commander gave her husband medicine from his own stores.
He had the rough islander as a regular guest on his flagship.

And Joshua was surely rough. Except in books with "a very limited vocabulary," a clergyman friend wrote in 1861, "he would soon be reading in what, to him, was quite an unknown tongue." His rustic speech was studded with odd verbal tics and tortured syntax, with a-goin's and a-tellin's. His appearance must have been frightful—the British found Tangier a place of "wretched poverty." But he could read the water and nature in general. He listened carefully, spoke the truth, and displayed a wry sense of humor. He was good company.

In August 1814, the fleet weighed anchor to carry four thousand Redcoat troops up the Potomac to Washington, where they sacked the capital and burned the White House. Once back on Tangier, they readied for their next presumed victory: an assault on Baltimore, at the time America's third-largest city, roughly 110 miles northwest of the island.

"I told them they had better let it alone," Thomas recounted many years later; "they might be mistaken in their calculations; for the Baltimoreans would resist them, and would fight hard for their city and their homes.

"'Oh!' said they, 'we can take it easily.'"

In early September, as the British prepared to shove off, Thomas was informed he was to "exhort the soldiers." And so Joshua Thomas stepped onto a small platform before the gathered Redcoat thousands, officers flanking him left and right, and started to shout. He talked about "what made this once good, happy world, so full of evil and misery as it now is; and what brings ruin on men, soul and body," he recalled. "Sin, I said, done all this." He reminded them "of the great wickedness of war, and that God said, 'Thou shall not kill!'"

Then he got down to business. "I told them it was given me from the Almighty that they could not take Baltimore, and would not succeed in their expedition," he said. "I exhorted them to prepare for death, for many of them would in all likelihood die soon, and I should see them no more till we met at the sound of the great trumpet before our final Judge."

This nervy forecast must have seemed far-fetched to his listeners, who'd prevailed in every encounter they'd had with the Americans, at least in this go-round. But they sailed off to Baltimore and proved Joshua Thomas right. The city's chief defense, Fort McHenry, shrugged off the Royal Navy's bombardment. Redcoats on the ground suffered heavy casualties. The defeated attackers limped back to Tangier.

The failed assault might have faded from memory like so much about the War of 1812, except that a young American lawyer and amateur poet named Francis Scott Key happened to witness the nighttime battle and celebrate it in verse. Later set to music, his poem became "The Star-Spangled Banner." Word of Thomas's bold exhortation spread even as the song gained popularity, and he achieved a stature unmatched by any Tangierman before or since.

Alas, no plaque marks the place where he delivered his warning. The site was deep underwater by 1900. Nowadays, it's probably half a mile out to sea. Which brings up one passage of Sugar Tom's book that rings more or less true. "When Captain John Smith first discovered the island it was then thickly set in pine timber of old growth, and the land was much higher than it is now," he wrote. "Whenever a ditch is cut through our lands now we are troubled with pine stumps. And as everybody knows where there is a pine stump standing, well rooted, there was a pine tree there at some time, and I know the land has sunk at least eight inches perpendicular since I was a boy."

He didn't get that exactly right. Sugar Tom was born in 1833 and his book appeared fifty-eight years later. In that time, the island didn't actually *sink* that much.

Even so, he was onto something.

EARTH HAS SEEN its oceans rise and fall over the eons, and the coastal plain of the Mid-Atlantic states has, at various points, been deep underwater or high above it. In the deepest cold of the last ice age, twenty-one thousand years ago, glaciers extended south from the Arctic to cover much of North America and Europe, and one of these glacial masses, the Laurentide ice sheet, covered virtually all of what would become Canada and the northern tier of the United States. Near the shore of Hudson Bay in present-day northern Quebec, the ice was two miles thick.

It's well established that these ice sheets pulverized and gouged the land in their paths, shaping it into new lakes, plains, and valleys, but they also left their mark on land they never touched. The Laurentide was so heavy that Earth's crust sagged beneath it and, in so doing, compressed the mantle, which underlays the crust and forms the bulk of the planet's mass. Part of the mantle is viscoelastic, meaning it's goopy, and when squeezed it acts as a gel is prone to do, flowing away from the point of compression. The pressure from the ice sheet propelled this plastic part of the mantle southward, and wherever it went the crust above had to make room. The ground there bulged, and this uplift was especially pronounced in what is now America's Mid-Atlantic region.

At the same time the ground was lifting, the oceans fell: With so much water locked up in glaciers, seas around the world dropped to roughly four hundred feet lower than they are today. The East Coast as we know it was well inland; the ancient shoreline followed the edge of the continental shelf, which lies forty to seventy miles east of today's Mid-Atlantic beaches and barrier islands.

So most of what we know as the Chesapeake Bay was dry land—a valley, down the middle of which ran the lower Susquehanna River. Tangier wasn't an island. Neither were its modern neighbors to the north, Maryland's Smith, South Marsh, and Bloodsworth islands, and what little remains of Holland. All were part of a continuous ridge that stretched south from the Eastern Shore and formed the eastern flank of the Susquehanna's valley. The future Tangier lay at or near the ridge's southern tip. The possibility exists, therefore, that the arrowheads that Sugar Tom held in his hand, as well as some found by Carol Moore and her fellow proggers, were fashioned by Indians who didn't paddle across the bay to hunt on what's now Tangier Island. They might well have been on foot.

Then, roughly twelve thousand years ago, the ice age ended. The glaciers melted away, and sea levels rose around the globe. The continental shelf off the East Coast slipped beneath the waves. The Susquehanna's lower valley was flooded, forming the long, slender Chesapeake. And the rising water made a peninsula of Tangier's ridge.

To the east, two other rivers flooded to become arms of the bay. What had been the lower Nanticoke River became Tangier Sound, four or five miles wide, lying to Tangier's east and Watts Island's west. The lower Pocomoke River became Pocomoke Sound, which runs between Watts and the Eastern Shore.

The changes wrought by the melting glaciers didn't end there. For with the great weight of the ice lifted from the crust in the far north, the ground there began to rebound and its displaced mantle to ooze back home. At the same time, the bulging ground to the south began to slump. As the crust beneath the bay settled, the Chesapeake inundated the lowest-lying portions of the two peninsulas, then chipped away at the surviving uplands until they became chains of distinct islands. That process, which scientists term "glacial isostatic adjustment," is still under way. The mantle continues its seepage, seeking equilibrium beneath the crust, and the Mid-Atlantic's ice age bulge, subsiding for thousands of years, is expected to continue doing so for thousands more.

Sugar Tom had never heard of glacial isostatic adjustment and lacked the tools to accurately measure it, but he sussed out its essential effect: His island was sinking, and it is still sinking today. The annual rate seems tiny—scientists estimate that subsidence measures about 1.6 millimeters per year at the Blackwater National Wildlife Refuge, not far away on Maryland's Eastern Shore, and doubtless it's very near that at Tangier. That amounts to just an inch every sixteen years.

But even that small loss makes a difference in an island's relationship to the surrounding bay. It amplifies the erosive power of tides and wind-driven waves. Every year sees the Chesapeake soak a little more upland into marsh and drown a little more marsh into open water. Every year sees the shoreline more salty, soggy, and vulnerable to the wiles of nature.

Assuming the rate of subsidence was the same in the nineteenth century as it is today (and it was likely close), Sugar Tom's island sank by about 3.6 inches between his birth and the time he wrote his book. That said, his eight-inch figure wasn't an exaggeration, because even as Tangier sank, the bay around it was rising—something an oysterman of 1891 would not have known. Ocean temperatures were rising, which increased water volume. Ice caps and glaciers were melting, which added new water to the seas. For most of the twentieth century, sea-level rise around the world averaged about 1.7 millimeters per year. Because it has accelerated since the mid-nineteenth century, it probably wasn't as rapid in Sugar Tom's day. Even so, the combination of ground sinking and sea level rising in those fifty-eight years would have approached eight inches.

He may have mangled the island's human history, but he detected and more or less described an ongoing and relentless process that would, a century later, threaten Tangier's end.

BY THE TIME SUGAR TOM published his history, 283 years had passed since John Smith's sighting of Russels Isles. We can only guess what the bay and its islands looked like in the explorer's day, but we do know that the relative sea level was about three feet lower than it was in 1891, and that the Tangier Sugar Tom knew was thus a vestige of its earlier self.

The first carefully surveyed map of the island—that is, drawn with sufficient context and detail that it can be accurately compared with more modern renditions—didn't come along until 1850. But earlier maps do provide general impressions, and the sum of them is this: What Sugar Tom knew as wetland had been dry ground in the seventeenth century, and much that lay underwater likewise had broken the bay's surface, and much less water separated Tangier and Watts Island from their neighbors. John Smith's map from 1612, though so vague that islands are little more than blobs, shows Russels Isles as two side-by-side chains so closely spaced that their origins as peninsulas (or "necks," as they're called in the Chesapeake) seem obvious.

The 1850 map, the work of the federal government's Coast Survey, offers startling insights to anyone familiar with the middle Chesapeake. It depicts a Tangier almost unrecognizably bigger than we know it to be, extending about half a mile farther to the west; the 76th meridian, which on the map runs through the island from north to south, just west of center, today crosses only water, well off the western shore. The West Ridge isn't that much west at all on the 1850 chart, but smack in Tangier's middle.

Uppards and Tangier proper are a single big island. West of Canaan, Uppards extends almost a mile north of the present shore, in a thick peninsula that meets and all but connects with a big island, Goose, at its north end; a trifling few feet of water lie between them. That peninsula has disappeared entirely. Goose Island survives as a squiggle of low sand and marsh, much withered and frequently overwashed.

Canaan appears on the old map, and the chart shows Uppards dotted with other occupied high ground, too: Aces, an oval of upland just to Canaan's west; Rubentown, a long ridge that lay well inland to its southeast; and Persimmon Ridge, a narrow rectangle of homesteads still farther south and closest to Tangier proper. Nothing remains of the settlements, and little of the dry earth on which they stood.

Tangiermen ascribe those changes solely to wind-driven waves, not rising seas, and it's true enough that waves have claimed much of their island's vanished acreage. But erosion and sea-level rise are no either-or proposition: They are inextricably linked, for the bay's erosive power grows as its level climbs. Accelerating erosion is a symptom of a global phenomenon, not a product of local winds alone.

That's borne out in the old map's depiction of Tangier's tidal creeks, which appear as slender rivulets, most of them narrow enough to leap. The Big Gut, which now wriggles clear through the island, did not do so in 1850; it petered out in the marsh. Likewise, Canton Creek, now a wide and straight passage that splits that ridge from the rest of Tangier, is shown on the map as a thread of water that dies halfway through the wetlands. Why? Because the bay has risen in relation to the island, converting wetlands to open water.

The map testifies that about a half mile west of the West Ridge, another hamlet occupied a fourth ridge on Tangier proper. Oyster Creek it was called, and it lasted long enough that old Tangiermen can remember it today. When I stayed on the island in 2000, I spoke with Ooker's father, Will, who recalled the place. "They had pine trees grew over there," eighty-nine-year-old Will told me. "They used to play ball over there. Oyster Creek was a right big place."

Leon McMann, twenty years younger than Will Eskridge, told me he played in the tall grass over that way as a boy, when a couple of houses, not long abandoned, still marked the spot. Jerry Frank Pruitt, a Situation Room regular thirteen years younger than Leon, remembered playing there, too, though by that time the houses were gone, along with most of the trees.

Today, a navigation beacon rises from the bay one hundred yards offshore. The foundations of Oyster Creek's homes lie at its base, under eight feet of water.

OF ALL THE 1850 MAP'S REVELATIONS, the most dramatic might be its depiction of Tangier's now-amputated spit. It was simply massive—a mile-long rib of sand rooted to the island's southwest corner, curving south and east until, at its hook, it broadened to a sandy rectangle covering dozens of acres. The map shows a grove of trees at this widening: the old Methodist camp meeting grounds.

As word of Joshua Thomas's famed sermon on those grounds spread after the War of 1812, the summertime gatherings on the spit became huge affairs, drawing the faithful by the thousands. By 1820 steamboats were carrying campers from Norfolk and the big cities up the bay. "From every creek come vessels of all sizes . . . ," Henry A.

Wise, later Virginia's governor, wrote of his own visit, "loaded with people and provisions, until the island harbors are studded with shipping and a forest of masts, which gives the wharves and island the appearance of some considerable mart of commerce."

Renowned preachers and humble exhorters shouted to the masses from bowers erected among the ghosts of the old British redoubts. The prayerful sang, danced, wept, swooned—and, it was said, witnessed the Almighty's hand. In August 1824, one Miss Narcissa Crippin, nineteen years old and "a zealous Christian," according to a report in the *Norfolk Beacon*, was "so operated on by the Spirit of God that her face became too bright and shining for mortal eyes to gaze upon, without producing the most awful feelings to the beholders."

"It resembled the reflection of the sun upon a bright cloud," related an eyewitness. "The appearance of her face for the space of forty minutes was truly angelic, during which time she was silent, after which she woke and expressed her happy and heavenly feelings, when her dazzling countenance gradually faded and her face resumed its natural appearance."

Better documented were spirits of an unheavenly sort. A motley assortment of hangers-on, not much hungry for the Word, pegged their tents in the sand alongside those seeking salvation. "Here the ministers of the Church winning souls away from Satan, and there the sons and daughters of vanity sipping the siren draught of sensual pleasure in all the ways of wanton delight," Wise wrote of the 1828 gathering; "here, at night, the camp at rest, and all its suburbs drinking, fiddling, dancing, and doing worse, uproarious in shameful frolic until morning light. The night is far spent, and at early dawn the horn is blown. The tents rise again to repeat the last day's scenes and exercises, and the sinners sink away to sleep until the curtain of the night falls again."

By the time the map of 1850 was drawn, the camp meetings were so trammeled by "the spirit of traffic and Sabbath desecrations," in the words of the Reverend Charles P. Swain (later the pastor of the church that bears his name), that they'd become all but useless.

The islanders themselves didn't misbehave, "but others from abroad
brought their wares here for sale, from a watermelon to a boat, and
even whiskey was smuggled in, and, as is always the case, where it
gets in sense gets out, and trouble was constantly being caused."

One August before the Civil War, a gang of Eastern Shore
toughs, riled by talk that a camp preacher was exhorting the virtues
of abolitionism, launched an amphibious assault on the beach. "By
some means used of God the people were made aware of the coming of the mob," Swain reported, and "a band of courageous men so
defeated and banished the devil's cowards as to make them jump into
mud holes and creeks to escape the genteel thrashing that every one
of them received."

Even so, Tangier's passion for the gatherings cooled. Joshua
Thomas died in 1853, and missing his eccentric but firm-handed
presence, the island hosted its last camp meeting four years later. Not
long after that, an Eastern Shoreman gained ownership of the spit
"and built a large boarding house with thirty-five rooms right on the
camp ground," Swain wrote. "The waves as if angry at this innova-
tion began to encroach upon the place until every vestige of the place,
pines and all, has been carried into the bay."

"Thus," Swain concluded, "has God kept unholy hands from de-
spoiling the place."

IN THE LATE SPRING 208 years after Joshua Thomas brought the Word
to Tangier, his spiritual descendants convene for two farewells, bitter-
sweet affairs that draw members of both churches and a good number of the unchurched as well. First comes the funeral of Henrietta
Wheatley, on a Sunday afternoon wrapped in a gloom befitting the
occasion. Rain falls cold and steady from a low ceiling of dark clouds,
ponding on yards and pavement, and the unseasonable chill that has
kept the crabs in deep water, beyond reach of the pots, continues to
grip the town.

I ride my bike to Swain Memorial from my rented rooms on the

West Ridge, a few hours after attending a morning service at New Testament that started with tugboater and church elder Kim "Socks" Parks opening the floor to prayer requests.

"Pray for relief from the flu situation," came from the pews.

"Oh, yes," Socks agreed, nodding. "That's terrible. Pray for the flu situation."

"Pray for Edna and her blood pressure."

"Let's pray for Edna," Socks said. "Her blood pressure's been fluctuating."

"Pray for Wendell," someone else suggested. "Wendell took a fall yesterday."

"Yes," Socks said. "Pray for Wendell. He took a fall to his crab house." Hearing no further requests, he added: "Pray for each other. Pray for our country and the election coming up. Pray for Israel." At which point Ooker, sitting in the front pew, hollered: "Pray for a heat wave!"

That earned a laugh from the congregation but no intercession from upstairs. Henrietta Wheatley's mourners pull up to the church in golf carts shrouded in waterproof tarps, and hurry in their coats for the shelter of Swain Memorial's porch. Inside, the sanctuary rapidly fills, those arriving earliest securing the rearmost pews, a long-standing Tangier habit. We occupy a soothing, convivial space, its high ceiling and walls sheathed in stamped tin overpainted in ivory, its interior illuminated by several large pendant lamps and the daylight washing through stained glass. Left of the altar, above and behind a raised perch for the choir, a large, carved-wood sign implores PLEASE OBEY THE HOLY SPIRIT, a request so disarmingly polite that I was charmed on first entering the church seventeen years ago. Elsewhere hang a big, gold-framed photograph of a rainbow crossing the Tangier sky to land right on the church's roof and a large black-and-white portrait of Charles P. Swain. Two lobes extend from the main sanctuary on its east and south sides—spaces used for smaller gatherings, such as the Sunday morning class meetings that precede regular worship, that can be sequestered from the rest of the room with heavy wooden shutters pulled from pockets in the ceiling. Today they're open, and the side rooms fill even before the rest of the church.

The casket is positioned before the altar and flanked by glass-shaded torchères and an array of memorial wreaths. The family files in—family in this case narrowly defined, for I'm one of the few people in the church who isn't kin to the deceased. The bereaved fill the sanctuary's front three pews as Nancy Creedle, Swain's organist since 1992 and the widow of a long-serving former pastor, plays a melancholy processional that incorporates her instrument's celestial special effects: a choir of angels, chords plucked on a harp, soft chimes.

Richie Pruitt, married to one of Henrietta's granddaughters, offers the eulogy. He opens with a short biography. Henrietta operated an island store for many years. She did not suffer fools well. She had two children, five grandchildren, six great-grandchildren.

"Henrietta gave her life to Christ at the '95 revival," Richie tells us. "I don't know how old she would have been in 1995. I'm sure someone will know it."

"Seventy," comes from the pews.

"Seventy," he says. "To give your life to Christ at seventy, and to have him accept it, is pretty good."

Pastor Flood replaces Richie at the pulpit. "Family and friends," he says, "what we have to understand this afternoon is that Heaven is a real place. It's not something that someone dreamed up. It's not a myth. And that's where Miss Henrietta is today. She is in that perfect house today.

"Heaven is also a place of reunion. Miss Henrietta is having the greatest reunion of all. She's in Heaven with her loved ones. And I can only imagine the reunion that is continuing there and that will go on for eternity."

So ends the brief service. We shuffle back outside to a hopeful medley of hymns from Nancy Creedle and follow the casket north through Meat Soup, the rain having eased to a sprinkle. Some walk.

Most climb into golf carts. The procession stops at a house near the road's end where, in the yard around back, a platoon of graves stands, and we look on from under umbrellas and the eaves of the house as Henrietta Wheatley is put in the muddy ground—and Tangier's buried population, already bigger than its living, grows by one.

A COUPLE OF WEEKS LATER we gather again at Swain Memorial for a different kind of farewell: the annual graduation ceremony of Tangier Combined School. Few island events are more eagerly anticipated, for all Tangiermen know all the graduates—as well as their parents, the circumstances of their births, private details of their home lives, their academic standing, their dating history, and whether they're right with the Lord—to a degree that would horrify most mainland teenagers. Tangier kids aren't allowed many secrets.

The flip side is that many island adults have played some role, large or small, in shaping the class of 2016's six boys and one girl. So on a Thursday evening in early June, Nancy Creedle launches into a medley of "Wind Beneath My Wings" and "Climb Ev'ry Mountain," among other inspirational favorites, to a Swain Memorial sanctuary packed with about three hundred people, nearly two-thirds of the population. Summer has arrived, at long last, and the church air-conditioning rattles and gasps in its battle with the heat. I'm sitting beside JoAnne Daley, herself a member of the class of 1968 and now the matriarch of the family that owns the grocery. "They used to have graduation over at the school," she tells me, while fanning herself with a program. "It was always hot. They'd have to leave the doors open, and the horseflies would get in."

The graduates march to the fore wearing dark blue gowns and mortarboards, and Nina Pruitt, the school's island-born principal and the only resident to hold a doctoral degree, takes the lectern. She points out a vase of flowers on a table below, explaining that in addition to the seven graduates, tonight's ceremony will honor Jordan Wesley Daley, who would have been a member of this class but died unborn in March 1998, when his mother suffered a fatal aneurysm.

I see many in the audience dabbing their eyes. JoAnne, Jordan Wesley's grandmother, sniffles beside me.

We hear from the new superintendent of Accomack County Public Schools, who urges the graduates to "get as much education as you can" and "always give to the community. Give to the old people." Next up: Jared Parks, a 1997 graduate, one of Tangier's two marine policemen, and the son of my landlady. He's a meaty fellow whose overpowering Tangier accent vanishes when he opens his mouth to sing Garth Brooks's "The River" to the accompaniment of a karaoke track. The song has a suitably maritime theme, with talk of rough waters and rapids and such; it peaks near the end with the good Lord at the helm, steering Garth to safety.

All of this is prelude. Nina Pruitt retakes the lectern to introduce the commencement speaker: the school's math teacher, a 1982 graduate, who a few years ago fell victim to an incurable neurological disorder that gradually crippled her, then woke one morning to find herself healed. "Do you believe in miracles?" the principal asks the church. "I present to you *our* miracle, Mrs. Trenna Moore."

Trenna, who is married to Lonnie Moore's younger brother, Tracy, strides easily to the mike. She is tall, slim, short-haired, no-nonsense. "You seven have heard your fair share of my two cents' worth these past five years," she tells the honorees. "I'm glad to be afforded one more chance to offer what I feel are important values as you begin the next phase of your life. But first, I would like to tell the audience assembled here a little about the seven of you." With that, she spends a few minutes talking about each of the graduates, enumerating their fine qualities, downplaying their flaws. "Austin came into eighth grade so quiet that I really could not learn a lot about him," she says of one student. "But in his eleventh grade year, along came Austin. It went from hardly knowing he was in your class to Austin having an opinion and discussing all subjects—home, school, and island gossip."

"Conner treated everyone equal during his high school years," she says of another. "He never just picked on a few. His jokes and mess-ing were to whoever was available. It was the same way with compliments. All have a chance to get a charming word from Conner, from the youngest student to the oldest person to enter the school's doors.

"I could always count on Conner to be truthful, even when it implicated him," she says. "Being known as an honest person is an attribute I admire."

So her speech goes, through all seven, until she shifts her tone. She quotes from the second book of Kings, about a leper, Naaman, who's told by the prophet Elisha that he'll be cured if he bathes seven times in the River Jordan. "It was the Word the Lord gave to me when I was told that I had an incurable disease," she tells us. "The doctor told me he would try to make my life as comfortable as possible for me, but there was no cure.

"This Word was comfort to me, because Naaman had an incurable disease and was not healed right away. God's Word was true then: After Naaman washed seven times in the Jordan River, he was healed. And in October 2014, I was healed. God's Word is true and trustworthy." She chokes back a sob. "The seven of you were a part of that journey, helping me without being asked. The morning I was healed all seven of you came to my classroom and hugged me. I will never forget that day.

"You see, jobs will come and go," she says. "Bosses and professors can let you down. Material possessions will become outdated or break. Your good health will fade in time. And sometimes even the ones you love most can disappoint. But you can always count on Christ to remain true."

I've never heard a public school commencement address quite like it. But then, this is a town with a cross on its municipal water tower.

Moving though Trenna Moore's remarks have been, the main event is still to come, for this is a graduation at which not just the valedictorian speaks, but every member of the class. One, the aforementioned Austin, thanks his parents, adding: "Without them, I literally would not be here." Another boy thanks both his birth father, who died when he was young, and his adoptive dad. A third, the "prophet of the class," offers predictions about each of his classmates: One will become a fighter pilot, and one will advance quickly in his work aboard tugboats, "in record time becoming a port captain." A third will master his computer classes in college, and another will become producer to pop singer Miley Cyrus. Hannah Crockett, at the top of her class, will become a "celebrity psychologist."

Laughter abounds, and tears as well, for the subtext to the ceremony is that Tangier, having raised these youngsters as a village, will soon say good-bye to all but one of them. Two of the graduates are bound for college and one for the military. Three will go to work for Vane Brothers, a Baltimore-based company that runs tugboats and barges up and down the East Coast. Most of them won't be back—or in any case, not for long. Tangier offers no opportunity to put a college degree to use; it used to be that a teaching job would open up once in a while, but with the school's falling enrollment, that isn't likely to happen again. Tugboaters might set up house here, but they tend to weary of the extra time that getting to and from the island adds to their two-week shifts. Those in uniform grow accustomed to the bigger, wider world and having a car, beer, and even privacy— which, scarce though it is in the military, might exceed that on Tangier. Besides, what is there for boys to do on the island but the work of their fathers, which they know to be hard, uncertain, and dangerous?

Nina Pruitt returns to the lectern. "As you leave us tonight," she tells the graduates, in what sounds like a bon voyage, "I ask one last thing of you. Take pride in your Tangier heritage. Be proud to be called a Tangierman.

"As a teenager you think of the place as too confining, like living under a microscope," she says. "But soon you will come to realize what a special place home really is. Take with you the strong traditions of hard work and devotion to family that are so evident on our island. These will take you far in life."

The graduates file out to Nancy Creedle's rendition of "Onward, Christian Soldiers" and form a receiving line on the church porch. We well-wishers slowly squeeze out of the sanctuary. As I near the doors I find myself beside Annette Charnock, a native who left for college and stayed away for decades, until she found love in her hometown and moved back nine years ago, and whose stepson is married to Carol Moore's daughter. Hair cropped in a no-fuss boyish style, Annette is bright-eyed, quick to laugh, and just a touch louder than most of her fellow islanders. "Oh, that was a tearjerker of a ceremony, weren't it?" she says, to no one and everyone around her. "Just when I stopped crying over one story, along came another."

Crab shanties line the main channel into Tangier harbor. (EARL SWIFT)

FIVE

T WAS ANNETTE CHARNOCK'S FUTURE HUSBAND WHOM I SOUGHT out one afternoon during my 2000 stay for insight into the blue crab. By that spring Edward Vaughn Charnock had been working the water for nearly forty years and was reputed to be one of the most capable crabbers on the island—an expert seaman, an aggressive potter who ranged far from the island in pursuit of hard crabs, and an honest businessman who overfilled his bushel baskets until the lids bulged, a practice islanders call "giving good measure."

I found him outside his house in Meat Soup, preparing his pots for the coming season. He clipped bars of zinc to their mesh to slow its corrosion in the bay's salt water, freshened the paint on his buoys, and replaced the frayed line that connected them to his traps, soaking each length in a muddy puddle before cinching its knots closed, for better grip. He looked like a character out of Coleridge, square-jawed and sun-chapped and gristly, a man who'd come to understand the sea and his quarry and who doubtless had much to share.

But no. Ed claimed ignorance. In all the time Tangiermen had devoted themselves to hunting the blue crab—and collectively, that amounted to thousands of years—what they'd learned about the creature, he told me, wouldn't fill a coffee cup. "Only two things known for sure," he said. "She'll run from you, and she'll bite."

I've since heard slight variations on this thought from other watermen, who lament how mysterious the animal remains despite generations of careful observation. And indeed, the island's crabbers will
disagree about aspects of the crab's life and habits and trade in some
pretty serious misinformation about them, too. That said, Tangiermen
know a great deal about the animal, starting with that bite: If circumstances require that you be pinched by a crab, see that it's the left claw
that gets you. At first glance the crab might appear symmetrical, but
its claws differ slightly. Both are lined with what resemble nubby teeth,
but those on the left claw are smaller and finer, and the claw itself is
slimmer and comes to a sharper point. That's because the left claw is
designed for cutting, while the right claw—stouter, with teeth more
reminiscent of molars—is built for crushing, and is far more powerful.

A Tangier crabber knows this from childhood, though he won't
likely pay much mind to the claws on a crab bound for a bushel basket. He won't devote much thought to the tender nuances of crab
courtship, either, though he uses his knowledge of the ritual to snare
female peelers. Flip a crab onto its back, and it presents a second clue
to its sex, in addition to the color of its claws. In the center of its belly
is a hinged flap, called an apron, under which its reproductive tackle
is located. On jimmies, it's a long, narrow panel that's often likened
to the Washington Monument, and on immature females, or she-
crabs, it's roughly triangular. Twelve to eighteen months after hatch-
ing, when a she-crab is nearing the molt at which she'll achieve sexual
maturity, her apron acquires a pink or bluish hue.

In Tangier parlance, this female peeler is a "doubler," and of great
interest to jimmies in the vicinity. She's attracted to them, too, for
a female mates only once, and only while a softshell. Because most
don't molt again after this key transition from she-crab to sook, the
coming days will provide her only shot at reproducing.

As invertebrate romances go, a crab's is rather touching. A jimmy
performs a dance for the doubler. If she deems him a worthy mate,
she backs beneath him, and he pulls her close to his belly with his

walking legs. For the next several days, he cradle-carries her wherever he goes.* When she's ready to molt, he forms a protective cage around her with his legs, and stands guard for the hours it takes her to work free of her old shell and fill out the wrinkled new one. Externally, the chief clue that she's reached adulthood can be found, again, on her belly: Her apron has changed shape, from a triangle to the Capitol dome. Once she's plumped up to her new, larger size, the jimmy flips her onto her back. His apron opens to expose a pair of long, wispy tubes called gonopods. Her new apron opens to expose a pair of receptacles called gonopores. They mate belly to belly for five to twelve hours. Afterward, the jimmy flips her back over and again cradles her for a couple of days, until her new shell has hardened. Then they part.

Crabbers know that as a general rule, peelers won't enter a pot where hard crabs are present. They're seeking refuge, and know that to molt in the presence of hard crabs is an invitation to death by cannibalism. If a pot contains both hard crabs and peelers, one can safely assume that the peelers entered it first. But peeler potters know, too, that a doubler feels a heightened sense of urgency to mate. So while the waters around Tangier are crowded with the prospective brides, it's common practice to toss a jimmy into a pot before setting it; a female peeler might sense his presence, the thinking goes, and head into the pot to get acquainted, her urges having overtaken her fears.

Most years, eligible females engage in one or more early-season runs past the island, and at such times the peeler catch can be almost ridiculously good—stories abound of Tangiermen finding dozens of doublers in a single pot or catching hundreds in a day of dipnetting. But the runs didn't happen last spring, and they haven't happened this year, either, and no one seems to know why.

That'll get a waterman talking about how much he doesn't know.

---

* Just for confusion's sake, this pairing is also called a "doubler" throughout most of the Chesapeake region. On Tangier, however, the term most often refers to the female peeler.

AFTER MATING, THE sook begins a long migration to spawn. It often takes her until the following year, after a winter's hibernation, to complete the journey. As she nears the salty water near the Chesapeake's mouth, she fertilizes her eggs with the sperm she's carrying. An orange sponge billows from her belly.

If there's an aspect of crab life about which some Tangiermen know as little as they claim, it would be that sponge. When I've wondered in conversation whether crabbers are hurting themselves by catching lemons—which are, after all, carrying the next generation of crabs—I've had crabbers insist that I need not worry, as those orange sponges are unfertilized eggs that will never hatch. This is untrue: The sponges cannot form without fertilization. A lemon in a crabber's basket will never see her eggs hatch, true enough, but that's because she's in the basket.

Assuming she lives to spawn, a sook can fertilize another batch of eggs with the leftover sperm inside her, which remains viable as long as she does. If she lives long enough, she can theoretically spawn three, four, even five times with the fruits of her single mating. Few sooks survive in the Chesapeake Bay to produce more than two sponges, however, and on another early morning at sea, I keep company with one of the reasons.

Ooker Eskridge guides the *Sreedevi* to a pot off the spit's east side, wondering, as he is wont to do, what's happened to the doubler runs of old. "They've moved off into deep water," he has theorized. "Used to be two crab runs every spring, with the females. Then it got to be that you only got one run in the spring. Then it got to be that we didn't even get that, hardly, except up near Smith Island.

"The crabs are shedding, but we didn't have any doubler run that we could find. It must have been off in deep water where you couldn't see it."

Two hours after daybreak, the water is slick calm, the air breezeless and leaden with humidity. The stillness feels weirdly ominous, for off to the south, twenty miles away, a curtain of black descends to the water and strobes with lightning. He leans for a buoy with his hook, pulls a pot aboard, and shakes its contents loose. Alongside several crabs, which Ooker plucks from the culling tray and tosses into their respective baskets, writhes a seahorse. He hands the creature to me. It's sheathed in a muddy slime and jackknifes in my palm. I admire it for several seconds, stroking its thorny skin with a fingertip, then lob it overboard.

Ooker's own courtship followed the course of many island pairings, which are only slightly less predictable than the blue crab's. He'd already dated a couple of mainland girls who visited in summer when, in the tenth grade, he turned his attention to Judith Irene McMann, a ninth grader. Neither was much of a mystery—they'd known each other their entire lives. She was quiet and pretty, with bright eyes and a dazzling smile. He was garrulous and long-haired; his eleventh-grade yearbook photo brings to mind a surfer, and his crinkle-eyed smile has the whiff of a stoner about it, but in reality he was a straight arrow. ("I've never smoked marijuana," he told me. "Never been drunk, either. Not even tipsy.") They were distant cousins—her great-great-great-grandparents were Ooker's great-great-grandparents. On Tangier, however, courting a relative is close to inevitable, and their family ties were looser than many.

They danced, cruised around in skiffs, visited the beach. They hung around with other teens at a couple of places with jukeboxes and pool tables, long since closed. They'd visit with each other in a "dating room" at the front of her house in Meat Soup. They'd go on long walks, taking care not to linger on the Heistin' Bridge, where—as many older Tangiermen will tell you—a girl's likely to "get a bad name."

He was eighteen when he married Irene in April 1977, two months before she finished high school. "A light comes on, and you just get the feeling, 'I think this is the one,'" he tells me, unable to further explain how the romance bloomed. "Irene played hard to get for a while. I tell her now, 'I know you wanted me. It was hard for you to do that.' And she'll say, 'Right.'"

By then, he was skippering his own boat—a wooden round-stern deadrise, the *Judith Irene*, powered by an inboard diesel engine. A deadrise is the classic Chesapeake Bay workboat, from thirty to fifty feet long with a broad beam and a simple, small cabin positioned up near the bow. That it had a round stern set the *Judith Irene* apart. Then, as now, most Tangier workboats had squared-off tails, or box sterns. But all deadrises share features suited to hauling pots aboard: a long, open weather deck aft of the cabin, stretching two-thirds of the boat's length and giving the crabber room for his catch; a steering console in the cabin and a second outside, on the starboard side, from which a man can control the boat while he leans over the side to tend the pots; and a swooping profile that culminates in a low freeboard at the stern, meaning that the sides of the hull drop to within a foot or two of the waterline, minimizing the vertical distance he'll have to lift his catch.

Ooker kept the *Judith Irene* for seven years, then traded it for the smaller, simpler *Sreedevi*, twenty feet long and cabinless—essentially a fiberglass tub with a point at one end and a ninety-horse outboard on the other. Unassuming though it is, it's Tangier's most photographed vessel. Thanks to Ooker's twenty-plus years as the island's mouth-piece, the little boat and its signature details—a Jesus fish and a Star of David hand-drawn in permanent marker on its steering console—have appeared in newspapers, magazines, and documentaries around the world.

WE'RE HALFWAY THROUGH a lucrative workday, with several bushels of hard crabs and a half bushel of peelers aboard. This, despite the appearance in the water of what Ooker calls "red moss," a feathery pink-brown algae that for the past several summers has clotted the water of Tangier and Pocomoke sounds. It fouls many a crab pot, its long strands snagging in the hexagonal mesh and sometimes forming a tangled mass so thick that the pot might as well be solid.

"It gets long, like brown hair," he tells me. He pulls up a pot bearded with the algae. "Growing fast," he murmurs. "The older the pot, it seems like, the easier it attaches. I made that pot, I'd guess, three years ago." Fashioning his own pots distinguishes Ooker from most Tangier watermen, who buy them ready-made. Of the 210 he has in the water, more than a third are replacements he made over the winter. He buys the heavy wire frame that forms each pot's base and sides and fashions the rest of the cube with a galvanized mesh that's sold as "saltwater netting." The effort costs about half as much as a store-bought model.

As we proceed down the row, even newer pots seem to have attracted the algae. All seem to be fast approaching the point where crabs will avoid them: A little red moss does no harm, Ooker tells me, but a lot will scare them off. Plus, he says, the algae adds to a pot's weight and increases its resistance as he pulls it through the water—and when you're pulling a couple hundred aboard by hand every day, that's an issue.

What's prompted the red moss's appearance? Well, add that to everything else watermen don't know. "Something in the water's causing it to grow," Ooker guesses, shrugging. That hypothesis, vague though it is, is shared by John Bull, who heads the Virginia Marine Resources Commission (VMRC), the state agency that regulates the bay's commercial crab, fish, and oyster harvests: "It seems to be a water quality issue," he told me.

Ooker pulls up another pot. Its line is draped with the stuff. "If it gets much worse, I'll pull a row and set them on the beach for a day or two," he says. "Put them in the oven." Inside the pot, a small perch lies gulping on its side, surrounded by hard crabs. Ooker dumps the load and picks up the fish, studying its belly, rubbing a gloved finger over it. "Lots of eggs," he says. "Lot of roe in him." He throws the fish back. In the next pot, a soft crab crouches among hard crabs. Ooker picks it out of the trap before dumping the others onto the tray. "He lucked out," he says, carrying the softshell to a live well filled with water at the bow and dropping it in. He adds: "I guess."

As he's culling a big jimmy, a sook grabs it and hangs on. Ooker gives the jimmy a firm shake. The sook loses the claw and drops to the tray, her dismembered limb clattering down beside her. Its pincers open and close of their own accord. "Look at that!" I cry.

"Oh, yeah, they'll keep moving," Ooker says. He picks up the claw, which continues to grab at the air. "Still got some bite in 'em."

We near the end of a row and draw within hailing distance of Leon McMann's boat. Leon, wolfing down his lunch in his tiny cabin, looks our way and waves. The *Betty Jane II* is small, as work-boats go—a design called a barcat, which shares a deadrise's grace-ful profile but is far less substantial a craft, with a snug open deck that rides much closer to the water. This is a boat made for protected shallows, rather than open seas—and not for potting but for crab-scraping, a method of catching peelers that has fallen out of fashion over the past twenty years. On a line off the stern, it pulls a device reminiscent of a lawn mower bag: a steel frame with a rope net trail-ing behind. Running slowly, at just two or three knots, Leon drags the scrape over the bottom, scooping anything resting there into the net. Every few minutes he hauls in the line, lifts the full scrape out of the water, and dumps the load onto a culling board.

We watch Leon finish his meal and shuffle back to the stern. He completes a lick, or run with the scrape deployed, and pulls up the line hand over hand. The bag is loaded with mud, seaweed, broken eelgrass, and a bundle of crabs. Almost everything goes back over the side: Unlike a potter, a scraper can't keep a by-catch—he's bound by law to return everything but peelers to the water—so most days, scraping doesn't net as big an overall catch.

But it can yield plenty enough peelers. Small as it is, a barcat can get into waters too shallow for a deadrise, so scrapers have some territory to themselves. And if the day isn't going well in one loca-tion, they need only pull their scrapes aboard and motor off to a new spot—which, compared with the daunting logistics of moving pots around, gives them an enviable agility. Scraping offers other advan-tages: It's cheaper to take up, as a couple of scrapes are far smaller investments than hundreds of pots. Its pace is less frenzied than potting. Scrapers have a few minutes during licks to breathe deep and look around. Plenty of time to cull carefully. And because they hug the shore, they're usually less exposed to disagreeable weather. On a pretty day, the work can seem almost relaxing.

Plus, there's the allure of a scavenger hunt about it, because there's no telling what might come up in a scrape: sharks, stingrays, terrapin, shrimp, and other creatures that won't venture into a pot; relics of bygone days, such as crockery and bricks, cutlery, pieces of boat and machinery, and, on one occasion, an unexploded bomb—which Leon, who caught it, recognized as "right dangerous." And every now and then a catch too big to lift, as I heard him describe in the Situation Room. "I caught a big ol' pole," he announced that afternoon. "It was probably this big around"—he spread his hands ten inches apart—"and ten feet long. There are all these big, strong bucks out there could pull that up easier than me, but I'm the one that got it."

"That must have been heavy," Bruce Gordy said.

"Heavy? It was like it was pulling back," Leon said. "A *big* pole. I believe it was laying on the bottom 150 year."

Now we watch Leon finish his cull as his scrape makes another curving lick along the bottom. "Not many doing it anymore, but I've always liked scraping," Ooker says. "I try to do it for a couple of weeks late in the summer. I like having all the gear you need right there on the boat. I like that you can go wherever you please. And you don't get as tired, because just standing up will wear you out, when you're potting—the water's rougher, and the boat moves around so much.

"The big disadvantage to it," he says, "is with scraping you're not catching anything until you throw the scrape overboard. With pots, you're catching crabs while you're at home, sleeping."

The row completed, we wave good-bye to Leon and head for Ooker's pots on the island's bay side. Rather than cut through the harbor, he opens the throttle and we skim north around Uppards, the boat's nose high, the flatwater smooth under the hull. On the island's far side, the first of his pots waits about two hundred yards from the boat channel's western entrance. As Ooker pulls it aboard,

I see that it's draped in red moss, and ask whether he thinks the algae scares off crabs because it helps them recognize a pot for a pot. "They *are* clever," he answers. "They're good at figuring something out." He dumps the pot, which contains a handful of hard crabs, including a big jimmy. Tangiermen have an expression about large crabs like him: "Don't take many to make a dozen."

"They're smarter than most fish," he says. Their canniness, combined with good eyesight, makes them formidable. Many a crabber has learned that using a holey glove will bring him pain, because a crab will soon recognize and exploit the weakness. "There can be a small hole in your crab pot, but the crabs will find it. They'll search and search and search until they find a way."

I notice a movement in the water off to starboard. *Callinectes sapidus* translates to "beautiful, savory swimmer," and now I see that the crab earns every part of the moniker: A jimmy, with a doubler clutched to its belly, is swimming sideways past the boat a few inches below the surface. Propelled solely by the jimmy's rear legs, the pair moves with surprising speed and grace.

COMES A THURSDAY in mid-June when much of the island again crowds into Swain Memorial, this time for the wedding of Lance Daley, scion of the family that owns the grocery store, to island native Erica Parks. In the week since graduation I've encountered Annette Charnock at church, at Daley & Son, and on the road in Meat Soup, prompting her to comment, usually with a cackle, that she can't seem to get away from me. As I approach Swain's porch, here she comes again, smiling and moving fast on the arm of an urbane-looking fellow in a sharp gray suit, expensive-looking tie, and dress shoes. It isn't until Annette introduces him as her husband that I recognize him as the waterman I interviewed sixteen years ago.

Once inside we split up, and I find my regular seat in a rear pew. A prerecorded rendition of Pachelbel's Canon in D issues from the speakers. Pastor John Flood and Duane Crockett, the New Testament elder, take up positions at the altar, along with the groom. Lance

Daley wears a gray tux that contrasts with his deep tan. His neck is tattooed with linked hearts, one labeled "J.W." for Jordan Wesley, the unborn member of the class of 2016, who would have been his little brother, and the other, "Mom," for Stephanie Crockett Daley, who died with her baby.

The bride crosses the church in a sequined, spaghetti-strapped ivory gown, hair piled high. Pastor Flood greets those gathered, then turns over the service to Duane, who teaches history at the school. He's also the groom's uncle—Stephanie Crockett Daley was his sister. Duane reads from the fifth chapter of Ephesians: "Wives, submit yourselves unto your own husbands, as unto the Lord. For the husband is the head of the wife, even as Christ is the head of the Church.

"In our society today, these Scriptures are not very popular," Duane allows. "It seems that we cannot get past that wives are to submit to their husbands. We get an image of a slave owner bossing someone around without regard to their feelings and desires." In his island delivery, "regard" comes out *regoward* and "desires" as *dizoyers*.

"Our world sees no problem in parents being responsible for their children or employees being loyal to their employer," the unmarried Duane says, "but the notion of a wife submitting to her husband seems archaic and ancient.

"These verses are not suggesting for a woman to become a doormat to anyone," he argues. "As I read and study these verses of Scripture, the bulk of the responsibility is placed on the man. He is told to love his wife as Jesus loved the church and gave his life for the church.

"Whenever a woman sees a man love her like that, I don't think she will have any problem in submitting—to her husband loving her so much that he is willing to die for her."

Pastor Flood announces that the two have written their own vows, and we all strain to hear them exchanged. I catch only a few words of Lance's: "Erica, you are the most amazing woman I've met," he tells his bride. "I can't wait to spend the rest of my life with you." The pastor calls for the rings, oversees their exchange, declares the couple legally bound. "Ladies and gentlemen," he says, opening his arms wide, "it is my pleasure to present to you Lance and Erica Daley, husband and wife." The church erupts in whoops and applause.

The reception is in the school's combination cafeteria and auditorium, fifty feet square with a high, exposed-truss ceiling and a cage over a big wall clock near the door. The room's edges are draped in a gauzy white fabric, through which white Christmas lights blink. A banquet table offers appetizers: tortilla chips and cheese sauce, a chocolate fountain and little cubes of pound cake, a veggie tray. I ladle myself a glass of nonalcoholic punch, the only liquid on hand. Everyday life can be challenging on a dry island, I reflect, but it's an entirely more rigorous test to get through a dry wedding. As I retreat from the table I again encounter Annette and Ed, who are sitting at an otherwise empty round table near the stage. "Mind if I sit with you guys?" I ask.

"You come sit right here," she answers.

Annette's own first wedding is well remembered on Tangier—first, because she was due some happiness after suffering a teenage tragedy. She'd been sweethearts with Warren Eskridge, and they planned to marry when he returned from Vietnam. "Warren had a scooter, a green Cushman," she's told me. "He put my initials on it when we started going steady: MAP, for Mary Annette Pruitt, on the gas tank. I thought I had arrived.

"I thought I was destined to spend my life on Tangier," she said. "I was going to get married and stay here, and I was very contented about that. And then he died at the end of my junior year."

Instead of staying put, Annette left in 1970 for what's now James Madison University, in Harrisonburg, Virginia. She met the first man she'd marry there, which brings up a second reason her wedding is firmly planted in the collective memory: The ceremony occurred while a photographer for *National Geographic* was on the island. The magazine's November 1973 issue featured a splashy image of her in her gown, mobbed by friends and family outside Swain. She was home from college at the time, and settled in the Shenandoah Valley with her husband after graduation.

Then, after thirty-six years—during which she taught first grade, raised two sons and a daughter, and divorced—friends on Tangier set her up with Ed Charnock, whose wife, Henrietta, had died in 2005. To hear Annette tell it, it was a tentative, almost tortured courtship, because Ed proved shy to the point of paralysis. But, evidently, she pulled him out of his shell: By their second or third face-to-face date, in December 2006, they were talking about marriage. "He said, 'Well, you know, with my job, it's got to be a certain time of the year,'" Annette recalled. "So we talked about the seasons. He said it could be after [winter] crab dredging and before spring crabbing, or we could wait until the end of the summer. And we both said we didn't want to wait that long. So that's what it was—we got married at the end of February."

Now I take a seat beside Ed and make conversation by asking him to identify islanders around the room. Ed's pushing seventy, but despite a face lined by sun and salt, he comes off as younger. He's a tall, powerful-looking character. His replies to my questions are friendly but minimalist, which brings to mind something Jerry Frank Pruitt told me in the Situation Room about Ed's father, Vaughn Charnock. He'd launch into a story, "get halfway through, then stop and say, 'There ain't no need to talk any more about it.' People would say, 'Well, Vaughn, aren't you going to finish your story?' And he'd shake his head. 'Ain't no need to talk more.'"

A disc jockey from the mainland has set up in the room's northwest corner, in front of a freestanding display of flashing lights. As Coldplay thumps from his array of speakers, we queue up to go through the dinner buffet of lasagna, Caesar salad, bread, pudding, and sweet tea. It's only when we're nearing the food that I realize I've seated myself at the grandparents' table—now, suddenly, I grasp that Erica is Ed Charnock's granddaughter—and I see in the staring eyes of many in the room that I've coat-tailed my way to an early shot at the lasagna. I'm relieved to get back to the table and resume talking with Ed. He remains a man of few words, but he smiles readily and has a dry, self-deprecating sense of humor. Annette compensates for his bashfulness. The meal passes quickly.

The speeches follow, all of them short and unabashedly sincere: "Me and Lance have been best friends all the way since kindergarten," the bulky young best man tells us. "And I just want him and Erica to know that if they ever need anything, I'll be there for them." A groomsman catches the garter. A bridesmaid catches the bouquet. Lance dances with his grandmother. The bride and her father dance to Aerosmith. The cake, shaped like a ziggurat, is excavated from the top down.

The DJ spins several dance tunes for the younger islanders before announcing that "we're going to switch things around, generation-wise," and "Mony Mony" by Tommy James and the Shondells blasts from the speakers. A squeal goes up, and a crowd of middle-aged island women materializes on the dance floor, Annette in its center and twisting with abandon. "Look at Netty!" someone hollers.

I look over at Ed, who like every waterman in the room has kept his seat. He's leaning back in his chair, face windburned red, watching his wife with an expression of incomprehension and delight.

PART TWO

THE LORD
TELLS THE WATER

Ooker holds a soft crab, top, and the empty shed from which it has just emerged. When its new shell fills out and hardens, it will dwarf the old. (EARL SWIFT)

SIX

L ATE ONE STEAMY MORNING I STEP ONTO THE DECK TO FIND the heat wilting, the sun intense, and I'm thankful that I'm not aboard one of the deadrises I can see out on the bay, working their pots. While watching them I sense a movement out of the corner of my eye and focus my attention there—just in time to see a finned arc break the surface, and a moment later and a short distance away, another. It's a pod of bottlenose dolphins, swimming at a languid pace up the island's western shore.

Dolphins are routine sights farther down the Chesapeake—when I lived on the beach in Norfolk in the 1990s, I saw them swimming just beyond the surf several times a week—but in recent years scientists have come to find that great numbers of the brainy, social creatures are spending time in the bay's upper reaches, too. Pods have been sighted up the Potomac River, farther north on Maryland's western shore at Solomons Island, even up near the Bay Bridge above Annapolis. One scientist figures there might be as many as a thousand entering the bay each summer. Another needed just two summers to identify (and name) five hundred individual dolphins from a boat in the Potomac's mouth.

Whether this is a new phenomenon or the product of increased human awareness remains a question, but some experts suspect the creatures have adopted the bay as a summering place, perhaps lured by its abundance of croakers and rockfish. Others hypothesize that the dolphins' regular food stocks are moving in response to the bay's warming—which it appears to be doing slowly but steadily, to the tune of more than 2.5 degrees Fahrenheit since 1960—and the dolphins are following the food. Exciting as they are to see, in other words, the dolphins might signal little-understood shifts in the bay's ecosystem, and those shifts are not likely to be for the better.

Whatever the case, I follow the pod through binoculars for several minutes before deciding that I want a closer look. The best place to get one is the slab, a concrete wharf just beyond the dump. I jump on my bike and race up the West Ridge and around the runway. It's been a while since the dump was picked up and barged away, and trash is piled high at the roadside: a washer, several discarded microwaves, and an old oil tank; empty propane canisters, a satellite TV dish, the skeleton of a motor scooter, and a stripped motorcycle frame; a couple of barbecue grills and several rusted lawn mowers, which on Tangier are invariably called grass-cutters; a queen-sized box spring; bald or lacerated tires of various size; played-out window air conditioners and space heaters; a house furnace, dented and gouged, and rolled and rusted chain-link fencing; pieces of toilets and bikes and golf carts; and a large cathode-ray tube from an old TV.

Beyond, the slab is dominated by enormous dumpsters filled to overflowing with smaller junk, and the concrete is littered with discarded oil bottles, ruined crab pots, and the plastic floats and metal cages used in oyster farming, a fledgling Tangier business. The rusting steel bulkhead at the slab's edge thus makes for an unsightly vantage point, but it offers a clear view west and north, and if the dolphins continue on their course they're bound to pass close by.

Unfortunately, there's not much of a breeze, and it isn't long before the greenheads find me. The Chesapeake's summertime scourge is just a half inch long, but judging from its bite, most of it is teeth. "Bad as a German shepherd," I've heard Leon McMann say, and, on another occasion, "Like a Doberman pinscher," and for the misery the insect delivers, those are only mild exaggerations. Its yen for blood is so pronounced that after tearing a divot out of an arm, leg, or neck, it will remain feeding on the bite, looking you in the eye, even as your hand descends to kill it. A single greenhead is a formidable adversary. A swarm is hellish. And here they come—one, now three, suddenly five—until slapping, cursing, I lose all interest in the dolphins and pedal hell-for-leather back out past the dump and down the West Ridge, desperate to get behind a screen.

WHEN I MENTION TO OOKER that I saw the dolphins, he tells me that he's seen whales venture this far up the bay—he saw one, he says, down by the offshore Tangier Light, and watermen saw a couple of humpbacks this past spring while out tending their pots. And one summer, he adds, his son Woodpecker found a dolphin tangled in crab pot lines. "He freed him," he says of the hard-charging water-man. "So he has a little bit of me in him. The rest of him is Leon."

We're out at Ooker's crab shanty after a long morning on the water. Animal rescue has figured in our conversation throughout the day: Hours ago, after dumping a pot onto the *Sreedevi*'s culling board, Ooker picked a jimmy from the tangle and announced it was blind. "See his eyes?" he asked me. "They're black." And so they were—the crab's stalked eyes, normally the grayish hue of the animal's belly, looked like burnt match heads. He waved a hand in front of its face. It didn't react. "I always throw them back," he said. "I figure if he's struggled this long and made it this far, I'm going to give him a break.

"My wife will hear that and say, 'You did *what*?'" he continued. "James will tell me, 'I don't believe you.' They throw a lot of terms at me. 'Weird.' 'Unique.' What's 'eccentric'—does that work, too?"

"It does," I said.

"James, he'll say, 'How do you know they're blind? I've never no-ticed a blind crab.' I say, 'Well, you're catching them, too.'" Ooker chucked the crab over the side. "I'll do it even with a big jimmy," he said. "I'm gonna let him go."

Once, he told me, he was crabbing off the P'int and came across an injured sea turtle. "It had a big gash on its head. It was having a hard time staying on the surface—it kept going down, and I could see it was struggling to get its head back up out of the water. So I pulled it aboard the boat." Once ashore, he called the Virginia Aquarium and Marine Science Center in Virginia Beach, which is home to a rescue lab. "They said, 'Well, we're going to have to call the feds,'" Ooker recalled. "'It's against federal law to be in possession of a sea turtle.'

"They came and got the turtle, but I got a call from NOAA," he said, referring to the National Oceanic and Atmospheric Administration. "This guy tells me, 'We'll let you off this time.' I said, 'Oh, you'll let me off, huh?'" He glared at me and rubbed at his mustache with his wrist. "Can you believe that? Like I'd done something wrong. If I hadn't pulled that turtle aboard my boat, it'd be dead.

"So anyway, he says, 'Yes, this time. But if this ever happens again, you can't touch the turtle. Just let us know where it is.' I said, 'If this ever happens again, you'll not be hearing from me. I'm through with that. I don't care if the turtle calls out, "Ooker! Ooker! Help me, Ooker!" I'm turning my head.'"

I told him I doubted he'd really do that. He shrugged.

We're surrounded by animals at the shanty. A large seagull is eyeing us from its perch on top of a piling. "That fellow there, that's Yellow Bill," Ooker tells me. "He hangs around." He points out a smaller laughing gull, which he's named Summertime. Both have made a home at the shanty for several summers straight and often fly out to his boat while he's crabbing. "Normally, Yellow Bill would eat right out of my hand," he says, "but if he notices there's somebody else here, he gets a little leery."

Two cats prowl the deck around us, another is balanced on the edge of a shedding tank, and a fourth animal is lounging inside the shed. Ooker rattled off their names on my first introduction to the creatures in May, adding, "They're a conservative group." Six weeks later, I know them well: Sam Alito, a fluffy smoke-gray cat, sturdily built and hungry for attention; big, thickset John Roberts, sheathed in glistening black fur; Condi Rice, also black, but slightly smaller and missing the tip of an ear; and Ann Coulter, runtish and starved looking, her black coat patchy.

"They've been out here about twelve years," Ooker says, shooing John Roberts from the tank. "There was a storm, a nasty one, and I was out here. There was a tree stump adrift with these kittens on it, so I went out and got them." He fishes several discarded sheds from a tank with a small hand net and flicks them into the water below. I watch them fall. With a splash they join hundreds of empty exoskeletons on the harbor's bottom. "My wife wouldn't let me bring them home," he says, "so they've been out here ever since. They've grown up with wood under their feet."

IT'S AT HIS shanty that Ooker demonstrates most impressively the deep knowledge and understanding of the blue crab that watermen gain over the span of their careers. Out back of the hut stretch three rows of tanks, plywood tubs four feet wide, eight long, and containing six inches of water pumped from the creek below and draining back into it. Ooker has eighteen tanks in all, though at this point in the season, he's running only half.

Years ago, before peeler crabbers took to using pumps to circulate water over their catch, their tanks were built with slatted sides and floated in the creek itself, and the natural action of the currents and tides kept the water moving within them. "Floats," they were called then, and even today some older watermen hang on to the term. A crabber typically arranged his floats in a chain or an open square and bustered up his catch from a small boat. It was hard on the back, but there are some who say keeping the crabs overboard caused the animals less stress than modern tanks, and kept more of them from dying in captivity.

I watch as Ooker sorts through the peelers he caught today, depositing them in tanks according to the time remaining before they'll shed. He makes this judgment quickly, with an inspection of their swim fins. One edge of the limb changes color; it develops a whitish tinge at first, then turns pink, and, just before a crab molts, deepens into red. Or so he tells me. No matter how many crabs he shows me, I fail to detect the telltale red he insists is there and plain to see. Peering into the tanks, I cannot spot soft crabs among the peelers, either, something Ooker can do with a glance. "How do you do that?" I ask him. "What do you look for?"

He thinks for a moment. "They have that soft crab look about them," he says, but he's unable to offer any specifics, except that their color is slightly off. Other Tangier crabbers have trouble explaining the process as well. One afternoon Ooker was showing a wedding party around the shanty. The groom, Allen Parks, born and bred on Tangier, was pointing out crabs at various stages of their molt to his guests, who were mainlanders—and unaccustomed to island pronunciations. "That one's soft," Allen said, pointing to a just-busted crab, "and that one's *howard*."

How can you tell that's Howard? a wedding guest asked.

Just by looking at it, Allen told her.

But what about him tells you it's Howard?

The way it looks, Allen replied.

Do you give names to all the crabs? she asked.

What? No, no, Allen said. It's *howard*.

One rule of thumb, only so reliable, is that the soft crabs tend to be bigger. Peelers are far smaller than the hard crabs Ooker finds in his pots; the state requires that they be just 3¼ or 3½ inches wide, depending on the time of year, versus the 5-inch minimum for hard jimmies and sooks. It's not until they've molted, and filled out their new and still-soft shell, that they achieve a respectable size. "People who don't know what they're talking about will say, 'Well, why do you catch such small crabs, those peelers?'" Ooker says. "It's because you have to catch a small peeler to get a large soft crab."

If a peeler's already big, it's almost surely a male. After mating, jimmies continue to molt, getting bigger all the while. Sooks can but rarely do, Ooker says: In all his years on the water, he's caught only eight or nine already-adult female peelers. As doublers, they're thus often said to be approaching their "terminal" molt.

The tanks closest to the shanty are reserved for busters, those crabs in the process of actually backing out of their old shells. It's a delicate stage: In the wild, Ooker tells me, a molting crab rarely "hangs up" in its discarded exoskeleton, but it happens regularly in the tanks, and it can kill the crab. He scoops soft crabs out of the tank and carries them inside, to the big cooler, where he drops them in a bucket. On the radio, an Eastern Shore seafood store is offering number twos for twenty-one dollars a dozen. "Listen to that," Ooker says. "They're charging twenty-one a dozen, and we're getting forty a *bushel* for the same crabs."

Therein lies one of the benefits of going after peelers. "A male crab, I can sell him as a number two for forty dollars a bushel, or I can sell him as a soft crab for forty dollars a dozen," he says. "I'd rather do that." Yes, his workday is longer—he has to monitor his catch into the night—and yes, he has to pay someone to buster up his tanks while he's on the water. Still, he doesn't have to catch nearly as many crabs to make a living.

This raises a question that I put to Ooker: Seeing as how soft crabs are worth so much more, why doesn't he catch a whole pile of hard crabs—forty-seven bushels a day, like the hard crabbers with the biggest licenses are allowed to do—put them in the tanks, and simply wait for them to shed? A couple of crabbers could join forces, one going out to fish up the pots each day, the other to buster up the collected thousands of crabs, and they'd make a mint.

Alas, it wouldn't work. "Being confined, it changes the crabs," he says. "Peelers will slow down their molt once they're in a tank. If you bring them in too early, they won't shed at all. You can't raise crabs from when they're small like you can with fish."

Even with rank peelers within hours of molting, "the process slows down," he says. "You put a peeler in the shedding tank, and it might take four days to shed, when if it was out in the bay it'd take a day or a day and a half."

Such expertise didn't come to Tangiermen quickly. It developed over generations, with crabbers talking among themselves and handing their collected knowledge down to their sons. The island's been studying the blue crab for two full centuries.

Among its early crab researchers was Joshua Thomas's stepbrother, George Pruitt II, who in 1818 bought property at Uppards with his wife, Leah, another of Joseph Crockett's granddaughters. George II was said to be as different from his dissolute father as a man could be. "It was said that he prayed more than any man that ever lived on the Island," Sugar Tom advised in his history. "He was an old man when I was a boy, and I have seen him in his canoe crabbing or getting oysters or fishing many times, and I do not remember ever seeing him alone without being engaged in prayer."

Another family was already living on the island's north shore when the Pruitts settled nearby—a fellow named Job Parks and his wife, Rhoda, who was Leah's sister. From those households began the long lines of Pruitts and Parkses on Tangier, surnames that today outnumber that of the founding family. Eighty-one current Tangiermen are named Parks, and sixty-five, Pruitt. Crocketts number fifty.

But in 1820, when people had occupied the island for forty-two years, Tangier was home to only seventy-four of any name. And that number seemed destined to rise slowly, for island life demanded much. For example, in September 1821, "a fearful storm of wind and swelling of the tide" swamped and battered the place in what became known as the Great September Gust. "The wind blew a storm from the east and southeast" through the night, Sugar Tom reported, "when suddenly the wind changed to northwest and blew a perfect hurricane, and the tide being very high, the hurricane brought the salt water from the Chesapeake Bay over the Island until it covered the Island. Even the highest land had three feet of water over it."

That would be neither the last hurricane to smite Tangier nor the most destructive, but the fact that Sugar Tom wrote of it nearly seventy years later testifies that his forebears saw it as a memorable test. Tangiermen have long considered their island an anointed place, however, protected by the style of fervent prayer George II practiced, and they found that even the most daunting setbacks left them with the spirit to press on. So it was after the Great September Gust, when it seemed sure that the flood had poisoned their meager cropland with salt. They prayed, then "went to work covering their land with drift and sea ordure and every thing of the manure kind," Sugar Tom wrote, and the following year "raised the best crop they had ever raised in all their lives."

The corn and potatoes they grew were essential, for they did not yet rely exclusively on the water for their living. They fished with increasing expertise and managed to sell some of their catch. They netted the odd crab for their own tables, too. But they faced long odds in attempting either harvest on a commercial scale; seafood—crabs in particular—was tough to get to market using the sailboats and wagons of the day.

Oysters offered the best prospect for income. The waters around Tangier were studded with oyster rocks—millions of the creatures piled in huge mounds down on the bottom—but getting their catch aboard was backbreaking work. The islanders used tongs, which resembled posthole diggers fitted with rakes at their ends. Tangiermen favored tongs with handles sixteen to twenty feet long and fashioned from heavy timber, meaning they weighed plenty even without a load of oysters in the rakes. Imagine balancing at the edge of a small boat and shoving such a contraption to the bottom, working the handles to scoop up shells (only some of them housing oysters), and hoisting it hand over hand to the surface, taking care to keep the rakes closed all the while, then lifting it clear of the water and the gunwales to dump the load into the boat's bottom. Imagine doing it in an icy northwest wind, the bay stirred into three-foot waves, the spray off their crests freezing on the tongs and the boat and chapping your face and hands—a realistic expectation, for oystering is a cold-weather undertaking. And imagine that after hours of such brutal work you end the day with a few bushels of oysters, each of which—were you able to sell it—would bring just a few cents apiece. All in all, working the land seemed a better bet.

Then, around 1840, the outside world arrived in Tangier Sound in the form of oystermen from New York and New England, who'd already ransacked the oyster rocks of their home waters and came looking for a fresh supply. These Nordmen, as islanders called them, set to work on the Oak Hammock Rock, a giant oyster colony just a little northeast of Uppards. They tonged, same as the Tangiermen, but also had their sailboats rigged to pull iron dredges that scraped oysters off the bottom and lifted them aboard by the bushel. The strangers also paid good money to locals who brought oysters to them.

The Yankee oystermen recast the island's relationship to the bay. "As this branch of industry increased very fast, people immigrated to the Island," Sugar Tom wrote. "The young men on the Island could now earn more money in three days than they could in a month before. They married and so did those who immigrated, which soon increased the population."

Indeed, it did. The census of 1850 reported the population of "Tangier Islands," which included a few sparsely settled islets to the north, as 178. Ten years later, the number had leaped to 411. Working the water became even more profitable during the Civil War: American Methodists split on the question of slavery, and Tangier, aligned with the Northern splinter, found its sympathies lay with the Union. The island sat out the fighting that consumed the surrounding mainland and, according to tradition, attracted come-heres from both the North and South who wanted to do the same, some of Ooker's forebears among them.

And as the place grew crowded, and properties were carved into ever-smaller pieces to make room for the newcomers, the croplands that once occupied all of Black Dye and most of King Street and all of the West Ridge shrank, snip by snip, until only isolated farmlets remained. Tangier grew ever more dependent on the Chesapeake and ever better at harvesting its bounty.

ALL THESE YEARS later, generations removed from the experiences of those early settlers, a certain stubbornness pervades life on the island. For what makes Tangier dear to its people, what makes their concept of *home* so meaningful, is that it is still not an easy place— that it can, in fact, be frightfully hard.

Or at least uncomfortable. The weather is muggy, and the Situation Room has grown sweaty and close. "Getting warm in here," Bruce Gordy observes.

"It is that," Leon agrees. "It's warm."

"We ought to put that air conditioner in the window," suggests Allen Ray Crockett, a semiretired waterman who, though soon to turn eighty, still has a full head of thick black hair.

"Yeah," Bruce says. "It's about time."

Our surroundings are decrepit. The old birthing room is sheathed in pale green tile that climbs four feet up the walls, a fair percentage of it missing, broken, or cracked. Higher up, on peeling wallpaper, members have tacked or taped unframed pictures: A photograph, printed on typing paper, of Bruce Gordy cradling a twenty-inch rockfish; a large Northrop Grumman publicity photo of the aircraft carrier *Harry S. Truman* under way; black-and-white photos of Babe Ruth and Lou Gehrig as teammates and of Ruth delivering his retirement speech at Yankee Stadium's home plate; a color photo of half-naked African babies, evidently snapped by a missionary; and a 1940s image of Tangier's county dock and harbor. The coffee maker rests on an old medical cabinet in a back corner. Fruit flies circle a trash can stuffed with discarded foam coffee cups.

The men take up the subject of a Tangierman who's lost his crabbing license for violating the state's fisheries laws. It's unstated what his offense was, but it's his third, and the VMRC has suspended him for the rest of the year. Leon is scandalized by the development. "That man can't work!" he gasps. "What's he going to do? They've taken away his living!

"What's the difference between that and getting three tickets from driving, and the judge says, 'You can't go to work no more'?

It's the same thing," he says, building up a head of steam. "There's something wrong with that part of the law. What am I going to do if I can't work on the water? What am I gonna *do*? I ain't got a Colonel Sanders fried chicken place I can go to for a job." It's a rule, he says, made up "by young people down in Newport News," where the commission is based, who know nothing about the water. Reminds him of a regulation the state considered a few years ago, citing concern about overfishing. It called for a halt to peeler scraping but left peeler potting untouched. Leon shakes his head in disgust. The scrapers "all had licenses for peeler pots that would catch four times as many crabs," he says.

"That's the truth, ain't it?" George "Cook" Cannon says. He's an in-law of Carol Moore's—his wife, Jody, is Lonnie Moore's older sister—and he lives on the West Ridge near the presumed site of Joshua Thomas's homestead. "At least four times as many."

Leon, wide-eyed: "It's like putting me and four or five others in here in charge of farming. It don't make no sense, none at all. What do I know about farming?"

Bruce: "Just because you're a marine biologist doesn't mean you know anything about the water."

Leon: "But that's what they look at, what letters they got after their names." He pauses to take a breath. "The whole problem is the people making the rules don't know what they're doing."

Bruce takes a drag off his e-cigarette. "All they need to do is pick the brains of the people here. Has anyone here ever been interviewed by those people?" Everyone in the room shakes his head.

Jerry Frank Pruitt, sitting nearest the door, speaks up. "The bottom line is that they want the Chesapeake Bay for recreation."

"No," Bruce says, "they want it to be for 'future generations'"—implication being the present generation doesn't count for much.

Jerry Frank brings the conversation back to the waterman who lost his license. It's his understanding, he says, that the fellow flouted the regulations. He was summoned to appear before the commission but decided not to show up.

"Oh, well," Leon says, "you can't do *that*."

"No, you can't," Jerry Frank says. "You can't do that. It's disrespecting them." He speaks slowly and quietly, as if debating each word. He owns a boatyard up by the dump. It's largely dormant these days, a victim of Tangier's shrinking population and an attendant decline in available boats to work on, but for forty years, Jerry Frank hand-built wooden deadrises said to be among the finest on the bay. He's accustomed to measuring twice before cutting.

"I know one thing," says Ernest Ed Parks, retired from a career skippering oceangoing tugs and pulling barges. "I'll put in with anybody aboard of a boat."

That's all well and good, Jerry Frank replies, but a man can have the best intentions, and strive to do right in his life and work, and still be in the wrong now and then. A while back, he says, he put a road through the marsh up by his boatyard. Intended no harm by it. "But I got a very threatening letter from the Corps [of Engineers] in Norfolk, telling me I was in big trouble and what all."

He wrote a letter in reply to the colonel in charge. "I dictated it. I'm not an educated man, schooled only through the eighth grade, and so I dictated this letter saying how sorry I was, saying that I had not realized I was breaking the law and that I knew that ignorance of the law was no excuse, but that I apologized and promised I would never do it again. And I asked for his forgiveness."

"That's what you have to do," Jerry Frank says. "You have to. Because the rules are the rules, and it's their job to hold you to them, and if you don't hold to them—well, whose fault is that?"

We sit silent, digesting his words. They make good sense to me, though I know they might not reflect the views of everyone in the room, given Tangier's long, fractious history with rules and officialdom.

Leon shifts in his seat. "It gets much hotter than this, we won't be able to sit in here," he says.

"Well," Cook Cannon says, "why don't we put in the air conditioner?"

"That's a good idea," Bruce says.

No one moves or speaks for several seconds.

"Where is it?" Cook asks. "Where's the air conditioner?"

"Out there," several members say, pointing toward the darkened hall beyond the door. Cook, who sports a feral gray beard, leaps to his feet with a crazed gleam in his eye. "It's time we got the air conditioner in here," he declares. He disappears into the hall and reappears half a minute later with a window unit in his arms.

"We talk about it," Bruce says. "He just *does*."

Cook grunts the unit across the room as Bruce forces the window open. Cook slides the appliance into position, pulls down the sash to hold it in place, unfolds the accordion seals on either side, and plugs in the cord. Chilled air washes into the room.

"Look at that," Bruce says. "Four and a half minutes, it took him."

Gravestones, bricks, and other relics of a vanished hamlet litter the beach at Canaan, June 2016. (Earl Swift)

SEVEN

F IFTY YARDS BEYOND UPPARDS, COOK CANNON LEANS THE *Noel C.* to starboard and we arc into the shimmering, flat strait between the Canaan shore and Goose Island, a junction of latitude and longitude that was once solid ground. Water has replaced the finger of land that once jutted northward from Tangier—a thin vestige of which survived on a government map from 1917 but had vanished when it was updated twenty-six years later. Off to port, Goose—formerly all but attached to that finger—survives as a wafer of marsh barely clearing the water.

We're headed to Crisfield, where Cook has to pick up some building supplies. Because he parks the *Noel C.* toward the west end of the boat channel, it's his habit to forsake the shorter but molasses-slow putter through the harbor in favor of cutting west into the bay and howling around Uppards at speeds that call to mind life's frailty. The diesel roars beneath the deck, so conversation means shouting. But it's a cloudless early morning, the air is crisp and refreshing, and the bay to our south is dotted with deadrises gleaming bright white in the low-angled light. It's a fine time to be out in a boat with an old salt who's navigated these waters for most of his seventy-two years.

Cook nods to his left. "That's Goose Harbor," he yells, using a term that historically applied to the water just east of the island, but which Tangiermen sometimes use for the island itself. "When I was a boy, this gap between there and Uppards was about three hundred yards across." I peer across an expanse of water more than four times as wide. Our movement shifts our perspective on the island, which I now see has been split in two and whittled to splinters. "Used to go over there all the time," Cook tells me. "And you could wade across this gap. I used to wade across it in knee boots, not waders."

"You could *walk* there?" I ask. It seems a tall tale.

"Yes, you could," he says. "Look now. Lot of water here. And deep water, all the way up to the bank." He turns the boat toward shore, and we hug Uppards's crumbling peat edge. On his depth gauge, I see that the bottom is ten feet down.

Later, I'll examine a navigation chart published by the U.S. Coast Survey in 1863 and find that as unlikely as Cook's assertion seems, it was once possible to walk a much greater distance—from Tangier north to Smith Island, six miles away across the state line—without encountering water more than a foot deep. That chart, shaded to reflect the water's depth, clearly shows the ghostly remains of the long peninsula that once contained both Tangier and Smith.

Cook steers the boat to the northeast, and we cross water that on the old maps was studded with an archipelago of small islets, holdouts of that sunken peninsula. Few remain. We pass Queen's Ridge, just east of Goose, which was home to at least one extended family back in Joseph Crockett's day. Now it has barely enough ground for a campsite. We cruise past a point marked on old maps as long and skinny Little Piney Island but find nothing there but water. The same goes for Reach Hammock, just above Little Piney: vanished. Farther north was a little square of land labeled Herring Island, which Tangiermen called *Hearn*. Ooker has told me he recalls stepping ashore there as a youngster. "Weren't very big, but you could walk around," he said. "My dad used to talk about the days when there was a store on it." Tangiermen still use Hearn as a reference point, but it's a phantom conjured from memory. There's nothing there.

What all these islands lacked in common, Cook hollers, was sand: Sand protects a shoreline. "Sand makes the water go up and over the bank. When the sand's gone and the water comes in, what it does instead is *this*." He smacks his fist into his palm. "What it does is undermine the shore, and it breaks off in big chunks."

Sand used to line Tangier's west side, he tells me. In his youth, the islanders didn't hike down to the spit to play on a beach; they made the far shorter journey to the point where West Ridge Road meets Hog Ridge today. Taking a lane that cut to the west, they walked past several houses and through marsh and high grass to a place they called Cow's Hole. "We used to play cowboys down there," Cook says. "There was that tall grass, that good-to-lie-down-in grass, and it was a good walk through there. I'd say as much as half a mile." The hike ended at a lovely beach. Whatever protection its sand provided to Tangier didn't last, for the beach itself proved no match for the bay. It washed away, the sand drifting south to the spit—until the bay dislodged it there, too, and carried it away from the island for good.

Cook steers us up along Smith's eastern flank. Smith is not one island but a cluster and, not counting the wildlife refuge that occupies the biggest of them, it's at least four times the size of Tangier. From its underside, three tentacles of land dangle southward, their tips crossing into Virginia. One used to have an island off its end called Shanks. It was settled in the eighteenth century and occupied through much of the nineteenth; Joseph Crockett's daughter Molly spent much of her life there. Nothing remains of it but its name, preserved in Shanks Creek, a tidal inlet to the east of where the island once lay.

We race past a crabbing area called the Peach Orchard, named for a point of land that once occupied the spot. Rooftops peaking over the marsh advertise Tylerton, one of Smith's three villages. Unconnected by road to the other two, it can be reached only by boat. The "yarnies" there, which is what Tangiermen call Smith Islanders, have much in common with their neighbors—a life sequestered from the mainstream, nearly indecipherable accents, and a home that has steadily shrunk over generations.

We cross the state line, marked by a widely spaced chain of white buoys. Up ahead Crisfield looms, its water tower and several big waterfront condo buildings glowing against the sky. Cook steers the *Noel C.* into the mouth of the Little Annemessex River, so wide that it seems we're entering a bay rather than a short creek. Off to port stretches a white sand beach, and on it stands a tall brick chimney. It's all that remains of a fish-processing plant that burned in 1932 and now serves as an aid to mariners as surely as a lighthouse. A good piece away on our other side is the river's low, marshy, and crumbling south bank. "When I was a boy, we had to stay to the left here, almost all the way to that chimney, to get into Crisfield," Cook tells me. "This over here"—he nods across the two hundred yards of water to our right—"was all land. It was really big, and now all of it has washed away. Just since I was a boy."

It's a reminder that the effects of rising seas are not limited to the Chesapeake's islands. It has fueled erosion that's stolen vast acreage from the bay's edges, from the Susquehanna to the Virginia Capes. Since the mid-nineteenth century, it has chewed away bluffs, erased beaches, and undermined forest. Up in Maryland, the Blackwater National Wildlife Refuge, created as a layover on the Atlantic Flyway for migratory waterfowl, saw five thousand of its nearly twenty-nine thousand acres of marsh turn to open water between 1938 and 2009. The loss is significant: The refuge contains one-third of the state's tidal wetlands.

The northernmost mainland community on Virginia's piece of the Eastern Shore, a low-lying watermen's village called Saxis, is flooded by even minor storms, and high water has carried off many of the crab houses that lined its waterfront for generations. With their destruction, Saxis lost its commercial heart. It's pervaded today by the unmistakable air of a ghost town in the making: The principal road snaking through town dead-ends at a business district consisting of a coffee shop, a tiki bar, and a cinder block crab-picking house.

Sixty miles to the south, at the Chesapeake's mouth in Hampton Roads—a metropolitan area of 1.7 million inhabitants that includes the state's two biggest cities, Virginia Beach and Norfolk—high tides routinely submerge neighborhood streets under water deep enough for motorboat traffic, and northeasters sometimes maroon entire sections of town. Photographs of cars stranded in water up past their windows have become a post-storm cliché in the *Virginian-Pilot.*

The region's relative sea level is rising even faster than Tangier's, because the much larger population there has tapped an underground aquifer for its water, and draining it has hastened the subsidence of the land above. Among the properties at risk is the mammoth Norfolk Naval Base, home to the Atlantic Fleet. Its aircraft carriers, guided missile destroyers, and submarines aren't at risk—they're built to float, after all—but the piers where they tie up, the pipes and utilities that supply them, and the streets their sailors travel to report for duty are in serious peril. The Hampton Roads Transportation Planning Organization, a panel of local, state, and federal officials and leading citizens tasked with adapting the region's highways to the looming crisis, reckoned in a May 2016 report that the bay could rise two feet there by 2045.

All of which is to say that Tangier Island is not alone in its struggle with the sea. It's just furthest along in the battle and the worst for it. The problems there will soon be felt elsewhere, just as Tangier is now vexed by challenges similar to those of Holland Island a century ago.

And we know how those turned out.

COOK THROTTLES BACK the engine and we chug past Crisfield's city dock, where the *Courtney Thomas* ties up. We veer through a narrow passage into the vast Somers Cove Marina, the biggest haven for boaters in this part of the Chesapeake, with more than five hundred slips, a big dockmaster's office, and a busy Coast Guard station lining its edges. Cook leaves the cabin for the stern steering station and, with typical Tangier nonchalance, spins the boat and backs it into a slip in a single fluid maneuver. It's all the more impressive because Cook's virtually blind in his left eye, thanks to a childhood accident, and as Leon has advised in the Situation Room: "If you ain't got both eyes, you got to be real particular if you're bringing a boat into the dock, because you can't tell how far away you are. You'll run into it."

We tie the boat in place and walk up the dock and across a parking lot to a black Ford F-150 that Cook uses when he's on duty. He's in his ninth year with the Chesapeake Bay Foundation, a nonprofit dedicated to protecting the bay and its drainage. It traces its beginnings to 1964, when a handful of Baltimore businessmen worried over lunch that people were overusing and underloving the bay, which was degrading its water quality, disrupting its wildlife, and spoiling its quiet beauty. The humble group they founded has grown into a noisy and effective advocate for restoring the Chesapeake's health, with a cadre of scientists, lawyers, and educators, backed by a committed army of volunteers.

Its mission has at times put the foundation at odds with Tangier's crabbers. Islanders will allow that the CBF has done a lot of good, particularly in raising public awareness about the bay's fragility. But in its past dealings with commercial watermen, the foundation could come across as high-handed and smarter-than-thou. It seemed quick to dismiss the wisdom that crabbers had gleaned in decades of working the water. Its fresh-scrubbed activists talked about the bay as if they knew all about it. Tangiermen *lived* in it. The activists, they argued, were often flat-out wrong.

The islanders, on the other hand, didn't always come off as wise or reasonable. They sometimes seemed to think they had a God-given right to plunder the Chesapeake as they saw fit, and that their heritage granted them immunity from laws and regulations. They denounced the science that CBF held dear if it didn't square with their more anecdotal learning. They invariably referred to the group by its motto—"Save the Bay"—and they often said it with a sneer. When relations were at their worst, in the mid- to late nineties, the island was pretty much at war with the outfit. Tourists pulled into a harbor lined with signs denouncing the CBF as an enemy to people who depended on the bay for their living.

These days the two are getting on better. The foundation has made a heartfelt effort to seek the advice and input of watermen—to treat them as valued players in the bay's culture and health—and to do a better job of explaining its research and aims. Tangiermen have tried to squelch their distrust of scientists and regulators and to treat Save the Bay as neighbors, especially the young staffers who spend summers on the P'int, at the Port Isobel education center. As they achieved a cautious détente, the foundation took to boating groups of students and teachers to Tangier to visit with crabbers and watch them work, giving both sides more time in each other's company. That helped, too. Ooker's crab shanty is a regular tour stop.

At first inspection, Cook Cannon might seem an odd fit with the organization. He is a man of strong opinion and very loud volume—even in private conversation, he tends to speak in a near shout—and while he's prone to smiles and devoutly religious, he also has a quick-flaring temper and little fear of an argument. Like most Tangier men of his generation, he quit school at fifteen to work the water with his stepfather, and he doesn't necessarily agree with the foundation's science. "The company I work for believes in global warming," he's told me. "They know how I feel about it, too. I don't believe in it. I think it's a load of crap."

But consider: When Cook was seventeen his stepdad was hurt aboard their boat, and the boy became his family's breadwinner. He chased work wherever he could find it—crabbing, naturally, but also building and remodeling houses, laying phone cable, honing his skills as a plumber, electrician, and mechanic. He spent six years in the engine room of an Army Corps of Engineers dredge. Then, in 1988, he landed a job as supervisor of the town's sewage treatment plant and trash incinerator, a $3.5 million facility that had quit working after just five years of operation. Few had faith that he'd be able to fix it. Gases from the plant's sludge tank had chewed through wiring and control panels and had so corroded steel catwalks and ladders that they couldn't support his weight. Failed pumps had spewed sewage ankle-deep on the concrete floor.

Mainland experts wrote off the mess as unsalvageable and talked about spending millions of dollars to replace it. Cook got to work. He cleaned out the sludge, rewired the controls, rebuilt the pumps, and returned the plant to full operation, mostly by studying each component and sussing out what made it work. "I'm not naturally smart," he said, "but I was determined."

The Tangier Town Council was so awed and grateful that it voted to grant him the job for life. All of which made him, twenty years later, a peerless candidate for the job of caring for the CBF's four educational facilities on the bay: a couple of big, old houses in Tylerton; a former hunting lodge on Fox Island; the big complex of dorms, docks, and a conference hall at Port Isobel; and a mainland outpost at Bishops Head, Maryland, near the Blackwater refuge.

WE ROLL OUT of Somers Cove and onto Crisfield's Main Street as Cook rails about Tangier's boat channel, which opens the harbor to westerly winds and agitates the water around the crab shanties. "You talk about a rough harbor," he says. "We're the only port on the whole bay that ain't got a good harbor. And we're the seafood capital *of the world*. You know that ain't right." Talk has bounced around for years, he says, about building a stone breakwater on the island's western shore to protect the channel. But every time it's appeared about to happen, either the money has fallen through or the Corps of Engineers, which oversees such projects, has opted to think on it a bit more. "They spend hundreds of thousands of

dollars to study it, and they never get around to actually *doing* anything," Cook roars. "I say it's time to stop listening to politicians
and take action ourselves—if it's right or if it's wrong." One grows
accustomed to hearing such talk from Cook, though that's all it is.
If Tangiermen could take action themselves, they'd have done so
years ago.

Crisfield is quiet this morning. We encounter few other vehicles
along the mile and a half to the hardware store and lumberyard
where Cook is to pick up a pile of posts he'll use to build osprey
nests on Smith Island. Cook's boss, a fit CBFer named Paul Willey, is waiting in the parking lot in a Patagonia ball cap and cargo
pants. He and Cook venture into the store and emerge with a load
of salt-treated four-by-fours. The three of us drive back to Somers
Cove, passing under Crisfield's water tower, the orange crabs on its
tank faded to a pale pink, and through a downtown of scattered restaurants, a shuttered bank, a stately but long-abandoned customs-
house, and an abundance of vacant lots. Main Street dead-ends out
at the end of the city dock, which is wide enough to accommodate
a lane of traffic, angled parking, and a covered waiting area beside
the mailboat tie-ups, and which is flanked by large condominium
blocks built on stilts at the water's edge. All's quiet down there,
too. Loudspeakers over at the Coast Guard station sound reveille
as we haul the wood across fifty yards of lawn and dock to the boat.
Sweating, we untie from the slip and motor out of the marina, passing a line of old brick crab-packing plants—relics of days when this
waterfront was far too occupied with the business of picking and
packing seafood to leave room for condos.

In the cabin, conversation centers on the meager number of peelers in the water off Tangier and the season's slow start for hard crabbers. "For the last three years," Cook says, "things been different. The
crabs have been acting different. I think they got smart. I think they
figured it out." Meaning, presumably, that the crustaceans have developed higher reasoning skills and are consciously outmaneuvering
the island's watermen.

Willey, a boat captain in his own right who has worked on and around the water for CBF since 1989, sounds like a doubter: "You think they got smart just in the last three years?"

"Let me tell you, every spring we've always had two runs," Cook says. "For the past three years, those ain't happened."

Willey cocks an eyebrow. "So you think that in the last three years, the crabs got it all figured out?"

Cook nods vigorously. "If you can train an alligator, you can train a crab."

Willey throws up his hands. "Cook, nobody's *training* the crabs!"

Out on Tangier Sound, the subject turns, inevitably, to erosion. Cook rants for a few minutes about the longed-for Tangier break-water, and Willey acknowledges that the Corps of Engineers is a tough organization to read. "One thing that's been a tough nut to crack, even though we have a good relationship, is getting them to tell us exactly what stage the breakwater is in," he tells me. "We try our best, but we always seem to be waiting for information."

The shriveled marshes of Smith Island slide past the boat's windows as Cook notes the sorry state of Uppards and the shoreline's retreat all around Tangier. Willey nods sympathetically. "The natural process of the bay is erosion and filling in," he says. "But it's tough when it affects critical pieces of the bay, places that shaped the bay— and, of course, people's homes."

Well, Cook says, one solution might be to take dredge spoils from elsewhere in the bay and blow them onto Tangier. Build the island up. Replace what the bay steals. Buy it more time. It worked at the P'int, he says—before the foundation moved onto the islet, it was owned by a fellow who dumped a mountain of fill there, then planted hundreds of trees, stabilizing what might otherwise have washed away. "That was marsh," he says, "and ain't that the prettiest place now?"

"But that's a slippery slope, Cook," Willey counters. "That'll give people an excuse to use fill to make uplands from marsh that they ought to just leave alone. When it's going away, you want to lash out at everybody, but sometimes you have to just accept that this is the natural course." The CBF's belief in that principle likely will doom its Fox Island education center, Willey tells us. The center occupies a rambling, wood-frame hunting lodge built in 1929, when the land on which it stood was still called Great Fox Island—to distinguish it from Little Fox, off to the south. Since then, Little Fox has disappeared completely, and its big sister has eroded to a whisper of marsh. The lodge is unprotected, wide open to the weather, its demise only a matter of time. "And it's a shame," Willey says, "because Fox Island is such a powerful place. Imagine being a kid from Baltimore and being out on Fox Island, in a place so remote, with no lights around. It's fantastic."

Later, as Cook and I cruise back to Tangier, having left Willey in Tylerton, we can see the Fox Island lodge a few miles off to port, a big white rectangle that seems to float on the water. The clubhouse, as mariners know it, is such a familiar sight that it's difficult to imagine this part of the bay without it. But the day is coming—and soon.

Erosion devoured not only Fox's uplands and marsh, but the sandbars that for centuries cushioned it from the worst a storm might dish out. Within the memory of older Tangiermen, the shallows south of Fox were shallow indeed. At low tide, only a film of water covered the flats of sand and mud all the way down to Watts Island. Jerry Frank Pruitt, born in 1944, has told me that he can remember "seeing the tree stumps under the water" down that way. "They were big old round stumps, big trees.

"It was so shallow, all you could get through there was a skiff, at low tide," he said. "That's how it was when I was a teenager."

No longer. Winds and currents have scoured the bay's floor, robbing Fox of the speed bumps that softened the effects of wind and tide. A single hard blow will finish the place. The clubhouse and Fox itself are, as Tangiermen say, "going away from here in a hurry."

THIS WOULD HAVE been unimaginable when the Fox Islands were inhabited, and Watts, too, and thanks to the oyster business, Tangier's head count was growing by leaps. By the outbreak of the Civil War the wintertime oyster harvest was Tangier's "main dependence," Adam Wallace wrote in his 1861 biography of Joshua Thomas, *The Parson of the Islands*. Even then it was "conjectured that the supply, in the Chesapeake waters, must soon fail, in view of the immense quantities taken up annually, and the increasing facilities with which (the oysters) are obtained."

War's end only intensified the oyster harvest, with boosts from several simultaneous developments. In 1866, the Eastern Shore Railroad pushed a line into the small, sleepy waterfront village of Somers Cove, and instantly the oyster trade had a way to get its catch to the big cities of the East. Shucking and packing houses sprouted along the water's edge by the dozens, and in 1872 the fast-growing burg renamed itself Crisfield, after the railroad's president. Enterprising townsfolk filled the marshes at water's edge with mountains of oyster shells discarded by the industry; over time they created a peninsula jutting into the Little Annemessex that became the town's business district. Crisfield was literally built on the oyster.

Canning plants followed not far behind the railroad, and soon diners a thousand miles from the sea could enjoy the fruits of a Tangierman's labor. The more available the oysters became, the greater the public's demand for them grew, and Crisfield boomed. Its register of sailing vessels grew to be the biggest in the country. Even today, older Tangiermen can remember it as a town that supplied virtually all their needs. Its business district boasted department stores and five-and-dimes, along with dress shops, shoe stores, places that outfitted kids for school, and a slew of restaurants, some catering to families, others to watermen. It had an opera house and, later, movie theaters. Grand churches. Big neoclassical and Romanesque banks equal to those anywhere. And rightly so, because the oyster catch landing at Crisfield ran into the millions of bushels a year, which brought money and lots of it.

That would soon enough cause trouble, as money will, between the watermen of Virginia and Maryland—and, in particular, the populations of Tangier and Smith islands. But in the meantime, the Civil War's end brought a more immediate crisis. Cholera arrived in October 1866. It was Tangier's first epidemic, and although there'd later be outbreaks of measles, tuberculosis, and smallpox, none exacted so high a toll or excited such panic. "The first case occurred on Tangier Island on the 10th instant," the *New York Times* reported on the twenty-seventh, "and from that date to the 21st, there were thirteen deaths." The paper advised that most cases "could be traced to some imprudence in diet," adding: "One or two ate watermellons [*sic*] and many others oysters."

Oysters could, in fact, harbor the disease, but the real culprit was more likely the growing settlement's casual approach to sanitation. Cholera is caused by a microbe that lurks in raw sewage and, when swallowed, attaches itself to the walls of the small intestine. The consequences are grim: gushes of watery diarrhea that lead to potentially fatal dehydration and that can, in turn, infect those tending to the patient. Easily treated today, cholera killed millions worldwide in the nineteenth century, including former president James K. Polk.

In 1866 Tangier's households relied on privies, which were not set astride deep holes, as was the style on the mainland—the water table was too high for that—but over narrow "sewer ditches" that were theoretically flushed by the give and take of the tides. Never mind that these ditches emptied into larger canals that served as the island's transportation grid. In those days, the roads were even narrower than they are today and suited only to foot traffic. To move cargo—wood for stoves, blocks of ice, furniture—islanders on the Main Ridge used small skiffs they'd "shove" with poles into the Big Gut, then turn up ditches they'd dug into their backyards, which effectively served as their driveways but also devolved quickly into open-air cesspools. Children routinely played in the waterways. Watermen invariably spent time in the same water, working on boats or rinsing off gear. It was all too easy to get an accidental mouthful of ditch water or to scoop drinking water from a cistern with unwashed hands. Or, yes, to eat an oyster exposed to sewage and infected with the microbe.

"The people began to die very fast," Sugar Tom recalled. "We did not know what it was until the physicians told us, and as many as six adults would die in twenty-four hours. I could hear the voice of weeping all night." The outbreak apparently led to the island's evacuation—Sugar Tom wrote that "nearly all the people left"—though how long they stayed away, how many decided to remain on the mainland, and how many succumbed have all been lost to memory.

The wonder is that the disease didn't return, for island homes continued to rely on privies and sewer ditches even as the population mushroomed. Most didn't have indoor plumbing until after World War II, and even then it didn't connect to a bona fide sewer system; indoor toilets flushed into septic tanks or, just as often, pipes that crossed the yards and emptied into open water. Bruce Gordy's wife, Peggy, told me that none of her homes had indoor plumbing until she and Bruce rented a place in Meat Soup in the late sixties. Their modern toilet dumped directly into the main harbor.

The situation wasn't remedied until Tangier installed water and sewer lines in the early 1980s, meaning that everyone in the Situation Room has had experience with outhouses. "Where those toilets used to be, the marsh would be this high," Jerry Frank Pruitt told me one afternoon, holding his hand over his head, "and everywhere else it would be this high." He dropped his hand to waist level.

"Yeah. Good fertilizer," Ooker said. "It got high around the sewer ditches, too."

"These ditches," I interrupted. "Did you guys go swimming there?"

Ooker nodded. "You'd run into a floater every once in a while."
"Have to do a reverse dog paddle," Allen Ray Crockett recalled.
They both laughed. "Yeah," Ooker said. "Exactly."

ON AN AFTERNOON four or five generations beyond the cholera, I
climb into Carol Moore's skiff to thud up the west side of Uppards to
the ghost of Canaan. I've prepared for the trip by donning long pants,
a long-sleeved shirt, and drenching myself in DEET. "I can smell
your insect repellent from here," Carol yells from the boat's stern,
where she's guiding the tiller. "It won't do any good. It won't even
slow them down."

At Canaan, several tombstones lie on their backs near the surf
line, along with tiny, scattered relics of the village: bricks worn porous
and smooth, rusted cogs and sprigs of iron that once made machines,
slivers of ancient lumber, glass.

"I'm going to walk this way." Carol points to the east. "You want
to hang around here and see what you find? Or you could walk down
the west side." As she speaks, flies home in on the fresh meat come
ashore and orbit around us. "Look close, and you might find some
arrowheads," she says. "I'll meet you back here."

She strolls away. A fly smacks into my cheek, but it doesn't land.
Others buzz within an inch of my ears, but the DEET, for the moment, appears to be working. I wander over to the stones and stop at
the tall, worn slab of marble dedicated to Polly J. Parks, who died on
December 4, 1913. Canaan was a busy village then, home to more
than one hundred people. Its docks jutted from a waterfront now
submerged under the strait that Cook and I traversed in his boat. In
Polly Parks's day, the long finger of land linking Uppards to Goose
Island, withered but intact, still guarded the Canaan shoreline from
westerlies.

I brush sand away from her stone, which is cracked across its
middle. Above the crack are the dates bookending her life. The in-
scription below the crack reads:

Here in the silent graveyard,
'Neath the sod and dew;
Never one moment forgotten
In sorrow we think of you.

What the stone does not reveal: That friends and family called her Dollie. That she was born a Pruitt and was directly descended from George Pruitt II, Joshua Thomas's stepbrother, and Leah Evans, Joseph Crockett's granddaughter. That she grew up in a crowded house, one of twelve children. That she married Harry Parks, a direct descendant of Job Parks and Leah's sister, Rhoda, the first settlers at Canaan. That her kin survive: Polly's great-great-nephew is Jerry Frank Pruitt, and her grandson Stewart married Jerry Frank's sister Connie. That after Polly's death, Harry remarried and outlived her by thirty-nine years.

Now her headstone lies broken and overwashed by the tide. The place where she raised her seven children, who would go on to seed much of Tangier's living population, has been swallowed by the rising bay, leaving behind nothing but shards. A palpable melancholy hangs over Canaan, not only because it offers such scant evidence of the generations who lived here, but because it offers a glimpse of what may come.

Carol returns, having found little in her wanderings. It's a lovely, windless afternoon, and the sun remains high in the sky, so we elect to cross the strait to Goose Island. We pick our way through the shallows around its edges, running aground time and again, Carol reversing course to keep the outboard's propeller from chewing too deep into the bottom, and finally make landfall on the islet's southeast tip. As small as Goose appeared from Cook's boat, it seems even less substantial as we tread its edges. It rises to a foot, maybe two, above the bay. Wild petunias sprout from the low dunes. The island's interior is a bog of stagnant tidal pools, surfaces veneered in scum, minnows busy below.

We prog the water's edge for a few minutes as shorebirds circle and swoop overhead, calling and crying. Then, having searched what little ground Goose has to offer, we push off for Queen's Ridge, a few yards to the east. This once-inhabited place is now little more than a narrow tuft of marsh grass. No point in searching for remnants of its past. There's nowhere left to look.

Ooker fishing up his peeler pots, June 2016. (EARL SWIFT)

EIGHT

———

TANGIER FOUND ITSELF BETTER CONNECTED TO THE WORLD in the decades following the Civil War. In 1867, steamboats began scheduled service around the Chesapeake, and in 1884 the island became a regular stop. Before long, four ships a week called there with passengers and goods. The harbor didn't exist at the time—the waters off Meat Soup were too shallow for anything bigger than a skiff, and the creek between Canton and the P'int was just inches deep at low tide. So the steamboats pulled into the haven formed by the curling spit—a pocket of deep, protected water that quickly earned the name Steamboat Harbor—and dropped anchor. Tangiermen poled and sailed small boats out to meet them.

In time, the steamship company built a wooden wharf in the harbor's middle, which made loading and unloading a little easier. But just a little. For more than thirty years, anyone or anything steaming into Tangier had to be ferried ashore. Still, for the first time islanders had easy access to mainland newspapers and groceries, and drummers could bring samples of dry goods and patent medicines to a population hungry for big-city merchandise.

They had the money to pay for it. The oystering boom showed no signs of letting up. The business became so big, and competition so fierce, that it wasn't long before friction developed between Virginia and Maryland as to exactly where their shared border crossed the Chesapeake. The line was well established on the bay's west side, where it hugged the south shore of the Potomac River, so that the river itself lay in Maryland. It was more or less settled on the Eastern Shore, too. But on the water in between, it got murky—a situation that the bay's people had not much worried about for two hundred years, because they'd never had a reason to.

They had one now. By law, oystermen could work only the waters of their own states, making the line's precise location a matter worth millions of dollars. The description of the line dated to a seventeenth-century land grant, which used points of reference that each state interpreted to its own advantage. Commissioners from both met to settle the question. They failed to reach an agreement, and the border wound up in arbitration.

The arbitrators, from Pennsylvania and Georgia, had their work cut out for them. They had to rely on centuries-old documents and John Smith's map of the Chesapeake, the only full depiction of the bay at the time the border was initially established. "Not only the names, but the places themselves have been much changed," they wrote. "Considerable islands are believed to have been washed away or divided by the force of the waters. Headlands which stretched far out into the bay have disappeared, and the shore is deeply indented where in former times the water line was straight, or curved in the other direction." But in a pragmatic 1877 decision, the panel established a zigzagging line that clipped the southernmost mile of Smith Island from Maryland, and granted Virginia dominion over the productive oyster rocks there and at various other points claimed by its neighbor.

The oyster industry was near its peak at that point. In 1884, Chesapeake watermen pulled fifteen million bushels of oysters from the bay, a level of pillage that boggles the modern mind. Even some of those watermen, an eternally hopeful lot, had to wonder how long the bay could keep giving.

That worry, quickly realized, explains why the peace ushered by the border agreement was short-lived. After the 1884 peak the catch began to slump. Oystermen responded by seeking their quarry with increased recklessness, and within a few years Virginians were disregarding the state line. Angry Marylanders met the poachers with gunfire, though not long after, they were sneaking across the line themselves. Both states created "oyster navies" to patrol the border, and their officers were soon fighting all-out sea battles with outlaw watermen. One day in the mid-1890s, Virginia oyster police found themselves in a desperate gunfight with Maryland watermen, who outnumbered them thirty boats to two. The following month, one of the Virginia boats came under fire from a small army of Marylanders hunkered behind breastworks on Smith Island. By then, the decline in annual harvests was pronounced.

Tangier had been transformed by the oyster. In December 1890 it was bigger, healthier, and better off than ever before, with nearly nine hundred residents, four schools, two hotels, seven stores, three fish factories, and fifty-two dredging boats tied up at the waterfront. Its people had no intention of surrendering this newfound wealth. So as it became obvious that the oyster business was on the wane, they turned their attention to a new fishery, abundant and familiar but largely untapped: the blue crab. By June 1903, the *New York Times* could report, "There is no slack time with the Tangier Islanders, for oysters take the place of crabs and crabs take the place of oysters, and there is no season when something worth having is not to be taken from Tangier or Pocomoke Sound, or the waters north or south.

"The Tangier Islander is in an almost ideal position to get the most in a simple material way out of life. He draws his living from the sea, and the shore affords him only occasional standing room and a place to sleep."

AN HOUR BEFORE ONE JUNE SUNUP I hitch a ride out to Ooker's crab shanty with Donald Thorne Jr., a.k.a. Thornie, first cousin to Carol Moore. I find the mayor packing soft crabs for market while listening to an Eastern Shore farm report on the radio. A crabber's life resembles a farmer's in manifold ways, he tells me. His livelihood is dependent on the weather, for one thing, and on other natural forces beyond his control: the bay's temperature and salinity, the health of its underwater grasses, the preponderance or dearth of fish that prey on juvenile crabs. His income, like a farmer's, turns on whatever price the market sets for his product, and he doesn't have much say in that, either. Prices typically start high in the spring, when crabs are fresh from the mud and just starting their migration, and drop steadily as their numbers increase after Memorial Day, reaching a seasonal low in late July or August, when the animals are at their greatest abundance. Beyond that general rule, however, a Tangierman cannot count on much. "You find out what the price is when you come in with your catch," Ooker says. "Sometimes they won't even put a price on your ticket—they'll just tell you it will be higher or lower than the day before."

This is a reality that frustrates pretty near every waterman on the island. Few are as vocal on the subject as Leon McMann, who's had seventy-one years of full-time work on the water to stew about it. "They pay what they want," he told me. "Ever since we been an island, we've never known what we were getting paid until after we caught it."

Unlike some farmers, a crabber isn't eligible for a government subsidy. He won't get a check from Washington to sit out the season in order to stabilize the commodity's price. And he can't stockpile his catch until the market favors a sale. "The stuff we catch, you got to get rid of it," Leon pointed out. "If you got live crabs and you got a boatful of 'em, what are you going to do with 'em if you ain't gonna sell to the man?"

So this morning Ooker is preparing a shipment of soft crabs to New York City's New Fulton Fish Market, which is offering a higher price than the crab house in Crisfield with which most of Tangier's peeler crabbers do business. The decision has its drawbacks: He has to buy heavy cardboard shipping boxes, which cost ten dollars apiece. He has to pay the cost of shipping his softshells from Tangier to the

Eastern Shore, then 260-odd miles by truck to New York City. And packing them for the journey requires a surfeit of labels and paperwork, which can be burdensome to a man who already spends eighteen hours a day on the job.

But New York prices more than make up for it most days, and the people up there offer another incentive: They'll defer to Ooker's expertise in grading his softshells, which the Crisfield buyers will not do. Like hard crabs, soft crabs are priced by size. The biggest of them, bona fide monsters that measure more than five and a half inches across and singly fill a dinner plate, are called whales. Those just a touch smaller are jumbos. Next come primes, then hotels—a name that not even Leon can explain—and, finally, mediums.

Ooker walks to an industrial cooler that occupies the shanty's middle and extracts a yellow plastic bucket containing the crabs that molted overnight. The first one he pulls from the container measures a full eight inches across. "That's a nice soft crab," he says. "They'll like him in New York." It's a whale of a whale, without question. Sizing most specimens is not nearly so obvious, however. "Crisfield, they grade the crabs, and they grade them smaller and usually pay less for them," Ooker says. "Crisfield, too, sometimes they care about how many legs are on the crab. Of course, you have to have a reasonable amount, but they'll get picky if you have a couple of legs missing." He knows that most diners won't open a soft-shell crab sandwich to ensure that the deep-fried creature within has all ten appendages; the legs yield crunch but little meat anyway, so a soft crab shy of one or two shouldn't be much of a worry. Unless it is.

Ooker is packing mostly whales and jumbos this morning. On the shanty's worn plywood floor sit several new shipping boxes, each of which accommodates three stacking cardboard trays. How many crabs fit in a tray follows a standard based on their grade: two dozen whales, three dozen jumbos, three or four dozen primes. He starts a tray of jumbos by dipping a page from an old edition of *USA Today* in a bucket of water, folding it to fit the tray, then plucking a crab from the yellow bucket and nestling it on the wet newsprint. The crab sits motionless, claws folded neatly under its chin, bubbles forming around its mouth. This is not a sign that the animal is in distress: Kept cool and damp, a crab can live for days out of the water. Ooker fishes for a second and sets it just behind the first, so that its face and claws rest on the first crab's back. A third crab is likewise positioned slightly overlapping the second, and so on—until nine crabs are arranged in a column comprising a quarter of the tray's width. He starts a second row.

The crabs lie still. In contrast to their feistiness as hard-shells, they're weak and keenly aware of their vulnerability and looking to avoid trouble. Ooker moves quickly, taking no more than a couple of seconds to grab a crab, grade it with a glance, and pack it into the appropriate tray. When he's filled one, he soaks another newspaper page and spreads it over the crabs within, then stacks the tray with others of the same grade.

"It used to be that you had to use dead seagrass to set the crabs in," he tells me. "Even for the local markets, you had to pack them in grass and paper and keep ice on them even in the shanty." That was before the island's crabbers got the big coolers and freezers they rely on today—a technological upgrade that didn't reach the shanties until the seventies. "Now they accept this," he says. "I guess the crabs look better in seagrass, but you're not going to fool with it if you don't have to. The newspaper keeps 'em wet, which is all you need."

Ooker was just starting high school when he and his oldest brother, Ira, built this shanty, on the site of an earlier crab house their father had used. It is a sturdy but unassuming structure, framed in two-by-fours, clad in plywood, all of its bones exposed and aged to a deep red brown. The center of the floor is occupied by the cooler, an enormous stainless steel model seven feet tall and six wide. Large freezer chests line three walls. A door is cut into one corner, and just inside it is the worktable at which Ooker does his packing. Piled on practically every horizontal surface are the tools and supplies necessary to life on the water: motor oil, caulk, paint and varnish, thousand-foot spools of nylon cord, wrenches and hammers and jars of screws. Old foam crab pot buoys. Cat food.

Ooker finishes a tray. He's assembled two of whales, five of jumbos, one of primes. He quickly puts together a ninth and final tray, split between jumbos and primes, then stacks the trays inside the shipping boxes, and onto each box pushes a snug-fitting lid emblazoned with the likeness of a blue crab and the words LIVE SOFT SHELL CRABS. He writes "JMS" in permanent marker in a corner of each lid—for the JMS Seasonal Seafood Corporation, the shipment's destination within the New Fulton Fish Market—and in another prints "J. Eskridge, Tangier, VA." He draws a Jesus fish under his name.

The farm report has given way to country music. The lyrics "You had a Corona, and I was drinking Bud Light" thump on a gush of static from the radio's overpowered speaker as Ooker binds the boxes in black string, then carries them out to the boat. JMS is paying him an average of $230 a box for his crabs these days, so the morning's chores will gross just shy of $700. It's already a good day, and he hasn't yet pulled up a pot.

AT 7:40 A.M. we cross the harbor to the mailboat dock, where a crewman from the *Courtney Thomas* tells us the big boat's port engine won't start, so the afternoon ferry—Mark Haynie's *Sharon Kay III*—will make the morning run to Crisfield. Ed Parks, maternal grandfather of mailboat skipper Brett Thomas, has pulled up to the dock in a skiff. Beside him are two boxes of swelling toads, as Tangiermen call the spiked and self-inflating northern puffers. Ugly though the fish are, they're undeniably delicious. Skinned and deep-fried, they enjoy growing popularity as "sugar toads," a restaurant appetizer. Even so, the eighty-year-old Ed is one of a few Tangier crabbers who fish for them.

"Brisk wind out there today," he says as we come alongside.

"Yeah," Ooker replies, and sighs. "It ain't gonna be blowin' none this summer."

We sit, rocking in our respective boats, and wait for Mark Haynie to arrive. Ed is one of two Ed Parkses on the island—a not uncommon situation with so few surnames spread among so many people. There are also two Denny Crocketts, two Jackie McCreadys, two Michael Parkses, and quite a few fathers, like Ooker, who share names with their sons. In days of old, when the island's population was bigger and the duplications were more numerous, islanders might distinguish two same-named neighbors with genealogy, by appending a father's name to one or both (so it is that my landlady, Cindy Parks, has an ancestor who was known as "Elisha of Zacharia Crockett"). Today, they do it with references to occupations, physical characteristics, or middle names. The two McCreadys, for instance, are referred to as Jackie Haskins or Jackie Burton, if the context doesn't pinpoint one or the other; Jackie H. is sometimes referred to simply as "Haskins."

Denny Crockett, the retired Tangier School principal, is "Principal Denny," versus Denny Crockett the electrical co-op worker, who—being twin to his brother, Donnie—is "Twin Denny." And the Ed Parks talking with us this morning is "Short Ed," as opposed to "Colonel" Ed Parks, a former mayor who now runs the town museum. Short Ed dislikes that label, and everyone on the island knows better than to use it in his presence; he's only rarely called to answer to it. Some Tangiermen with alternate forms of address, on the other hand, are rarely called by their given names. Ooker is always Ooker. Christine Charnock, who became the oldest woman on the island with Henrietta Wheatley's death, is always Teany. Carol Moore's brother, David Charles Pruitt, is always Tweet, and her father was Flapper to everyone but his family. George Pruitt is always Hoot, for reasons unclear, except among those Tangiermen who know him better as Monk, short for "monkey" and a reference to his childhood talent for climbing trees. Many of his neighbors, asked his real name, will tell you it's Frankie, though that's actually his middle name. Meanwhile, Frankie Crockett—that *is* his name—is always Tabby. Ed Charnock is Eddie Jacks, after a character on the old *Peyton Place*

TV drama. The island's dead include Chowder and Sea Biscuit, Popcorn and Spaniard, Ponk and Spurge, Puge and Miff. Puff Cheeks. Kisses.

Some nicknames are easily traced to specific traits or incidents. Shithouse Al, for instance, "led a gang who thought they'd go around knocking over toilets," according to Jerry Frank Pruitt. And Half-Ass Buck was a come-here who "lost a piece of a cheek in an accident or something," as Ooker put it at the Situation Room one afternoon.

"Sawmill," Leon chimed in.

"Yeah," Ooker said, "a sawmill."

"He came here and married Elmer's daughter," Leon recalled, referring to the late Elmer Crockett. "He saved Elmer's life, too."

"How did that happen?" I asked.

"Elmer hit a wire with his head," Leon said. "I don't know how it didn't kill him. Half-Ass Buck throwed him away from the wire, and that's how he got him away."

"What, like an electric wire?"

Jerry Frank nodded. "Burned a hole," he said, forming a circle with his hands the size of an orange. "It put a big hole in his head. I sat behind him in church, and he always had a bandage over it. They'd have to go in and drain it now and then."

That was Half-Ass Buck. The origins of many island nicknames aren't as well remembered, by accident or design. Both Allen Ray Crockett and his son Mark are known as "Mooney," after an ancestor who had the same handle, though neither can explain the name's genesis. Kim Parks prefers Socks to his given name, but he won't share how he earned the moniker. "I ain't going to discuss it," he told me. "It ain't Christian." Homer Williams, who owned a grocery and the island's first TV, was known as Dobbins. "I don't know why they called him that," Leon told me in the Situation Room. "He had one leg shorter than the other one."

"He had no neck," Bruce Gordy added.

Leon: "Think he was in a car accident or something."

"Did he get the name after his accident?" I asked.

"No, I don't think so," Leon replied. "Think he was always Dobbins."

During another Situation Room gathering, I asked Richard Pruitt why his father, the late Carlton Pruitt, was called Chiney. "I couldn't tell you to save my life," he said, chuckling.

Jerry Frank spoke up. "Miss Maggie Walter gave him that name. He and John Lewis Parks were coming down the lane, and she called out, 'Here they come, the Spaniard and the Chinaman!'"

"Did John Lewis Parks look Hispanic or something?" I asked.

"No, not really," Jerry Frank replied.

"Did your dad look Asian?" I asked Richard. He shook his head.

Don't expect it to make sense, Leon told me. The late Miss Maggie was pretty crazy. "There was a streetlight on the corner outside her place," he said, "and people couldn't walk past there without her saying something. People would walk all the way around to avoid it."

One evening at Swain I asked Marlene McCready, wife of Jackie Haskins, how Tangier old-timer George "Hambone" Thomas came by his nickname. She was stumped. "How *did* Hambone get his nickname?" she mused, turning to Principal Denny's wife. "Glenna, do you know?" Glenna did not, so Marlene buttonholed Hoot Pruitt, who was passing by.

"I don't have any idea," he answered. "Ask some of these old people."

She turned to the much younger Principal Denny. "Do you know how Hambone got his nickname?"

Denny, settling into the pew behind us, shook his head. "I don't know."

Marlene resorted to theory. "I'm sure it goes back to when he was little and something his mom was cooking. And a hambone got put in front of him or something."

"It doesn't seem to take much," I commented. I nodded across the sanctuary to tugboater Bill "Shoot" Parks. "How did Bill Shoot get his?"

Marlene called across the church to him. "Bill Shoot? How did you get to be Bill Shoot? How'd you get your nickname?"

"I don't know," the ruddy-faced, crew-cutted Shoot replied. "Something that happened early. It's hard to shake it."

"Yeah," Denny agreed, "once you have it, you can't get rid of it."

Now, as we sit in Ooker's boat, Mark Haynie pulls up in the *Sharon Kay III*. He's known as "Poopdeck." Years ago he wore a slouchy yachtsman's cap, which reminded some of the hat favored by Popeye's father, Poopdeck Pappy, in the old comic strips and cartoons. He hasn't worn the cap in decades, but he's Poopdeck forever.

WE FISH UP OOKER'S POTS on the island's bay side under a low titanium sky. A breeze comes steady out of the northeast at twenty miles an hour, but we're in the island's lee, and this close to shore we're shielded from its effects. Just a little farther west, the protection dissipates: Out in the open bay a big sailboat is bucking the wind through seas turned olive green and frothy, and having a time of it.

Ooker pays little mind to the morning's ominous cast, however, because the pots are coming up loaded with peelers and number ones. On the first row of twenty-seven pots, he harvests forty-two peelers. "If you do one peeler to the pot, you're doing good," he tells me, swinging the boat around to start the second row. The first pot clatters aboard carrying a peeler, a number two, and a couple of undersized jimmies that he tosses overboard. A half-dozen crabs wait in the next. He studies one recently molted sook, decides it's a buckram—a soft crab that has started to harden up, worthless to both peeler and hard crabbers—and throws it back, along with two little juveniles. He keeps a peeler and a number two. In the third pot he finds a buster, a number one, a number two, and a lemon.

"Man," Ooker says, "I wish I had all my pots over on this side today." As it is, he has 110 arrayed in four rows here and the remaining 100 on the island's far side, most of them just off the spit. They're borrowed pots, Ooker having beached most of his own to let the sun burn away the red moss that choked their mesh, and they're ancient. Leon bought these back before he switched to scraping and eventually sold them to Woodpecker, who used them until he quit chasing peelers, which was two or three years ago. And as Ooker has experienced in past years, the old pots have become clogged with moss in very little time. It hasn't put off the crabs, however. The second row yields thirty-four peelers, and the third, thirty-eight. Ooker is both pleased and puzzled. "I'm sure they see it," he says of the algae. "A little of it don't seem to bother them, but a lot of it usually will."

He takes five minutes to wolf down half of a PB&J and a cellophane-wrapped snack cake, and is encouraging the last few drops from a bottle of Yoo-hoo when a clump of seagrass floats by. "People talk about the crabs being down," Ooker says. "We had a slow start because of the cold weather, but I don't think the crabs are down at all. Some people, they make conclusions without any backing for them.

"Some guy saw all the grass that you get at this time of year floating on the surface, and he said the scrapers were doing it—they were pulling the grass up from the bottom. Well, you go to places along the bay side, like back in Shanks Creek, and you see the same thing there. And nobody scrapes there." He snorts. "These are the healthiest grass beds in the bay, around here. If scraping was doing it, they'd be long gone."

We start the fourth row. A buster comes up, half out of its shell. Soft crabs, all whales, turn up in three straight pots. One comes up with a peeler crouched on top, outside of the trap. Ooker snares it with a deft stab of his hand. "A lot of peelers here this morning," he says. "I guess they're in the stumps." I peer overboard into the water, which here runs about eight feet deep, but can't see the remnants of any drowned trees.

The next pot comes up empty—a surprise, given the day's bountiful haul. "Sometimes a crab pot will come up with nothing in it, and what's happened is that you've thrown it down on a stump, and it's tilted," Ooker explains. He turns to eye the island's western shore, about one hundred yards away. "Hard to believe this was a wooded area."

Indeed, the next pot is loaded with crabs, and after pulling the fourth row's last pot, Ooker does a quick tally: sixty-seven peelers on that row. "That's as good as you get," he tells me. "Real good. I wish I had some more pots here." He pauses before adding: "You got to be careful, though. You put too many pots out, they'll catch on."

He guns the outboard and we swing into the boat channel. Beyond the harbor we come into the wind, which is blowing hard now, and into three-foot waves that seem to be headed every direction at once. The boat bucks and thumps through the chop as we approach the southeast corner of Uppards, where Ooker has set a few pots. When he cuts the motor to snag the first buoy, the boat starts to rock aggressively. I grab the steering console and hang on. "Breezy," Ooker observes.

A couple of crabbers pass us, headed into port. One, a scraper, holds out a fist, thumb pointed down. The other, a hard-potter, draws a finger across his throat. Ooker waves and hollers: "That's all I needed to see!" Just the same, we pull up five pots, looping around each to keep the boat's nose into the wind as Ooker hooks the floats. All are empty. The wind gusts to thirty miles per hour. Standing in the boat now requires knee bends straight out of an aerobics class. "Kyowking!" Ooker yells.

"Excuse me?"

"Wind like this," he says. "Kyowking. Which isn't worth it, if the pots are empty. I'm calling it a day."

"BAD," LEON DECLARES of such days. "Rough." He makes the observation to a crowded Situation Room. Cook, Jerry Frank, Richard Pruitt, Allen Ray Crockett, and Bruce Gordy are here, as well as John Wesley Charnock, the town cop. All are drinking coffee from Styrofoam cups that someone bought by the thousands on a trip to a bulk warehouse; they're stacked neck-high next to the room's trash can, which remains under siege from a spiral of fruit flies.

I'm sitting beside Allen Ray and notice a tattoo on his left forearm. It looks to be a split-tailed bird, but Allen Ray's skin is roughened and cracked and tanned the color of tobacco, and the tattoo is impossible to make out. I ask him: "What is that—a barn swallow?"

Allen Ray looks down at the image as if he's forgotten it was there. "Woodpecker, I think," he says. "It's been on there so long I don't even remember. Can barely see it."

"It's had time to fade," Leon notes. He adds, since the subject has turned to the effects of old age, "Walking don't agree with me. Once you stop walking and start riding, walking gets hard." This from a man who just spent eight hours on his feet, hauling a loaded scrape onto a rocking boat.

Aging is a topic Leon frequently puts on the table. He may do so with an anecdote, as he did when discussing his recliner on another afternoon: "You set that chair back, and it ain't long before you nod off. And you wake up and think, 'Is I supposed to be leaving for work?' You wake up that sudden, and you wonder, 'Has I been out already, or is I gettin' ready to go out?'"

"I do wake up from a nod and I get to thinking," he told us, grimacing. "That's a sign of failing, if you're an old man."

More often, he introduces the subject with a mild complaint. "The years get different, the older you get," he announced on another day.

"They're shorter?" someone guessed.

"There's more trouble in 'em," Leon said. "More aches and pains."

"Reminds me," Jerry Frank now says, "of Ray Crockett years ago," Ray being Leon's father-in-law. "He went to see Dr. Smoot, the ear, nose, and throat specialist." The story's a long one and ends with Ray telling the doctor: "Look, I'm eighty. *Everything* is down 50 percent."

Ooker strides in, and as he's mixing powdered creamer into his coffee he announces he's mad, which he pronounces *my-yid*. Seems he had an exchange with an islander who thinks it unfair that the town charges twenty-five dollars for golf cart registration tags. Tangier is the only Virginia locality where it's legal to drive without state license plates; instead, the island's 285 motorized vehicles, all but a few of them golf carts, are required to display town stickers. This citizen told the mayor he didn't believe he'd be paying the fee.

"I said, 'Well, give it a try. See what happens,'" Ooker says. "'But I'll tell you what I think will happen: You'll put your cart up for the year.' He said, 'Well, it's the principle of the thing.' I said, 'You're right. It *is* the principle of the thing. You'll buy a tag or you'll park your cart.' But he just wouldn't let it go. Just kept at it. I said, 'Well, go ahead. Give it a try.'"

Leon evidently knows who this citizen is. "I'll bet when he walks to Canton and back he'll come up with that twenty-five dollars," he says. "Walking is hard."

Ooker busters up his peelers in the shedding tanks out back of his crab shanty. (EARL SWIFT)

NINE

M ARY STUART PARKS, A RED BANDANNA KNOTTED AROUND her head, surveys the kitchen at Fisherman's Corner. It is nine o'clock on a Saturday morning. The dining room opens in two hours, and Stuart, as she prefers to be called, faces a tumble of chores to ready the restaurant for the weekend's anticipated spike in tourists. She consults a list of the most pressing tasks to complete first: "5 lbs crab cakes. Thaw. Slaw. Squash casserole. Green beans. Bisque. Heat soup. Toast points. Hush pups."

Much to do, but—as she has learned over the seventeen years she's owned the place with Irene Eskridge, her first cousin and Ooker's wife—doable. Stuart crosses the room to a big industrial cooler that closely resembles the model Ooker keeps in his crab shanty. Lisa Crockett, first cousin to Ed Charnock and a ten-year veteran of the kitchen, is already there, and together they perform a quick inventory. It's rainy and cool outside, which will no doubt keep some tourists away, so the crab cakes they've made over the past couple of days—about fifty of them, incorporating ten pounds of crabmeat—will probably be enough. One item down.

Stuart pulls a cauldron of tomato-based vegetable soup from the cooler, ladles some of it into a large bowl, and stirs in fresh-picked crabmeat to create one of the restaurant's signature dishes, crab vegetable soup. It goes into a pot and onto the six-burner stove, part of a wide commercial range that occupies much of the room's back wall. The ventilation hood was hand-painted by Irene: HIS MERCIES ARE NEW EVERY MORNING.

As the women settle into their routines, their speed and energy rise. Lisa slides a dozen russet potatoes into the oven, while Stuart turns to a stainless steel island three feet away to tackle another specialty of the house, crab bisque. She combines evaporated milk, butter, flour—"No calories whatsoever," she assures me—and puts the mixture on low heat, working it with a spoon to make a roux that will give the bisque a satisfying thickness.

While Stuart stirs, Lisa pulls a large tray of toast points from the oven, an accompaniment to the Corner's popular crab dip. At the far end of the prep table, Ginna Giles, in her fourth year of cooking and waiting tables, mixes fresh-cut spring onions into the batter for crab hush puppies. She asks Stuart to taste it.

"It needs more hot stuff," the boss says. "Did you put Old Bay in it?"

"No," Ginna says.

"I'd put in some Old Bay and a little cheese."

The rest of the staff wanders in: Stuart's aunt, Dot Dize, the dishwasher; waitress Jennifer Bowden, an Eastern Shore native of Tangier descent, now married to an islander; and waitress Erica Daley, Ed Charnock's newlywed granddaughter. While Dot starts on the dishes, Jennifer and Erica make coffee and set up the salad station near the door to the dining room. Ginna prepares the squash casserole—today's featured side—and Stuart chops celery and halves shrimp for crab and shrimp salad. She interrupts the task to add crab and evaporated milk to the bisque and to pull a pot of green beans off the stove.

Shortly before eleven, Irene arrives to run the register. She and Stuart take turns overseeing the kitchen and dining room, while dividing the other duties of ownership. "Irene handles more of the administrative stuff," Stuart says. "We complement each other.

"We don't argue. We couldn't have worked all these years to-

THE LORD TELLS THE WATER

gether if we did. We try to run it as a God-centered business, as a Christian business."

It's also a family business. The restaurant gets its soft crabs from Ooker and Andy Parks, Stuart's husband, one of Tangier's few remaining scrapers. Ooker has told me he supplies six or seven dozen a day at the season's height. He often eats lunch in the dining room, witnessing firsthand the product of his long hours in the *Sreedevi* and his crab shanty. Here, on the plates of the tourists around him, his peelers' journey from bay to basket to bustering comes to an end.

The final chapter of that journey is not for the squeamish. Stuart, or whoever's cooking, places the living soft crab on its back and pulls away its apron. She takes the animal in hand and cuts off its face—including its eyes, antennae, and complicated mouth—with scissors. With the crab's front end excised, she can lift its upper shell to expose its innards and plucks out two clusters of long, spongy "dead man's fingers," the crab's gills. Contrary to pervasive belief, they're not poisonous, but their taste and texture are disagreeable, so out they come.

Stuart also scoops out the crab's bright yellow hepatopancreas, a mushy organ that filters the creature's blood. It has a sharp, somewhat bitter flavor, which contrasts with the crab's sweet and succulent meat. On the mainland, many restaurants leave this "mustard" in place, and many diners seem to love it. On Tangier, however, it's denigrated as "yellow mess" and removed. Though it's a decision based on taste, not safety, there's strong argument for siding with the island, for if a crab ventures into polluted water and absorbs its toxins, the hepatopancreas is where they accumulate.

Stuart then dunks the animal in batter, breads it, and either deep-fries or sautés it. If it's to be eaten on a sandwich, frying is the way to go, because a sautéed crab will make the bread soggy. True to tradition, the bread is always untoasted white. No need to dress up the proceedings with a bun—it will only get in the way of the crab's innate deliciosity. No, the less bread, the better; it serves only to ease the crab's journey from plate to mouth.

FEW RESTAURANTS—VERY FEW—can claim their crabs travel so direct a path from bay to plate as those served at Fisherman's Corner. Ooker sets his peeler pots within sight of the Tangier shore. He busters up his catch about five hundred yards from the restaurant's stove. Coming ashore, he carries his contributions to the menu across two hundred yards of Meat Soup from the *Sreedevi's* tie-up.

The soft crabs that Lorraine Marshall serves up at her namesake restaurant, about a dozen feet from Fisherman's Corner, are likewise the most local of products, supplied by Tangier crabbers. Order a soft-shell crab sandwich anywhere in bay country, and you can rest easy that you're getting a blue crab caught and bustered up on Tangier or elsewhere in the Chesapeake—or in the other American waters (North Carolina, Louisiana) where one can haul peelers aboard. But the farm-to-table experience you'll find in these two island restaurants sets them apart. No middlemen are involved. The provenance of their softshells is beyond question.

Down the road at the Chesapeake House, which serves a vast all-you-can-eat lunch and early supper built around its massive crab cakes, a patron can be equally assured that the crab on her plate originated nearby. The restaurant buys its crab from Lindy's Seafood, where Tangier hard-potters sell their catches. Lindy's steams and picks the crabs and delivers the meat to the Chesapeake House, ready for fashioning into cakes.

Off the island, however, bona fide Chesapeake Bay blue crab has become elusive—at least in terms of the steamed and picked crab-meat that goes into crab cakes, crab dip, and other regional specialties. That's even true in restaurants that trumpet their connection to the bay and its traditions—that decorate their walls with crab pots, buoys, and photos of watermen and deadrises, and that tout their timeless Chesapeake Bay recipes. All too often, there's nothing Chesapeake Bay about the crabs they put in their dishes. Many Mid-Atlantic restaurants, if not most, use pasteurized crabmeat from thousands of miles away. As for restaurants outside the region: assume their crab is imported.

Economics drives this state of affairs. The bay's crab population fluctuates—at times alarmingly—which over the past thirty years has prompted both Maryland and Virginia to tighten their harvest regulations to safeguard the species. Even in good years, the supply of fresh Chesapeake Bay blue crab is limited by the animal's life cycle, in that it's unavailable from December to March. So restaurants and their suppliers sought a substitute, and have found it in the Asian swimming crab—similar to the blue crab in many particulars, plentiful throughout the Philippines and Southeast Asia, and far cheaper to obtain. Most of the crab you'll find in the supermarket, whether canned or refrigerated, is imported, too.

Why should you care? First, the influx of imported crabmeat affects the bottom line for crabbers on Tangier, Smith Island, and the many small watermen's communities around the Chesapeake. Imported crab cuts into the earnings of the companies, like Lindy's, that buy, process, and market the bay's crabs. And not least, the Asian swimming crab tastes nothing like the Chesapeake Bay blue crab. To me, the homegrown crab is sweeter and more luscious than any other. I'd argue that its distinct flavor even sets it apart from the *Callinectes sapidus* fished up in Carolina and along the Gulf Coast—that within the same species, place affects taste.

I'm not alone. Aficionados can detect the difference readily. "I'm from Mississippi, so I grew up eating blue crabs down there. They're delicious," said Hampton Roads chef Sydney Meers, who has earned a reputation for imaginative dishes and flamboyant style at a succession of storied restaurants over the past thirty years—and whose sautéed softshell had a transformative effect on me when I first encountered it in the late eighties. "But when you get them up here, they're a different kind of flavor. Chesapeake Bay blue crabs are sweeter."

Such nuance is lost on many diners, particularly those who haven't grown up eating the bay's crabs. "I don't think you can get no crab anywhere else that'll beat this for the taste," Ed Charnock told me. "But I guess those restaurants don't care as long as it's cheap. People get a crab cake, and they don't know how it's supposed to taste, so they don't know the difference.

"I'll tell you one thing," he said. "It makes a difference to us."

Back at Fisherman's Corner, two parties of diners enter. An order reaches the kitchen: a crab cake sandwich and a side salad. The waitresses assemble the salad and hurry it out the door. Stuart warms a half inch of oil in a high-sided pan, and after waiting a few minutes to allow the customer to start on his greens, she sets two roughly spherical cakes into the pan. They sizzle loudly. She flips them to brown and caramelize the other side, but plucks them out in under a minute— the crab inside has already been steamed and is easily overcooked. From the pan they go into the microwave just long enough to heat through, and from there to the plate.

In walks Ooker, carrying a Ziploc bag of crabmeat, which he hands to Stuart. Fisherman's Corner has a menu side called "Chesapeake Bay Soft Crab Treasures," and here's the chief ingredient: pieces of soft crabs that have lost most of their legs and have thus become what watermen call "doorknobs." Rather than try to sell them damaged, Ooker cuts the soft crabs into meaty chunks.

Stuart pulls four from the bag, batters and breads them, and tosses them into the deep fryer. A couple minutes later she places them before me, warning, "Give them a minute. Let 'em cool." They're each the size of a hush puppy and unrecognizable as softshell—until, unable to wait more than a few seconds, I take a bite.

The shell gives way with a perfect snap. Inside is an unbroken lump of tender, sweet, and remarkably juicy meat. If there is a better food in all the world, I haven't encountered it. And the waterman who caught it, and who plucked it from a shedding tank just minutes ago, is loitering a few feet away.

"How are they?" Stuart asks.

"Amazing," I tell her.

"Well," she says, "they're certainly fresh."

FISHERMAN'S CORNER has operated under its present ownership since May 2000, when Stuart, Irene, and two other island women threw in together to buy the business. "None of us had any idea. We were just housewives," Stuart says. On opening day, "we cried, we were so scared. We had no training at all."

The other two partners eventually moved away, leaving the cousins in charge. Both Stuart, born in 1954, and Irene, born in 1959, grew up in Rayville, a snug cluster of houses in Meat Soup that was named for Irene's maternal grandfather. The two have known each other since childhood and remained close as adults—and even more so as business partners. "We worked together and found each other's strengths," Stuart says, looking around the kitchen. "It's changed a lot in here. Like, we originally had the dishwashing over here"—she points to a sink near the range—"and we were constantly running into each other. Every year we would learn how to make it all flow a little better.

"Tables of ten, it don't even faze us anymore. We used to be, 'Oh good Lord, how are we going to do this?' Now it's nothing."

This feat is all the more remarkable for the fact that none of the restaurant's founders came into the venture with a lot of capital; they were all married to watermen and hostage to crabbing's uncertain revenues. All had busy lives with family and church. Stuart's two sons were grown, but she was still raising a young daughter. Irene carried a titanic load, and not just because she was married to the island's emerging mouthpiece. The Eskridges' younger son, Joseph, was still in high school at Tangier Combined. And then there were the four girls.

Irene and Ooker had long talked about adopting. The subject moved to the fore after Irene suffered a miscarriage while carrying their third child. After learning that a domestic adoption might take years, they chose the expensive but speedier option of seeking a daughter in India. Sreedevi, four years old, arrived from an orphanage in the Indian city of Hyderabad in 1996. "Lice? She was covered with lice," Leon recalled. "Four of us went up to get her when she come in. We all had to get RID—we went to a store in Salisbury. We had lice all over us."

Leon's sentimental recollection notwithstanding, the entire town embraced the quiet little girl. Her assimilation went so well that two years later, the Eskridges sought to adopt a second daughter. They aimed for a second or third grader, so that Sreedevi would have a playmate. That brought them Devi, from the same orphanage, whose operators claimed that she was seven. She was more likely nine or ten.

The orphanage also told the Eskridges that Devi's mother had given her up for adoption and that both of her parents had since died. That was another fiction—she'd been kidnapped and sold to the orphanage. "It said on all my papers that my mom had me out of wedlock and couldn't keep me," Devi recalled nineteen years later. "I thought, 'That's a funny thing, because I was with her.'"

On the plane to Washington Dulles Airport, an Indian adoption agent coached her to say "Hello, Mother" and "Hello, Father." That was the only English she knew when she stepped onto U.S. soil. "I wasn't scared, but I was nervous," she told me. "I didn't want to be there. I was homesick. When the doors opened to the airport, they pointed to James and Irene and said, 'Those are your new parents.'"

Despite this traumatic introduction, the outgoing and adventurous Devi took to Tangier. For one thing, it seemed familiar. "When I lived in my home of Kakinada, the business was mainly fishermen," she said. "I would come into town to sell spices and would watch them come in with their boats and nets. And when I got to Tangier, I was surrounded by seawater, and most of the men are fishermen."

Irene homeschooled her in English, labeling everything in the house with Post-its—beds, doors, salt and pepper shakers, the fridge. Devi caught on fast, which isn't to say it was easy. "You take children from India who can't speak English," Principal Nina Pruitt observed, "and move them to Tangier, where some people think we can't speak English, and you're in for an adventure."

Once she had tentative command of the language, Devi explained to Ooker and Irene that she'd been stolen from her mother. Horrified, they contacted an Alabama couple who'd adopted two sisters from the same orphanage the same year, and with whom they regularly spoke. When the couple approached their girls with Devi's story, they learned

that the sisters had been stolen, too. The discoveries spurred investiga-
tions and the prosecution of orphanage officials in India. The Alabama
couple still actively campaigns to reform the adoption process.

Both the Eskridges and the Alabamans gave their children opportunities to return to India, to meet their surviving relatives, and
to choose where they wanted to live. All decided to remain in the
United States. Meanwhile, to distill a long and complicated story to
its essence, the Alabama couple sent their two girls to live with the
only people in America the youngsters knew from India, those being
Sreedevi and Devi. In 2003 the Eskridges took Bhagya, thirteen, and
Manjula, fifteen, into their already crowded household on the West
Ridge.

And so it was that Tangier became home to a tiny but celebrated
Asian population. "They had a good childhood. Oh, yeah," Leon said.
"They accepted them like crazy around here."

So, yes: Irene had her hands full when she cofounded the restaurant. And her life didn't get any less complicated for years. The oldest
of the girls, Manjula, graduated from Tangier Combined in 2005.
The youngest of her six children, Sreedevi, didn't graduate until 2010.

Nowadays, all the girls live on the mainland. Manjula is married
to Warren Eskridge, a barge captain and the son of Ooker's brother
Allen Dale. They have two children and live on Maryland's Eastern
Shore. Bhagya is a hairdresser in Birmingham, Alabama. Sreedevi is
a nanny and studying to be a teacher in Austin, Texas.

And Devi, who visits the island frequently, lived in Delaware,
Florida, and the Eastern Shore before moving to Lovingston, Virginia, just east of the Blue Ridge, where she works as a nanny. She's
an hour's drive from Ooker and Irene's son Joseph, who has lived in
Lynchburg since leaving Tangier for Liberty University. They often
attend church together.

WHEN NO CUSTOMERS ARE ABOUT, the kitchen crew at Fisherman's
Corner will decamp to a table out front, to thumb through a pile
of magazines and catalogs they keep on hand for such breaks. The

publications are smudged, their paper crackled and supple with age. One is a *Country Living* from 2006. Lisa Crockett studies an issue of *Southern Living* dating to 2010. "We look at 'em like we've never seen 'em," she tells me. She points out an ad. "They probably don't even make that anymore."

A call comes in for a to-go order—fried pickles, fried green tomatoes, hush puppies, and two cups of crab bisque—from workers at Daley & Son. The staff hustles into the kitchen. Lisa uses an ice cream scoop to dollop Ginna's crab hush puppy mix into the deep fryer.

Everyone who works at Fisherman's Corner is female. That's true, too, of the crew at Lorraine's. The relative dearth of testosterone on either premises might help explain how the competitors have coexisted on good terms for so many years. "It's not the greatest thing to have the two restaurants next door to each other," Stuart says, battering up pickles for the fryer. "You'll see tourists come up and look at one, then the other, trying to decide, and it's nerve-racking on a slow day.

"But we get along very well. We run out of fries, I get 'em from her. She runs out of something, she comes over here. That's the way it's always been, and God provides business for all of us."

The Chesapeake House, in business for nearly eighty years, was also founded by a woman, was operated for decades by her daughters, and is now managed principally by Glenna Crockett. She owns the place with her husband, Principal Denny. They employ an all-female staff.

Walk into most of Tangier's terrestrial workplaces, and you'll likewise find few men on the payroll. The gift shops, which do a healthy summer trade in souvenir ball caps, T-shirts, and sweatshirts, along with handmade arts and crafts and imported kitsch, are the province of women. They dominate the tour buggy operations, offer the island's only salon services, staff the museum's front desk. The Bay View Inn, the only year-round bed-and-breakfast, is owned by a married couple but overseen day to day by just the female half: Maureen Gott, a come-here from New Jersey.

Brett Thomas drives the mailboat, but it's his mother, Beth, who runs the business end of the operation. Terry Daley Jr. manages the grocery store with his son, Lance, but he answers to the matriarch of the family—his mother, JoAnne, who's owned the business since 1986.

Leadership of the island's public institutions is decidedly female, too. Since 2005, Nina Pruitt has served as principal at Tangier Combined School, where she oversees twelve teachers and nine staff. Only one, Duane Crockett, is a man. Inez Pruitt went to school on the mainland for five years, commuting on the mailboat for the first three and coming home only on weekends the last two, to earn her post as the island's physician's assistant. Her daughter, Anna Pruitt-Parks, a town council member since 2006, is emerging as one of the panel's strongest voices—and is also Tangier's always-on-call paramedic and the only full-time employee of the Tangier Volunteer Fire Department. The post office is run by women. And while Ooker is mayor, the details of administering the town's business have long been the task of female town managers.

With most of the island's able-bodied men away in boats most days, it's also left to Tangier women to disseminate information and conduct the social transactions that keep island life running smoothly. Their communications network is reliant on the landline telephones still present in almost all homes, and it's lightning fast. When the rooftop siren at the Tangier firehouse wails—signaling serious injury, illness, or, worst of all, fire, and sounding exactly like the London air-raid warnings in old movies, which makes its air-driven warbling even more ominous—phone calls instantly flash among the ridges, buzzing with questions and guesses as to what's going on. In minutes the entire island has the answer. Good news travels only a little slower.

I experienced the network's efficacy during my stay in 2000. I was quartered at the Bay View Inn, which at the time was run by Ed Charnock's sister, Shirley. As I ate breakfast one morning, Shirley asked me what I had planned for the day. I replied that I was going to the Main Ridge to see a woman who'd agreed to give me her collection of recorded Swain Memorial sermons, preserved on cassette tapes. Both churches make such recordings for the island's shut-ins; these days they're on CD. The churches burn through a lot of discs.

Shirley picked up the phone as I stepped outside. As I was going over on Wallace Road, I encountered an islander coming the other way, who said, "I hear you're going to see Mrs. So-and-So to get her tapes." As I reached King Street, another islander gave me directions to the woman's house without my asking for them. By the time I got halfway down the Main Ridge, everyone on Tangier knew where I was going and why.

I took a lesson from that. A visitor to Tangier had best assume he's the object of almost anthropological study and conduct himself accordingly. The islanders might complain that tourists eye them a bit too closely, but it cuts both ways.

Island men exchange information, certainly—at the Situation Room and in similar daily gatherings at the Tangier Oil Company, the combination fuel pier and marine hardware store known as the "oil dock." But they spend most of their days alone or in the company of one or two men aboard their boats. Their networking, limited to periods around and between the obligations of work, lacks the immediacy of the women's contacts. By the time the men get around to a subject, odds are good that it's been hashed over thoroughly by their wives and that the island's collective opinion on the topic has started to take shape.

That point bears repeating: Tangier women shepherd the island's thinking. More often than not, they're the catalysts for action as well. If money is needed to meet a need at school or church, it's the women who raise it. When islanders decided to send a video about Tangier to every member of Congress, it was Anna Pruitt-Parks who planned and led the campaign. Some veterans' graves lacked flagpoles, which islanders thought a shame—until Carol Moore spearheaded a drive to buy and erect them.

Bottom line: In just about every aspect of island life but two, working the water and running the churches, women are in charge.

And the churches are only ostensibly ruled by men. Influence within the congregations is largely wielded by women.

When I ran this observation past Devi Eskridge, she was quick to confirm it. "People will say it's men who are boss," she said. "But it's women. If Mom died tomorrow, my dad wouldn't know where his socks are."

The dependence goes beyond questions of wardrobe. Soon after Annette Charnock married Ed, she learned that he knew nothing of running a household or managing his finances. He'd never so much as written a check; his late wife, Henrietta, had overseen virtually every land-based aspect of their daily life, and his daughters had assumed the duty after she died. "I remember one day I was cooking dinner," Annette recalled, "and I said, 'Ed, would you mind running to the mail? I don't want to leave this.' And he said, 'I don't know how.' I said, 'What do you mean, you don't know how?' He said, 'I've never done it.'

"I said, 'You mean to tell me that you've never been to the post office, put a key in the mailbox and turned it, and taken the mail out?' He said, 'No.' And I said, 'Well, I think it's time for a lesson,' and we went on a field trip.

"He said, 'I feel like a child.' I said, '*Well*.'"

Back at Fisherman's Corner, I take a seat in the now-bustling dining room. Ooker sits across the table from me. "Who's boss on Tangier?" I ask him. "The men or the women? Who's the *real* boss?"

Five or six feet from our table, well within earshot, Irene is bent over paperwork behind the register. "Well," Ooker says in an unusually loud and authoritative tone, "I'd say that the women have their *input*." He half glances over his shoulder. Irene does not look up from her work.

"You receive that input," Ooker continues, at a slightly higher volume. "You consider the source." He throws another quick glance her way. "Some of it, after you consider the source, it don't go no further than that." He hazards another look.

Irene isn't listening to a word he says.

The shifting moods of Leon McMann, June 2017. (Earl Swift)

TEN

COME THE END OF JUNE SUMMER HAS SETTLED HEAVY OVER
Tangier. Temperatures soar well into the nineties most afternoons, the humidity rising in tandem, poaching the shadeless island. Absent a breeze, the flies launch from the marsh and come
hunting for blood. The roads empty.

The sticky heat can be even less bearable out in the boats, where
the labor of pulling scrapes and pots, combined with the sun's rays reflecting off the water, can render a man woozy and half blind. But the
effort pays well this time of the year, for the Chesapeake has warmed
as well, prompting the crabs to move from deep water to the shallows
and, with some encouragement, into the watermen's bushel baskets.
Hard-potters are catching their limits. Peelers abound.

A crabber has to keep his eyes on the forecast, however, because
with the heat comes instability. A morning might be kyowking, as
Ooker put it, and that very afternoon, dead calm. A wind might blow
hard out of the southwest for days straight and build seas that make
for rough going off the island's western shore and in the harbor, too.
A persistent nor'wester might drive the tides down. A southeasterly
might push them up.

To a point, Tangiermen treat wind as a nuisance, rather than a
threat. If they know to expect it, they'll usually be able to work in it—

and they're inclined to do so, for the crabbing season doesn't include makeup days. One Sunday evening, having made plans to join Ooker on his boat the next morning, I approach him after a New Testament service to ask when and where he wants to meet. He holds up a hand. "It speaks of wind in the morning," he says.

"It speaks of wind," I echo.

"Yes," he says. "It speaks of wind."

"So, are you going to go out?" I ask.

"Yeah, I'll go," he replies. "But you might want to hold off. If the wind's up, I don't think you're going to be too comfortable in the boat."

"Well, if you're going, I think I might like to tough it out," I tell him, "as long as it's not *shouting* of wind."

He suggests that we talk in the morning, in case I change my mind. When my alarm goes off at four, the wind is buffeting the darkened house, and the chimes hanging from the underside of my deck are banging around crazily. I step outside. A blow's coming steady out of the west at what must be thirty miles an hour. I can't see the water, but I can hear its hiss, along with the thumps of breakers hitting the shore beyond the airstrip. There's no way, I decide, that I'm spending the day in an open boat in water I can *hear*. I leave a message for Ooker telling him so, and go back to bed. When I wake again the sun's shining bright, but the wind is blowing as hard as ever. I step back outside. The bay is stippled in whitecaps. In their midst are several deadrises, their captains fishing their pots. Later, conversation in the Situation Room centers on how hard it "blowed" out there, but no one seems much put out by it.

No, it's the weather they don't see coming that can get a waterman in trouble, and the Chesapeake serves up plenty. A few days later, after a quick trip home to cut the grass and answer mail, I'm headed back to the island aboard Mark Crockett's *Joyce Marie II*, the summertime passenger ferry linking Tangier to Onancock, the nearest Virginia port. It's a sixteen-mile passage across both Pocomoke and Tangier sounds, and it can get rough with wind from pretty near any direction. Mark, regarded as an especially gifted boat handler on an island of experts, keeps a close eye on the forecast.

The outlook for today was calm and hot, and so it is. But as we near Watts Island, dividing the sounds, Mark calls my attention to a thunderhead towering over the western shore. It's the color of slate, thousands of feet high, takes the classic anvil form that advertises trouble—and it wasn't there a few minutes ago. We watch as it slides down the shore and over the water far to our south, moving at highway speed.

Storms like that, moving so fast they're on you before you know they're coming, are not infrequent on the Chesapeake. Some hot summers they sweep over the bay almost every afternoon, and though they pass quickly, often lasting just a few minutes, they can spawn blasts of hurricane-force wind, deadly cloud-to-ground lightning, and near-biblical rain. The morning after I ride over with Mark, a squall races across the bay north of Tangier and slams into Crisfield with winds estimated at eighty miles an hour—some reports even claim one hundred—and wrecks boats, topples trees, and snaps the stern line of the *Steven Thomas*, a big Tangier tour boat tied up at the waterfront. The line in question is an inch and a quarter in diameter.

"A squall can blow five miles an hour or can blow a hundred," as Leon advised in the Situation Room. "You never know what it'll do."

"I know one thing," Allen Ray said. "When I see one of 'em coming, I respect it."

Leon nodded sagely. "You have to."

"Just have to bring that bow around into the wind," Allen Ray said, "and try to hold it. And that's a job."

"It is that," Leon agreed. "And the trouble is it gets to blowing so hard you can't tell what direction it's a-goin'."

Some violent storms, especially in the spring, can sprout dark tentacles that pack a more selective punch. Ooker's told me that even a small waterspout, about shoulders' width across, can spin a forty-foot workboat, and anything larger would lay waste to it. One morning in 2015, Tangier witnessed a procession of thunderheads chug up the bay. Islanders counted nine waterspouts hanging from the clouds at once, and before the storm passed, nearly forty all told. On another occasion, Ooker said, a waterspout touched down in the creek not far from his crab shanty and moved ashore behind the grocery store. He was north of the school and saw that it was headed toward his house on the West Ridge, so he started that way. En route he passed a weeping willow, all of its branches standing straight up.

Full of water, the twister took its time. "It went across Wallace's Bridge and picked up a dinghy there," Ooker said, then lifted off the ground and ghosted over an aboveground pool he had in the yard. "We had some swim rings—some orange swim rings—in the pool, and it picked those rings up. Before I could get in the house it passed directly over me, and I could look up into it, and I saw crab boxes, baskets, a lot of stuff in it. Then it moved over the water, and I could still see those orange swim rings in it." He paused. "I didn't get those back."

I CAN ATTEST TO THE POWER of a summer squall and to just how frightening one can be. In the summer of 1994, I convinced my editors at the newspaper to buy a sea kayak and let me paddle it in a five-hundred-mile circle around the Chesapeake, filing stories and pictures as I went. I pushed off from Norfolk, paddled twenty-odd miles across the Chesapeake's mouth to the southern tip of the Eastern Shore, and started north from there, my boat loaded with food and camping gear.

Three days into the voyage I pulled into a wide break in the shoreline at the mouth of Mattawoman Creek and beached for the night on tiny Honeymoon Island, a lump of sand in the creek's middle sprouted with beach grass and a few water bushes. I set up my tent, broke out my stove, and cooked dinner. Then, as darkness approached, I crawled into my sleeping bag to read by headlamp before turning in. I was immersed in a book when, at about nine, I heard a low, long rumble of distant thunder. I paid it little heed. Not three minutes later I heard another snarl—this one much louder, and deeper, and closer. And just seconds after that a gale blasted the tent with sudden, extreme force, ripping up the stakes and prying up the floor and rolling the shelter onto its side before I had time to scream. I threw myself to the tent's windward side and stretched to pin down the corners with hands and feet, while from outside came the sounds of my cook set skittering away and the kayak sliding on the sand. I heard that for only a moment, though, because now came a deluge pounding the tent, and lightning in a flurry, bolts striking by the score, so close that the ground bounced under me, their blue-white strobes blinding through the tent's two layers of nylon, and the sound of this hellstorm—the roar of the wind and rain, the concussions of the thunder—blotting out my every thought except that I was about to die.

My tent had an aluminum frame. I was trapped in a cage of conductive metal that stood tallest of anything for a quarter mile in any direction. I was certain the lightning would find me. As fast and close as it came, it seemed impossible that it wouldn't. For twenty-five minutes I crouched inside the tent, wrestling the wind to keep its floor down, listening to the sky make sounds I'd never heard and haven't since—like great sheets of fabric ripping and fighter jets buzzing just overhead. And layered on top, the cacophony of the strikes. And then, as suddenly as it started, it stopped. After a few retreating rumbles, the creek fell quiet.

The floor of my weatherproof tent was under an inch of water. My sleeping bag was sodden and all my gear soaked. I was so spent that I hardly noticed: I have probably been more frightened in my life, just for a moment or two, but never have I been so terrified for so long. I bailed out the water as best as I could, collapsed on my wet bag, and slept like a boulder.

Mind you, I was on land. I can't imagine what it would be like to encounter such a storm on open water in a small boat. I hope to never find out.

HEAVY WEATHER ON THE BAY is not relegated to the warm months. Fierce storms crop up in the dead of winter, too, and can likewise come out of nowhere. Among Tangier's most storied tragedies took place in January 1896, when a squall out of the northwest pounced on William Henry Harrison Crockett, great-grandfather of Ooker and great-great-grandfather of Carol Moore, as he sailed home after delivering a load of oysters to Washington, D.C. Wind hit the captain's two-masted schooner with such force that the vessel keeled over, tossing Crockett and his three crewmen into the icy Tangier Sound. One of those three was Crockett's son-in-law, Tubman B. Pruitt, also Carol's great-grandfather and Ooker's great-uncle.

"She sank at once . . ." an Eastern Shore newspaper reported. "Captain Murphy of the police boat did not get half way to the sinking boat before it went down, and when he reached the scene of the disaster nothing was to be seen but the wild waste of roaring waters."

Tangier felt "its loss most keenly," as Swain's church board put it in a resolution a few days later. Especially that of Crockett, who was an exhorter and Sunday school superintendent. "In these men, it has lost highly esteemed and very worthy neighbors and citizens, whose places will be hard to fill, whose presence will be sadly missed and whose memory will be long cherished." Long cherished it has been: The men are still talked about more than 120 years later.

Another hard loss came in February 1914, when William Asbury Crockett, assistant keeper of the offshore Tangier lighthouse, encountered a sudden and powerful blast of wind in the midst of a larger winter storm. Crockett, who was Jerry Frank Pruitt's great-grandfather, was returning to the light from his daily run to the island for mail and groceries when the gust took him by surprise. Depending on who's telling the story, it either capsized the boat or jibed his sail, knocking Crockett overboard. Either way, he drowned.

The years since have seen a procession of islanders claimed by violent winter weather. James E. "Puck" Shores drowned in such fashion in November 1965. He was the father of Rudy Shores, a peeler crabber whose shanty sits alongside Ooker's and who's married to Ooker's sister-in-law. Harry Smith Parks, JoAnne Daley's brother, disappeared in April 1989, after he reported engine trouble while motoring his forty-foot deadrise, *Miss Annette*, home from the lower Eastern Shore.

And then there's a much more recent mystery, one that every Tangierman over twenty recalls with sadness and frustration. James Donald "Donnie" Crockett was seventy-seven, a lifelong waterman, and a man who kept his head in difficult circumstances. He lived in Meat Soup behind Leon's place. In a trim two-story house there, Donnie and his wife, Eldora, had raised four sons who followed him onto the water.

Eldora died in 1982, and most every evening after, Leon's wife, Betty Jane, would knock on her kitchen window as Donnie walked past and offer him leftovers from the supper she'd cooked. He later lost his oldest son, Don, to cancer. With his other boys grown, he now stayed busy by potting in the summer and oyster dredging in the winter, fixing bicycles, and tending to a large and growing family of cats.

Even on an island overrun with the animals, his brood was noteworthy. He kept more than twenty cats, had a name for each— biblical and weather references mostly, as in King David and Solomon, Foggy and Frostbite—and looked after them closely. Getting home to feed them was a daily priority. And it was one of his pets, a gray tabby named Spottie, that prompted Donnie to set out for Crisfield on the blustery morning of March 8, 2005.

It spoke loudly of wind that day, in the form of a fast-moving nor'wester that had prompted the National Weather Service to issue a gale warning. His sons pleaded with him to stay off the water. But Spottie needed to be spayed, so Donnie Crockett left Tangier in his forty-foot box-stern deadrise, *Eldora C.*, to deliver his cat to her veterinarian. The twelve miles to town, a trip he'd made thousands of times, was uneventful.

By the time he picked Spottie up for the trip home, the conditions had turned and were worsening by the minute. The temperature plummeted. Freezing winds raked the Little Annemessex; in fifteen minutes, they jumped from twenty miles per hour to forty-five. Snow blew sideways so thick that visibility was near zero. The storm was so fearsome that Leon, headed back to Tangier with Betty Jane after visiting their daughter Carolyn on the Eastern Shore, doubted that even the sixty-four-foot mailboat was safe. "I seen that rim to the northwest," he told me. "Just when we got to the mailboat, it came in. We went by the dockside, and I looked out and decided not to go."

Indeed, the mailboat's skipper—Rudy Thomas Jr., Brett Thomas's father—had second thoughts once he set out for Tangier. He had a good many passengers that day, among them Donnie's grandson Twin Denny and Twin Denny's wife, Danielle, and their two small boys, and Denny's sister-in-law Andrea, pregnant with another of Donnie's great-grandchildren. The seas they encountered ranged from four to six feet, and the big, steel-hulled mailboat heaved and tossed. Spray froze thick to the rails and lay heavy on the weather decks, compromising the boat's handling and balance. Rudy would have turned around, he said later, but feared coming broadside to the waves. When they reached the island after a white-knuckle hour, Rudy told his passengers, "I don't know who's dumber—you for coming aboard or me for leaving the dock."

Donnie, meanwhile, got back to his boat, hauling Spottie in a carrier. He had a sister in Crisfield. He could have stayed. Perhaps he worried his other cats would go hungry. In any event, he climbed aboard the *Eldora C.* and headed for home. "It was an awful time, blowing about fifty," Leon told me. "And he went right out in it."

He had company. Dorsey Crockett, another Tangierman, was eyeing the weather at the Crisfield waterfront when Donnie's boat chugged past, and he reacted in a manner common among watermen: "I figured if he could make it, so could I." He fired up his own dead-rise and tailed the *Eldora C.* out of the harbor, overtaking it a short way out. He and Donnie exchanged waves as he passed.

By this time, Tangier Sound was unfit for any vessel. The light station off the Tangier beach recorded winds of fifty-eight miles per hour. Once the boats left the shoreline's protection, six-foot waves crashed over the decks and froze to glass and wood. The men could see nothing through the whiteout of blowing snow.

"This is rough, ain't it?" Donnie radioed to Dorsey.

"Sure is," Dorsey replied.

Minutes later, the *Eldora C.* disappeared.

Dorsey Crockett, running blind, was navigating by radar when he saw the blip representing Donnie's boat vanish from the screen. "I tried raising him on the radio," he told the *Virginian-Pilot*, "and there was nothing." Less than a mile separated the boats, but going back was out of the question. "There weren't no letup," Dorsey said. "I ain't been out in no rougher."

Word reached Donnie's sons that their father was missing. They struck out into the storm to look for him. As they left the harbor, Lonnie Moore pulled in, having left Crisfield about forty-five minutes behind Donnie. "That was one of the roughest days I've ever been out in," Lonnie said. "We saw his son Will going out as we came in the creek, but we didn't know Donnie was missing." The men searched the crazed sound until dark. The Coast Guard joined in with search-and-rescue helicopters and a C-130 fitted with night-vision equipment. In the days that followed, Tangier watermen and boats from the Virginia Marine Resources Commission and Maryland State Police zigzagged over a 450-square-mile slice of the bay, on the lookout for any sign of the boat or man.

A diesel-powered deadrise has many loose parts that float, among them the big plywood lid that covers its engine box. Nothing came to the surface. For the *Eldora C.* to go down without marking its position, and without a Mayday from Donnie, suggested to his fellow watermen that the sinking was instantaneous, or nearly so, and that the boat settled upside down on the bottom. In the conversations that inevitably follow any tragedy on the water, Tangiermen traded theories as to what might have brought the end so fast. Donnie's boat was forty-odd years old and reputed to have a balky bilge pump; it could have simply taken on more water than the pump could bail. A rogue wave could have come over the low stern and overwhelmed the vessel. Or his oystering rig, which Donnie hadn't removed at the season's close, could be to blame. The device might have thrown the *Eldora C.* off-balance in those heavy seas, especially with ice adding to its weight. The boat might have flipped without warning.

For days, then weeks, then months, hope prevailed on Tangier that Donnie Crockett's body would be found, or that the bay would turn loose pieces of the boat so that searchers would know where to look. History, or at least tradition, was on their side. Months passed before Puck Shores's body washed up, but so it did, the spring after his boat sank. Harry Parks's body was found more than three weeks after he went down, clear across the Chesapeake.

When William Henry Harrison Crockett drowned in 1896, his body went missing, too. Five months passed, the story goes, until his distraught widow paid a call on Charles P. Swain, the Methodist pastor, and told him she feared her husband would never receive a decent burial. The pastor told her he had been praying on the matter and that he was sure the captain would come home soon. Within days, the tides carried his body up the ditch right behind his house.

Donnie Crockett's family was not so fortunate. The only trace ever found of him was a life ring from the *Eldora C.* that washed up on Watts Island.

WHEN WILLIAM HENRY HARRISON CROCKETT DIED, the Methodist congregation that he'd so faithfully served was housed in a small, plain building in Meat Soup, snug every Sunday and stuffed to overflowing for weddings, funerals, and holidays. Pastor Swain suggested that the growing and godly town required a bigger house of worship, and so began a campaign to build one. The new church, finished in 1899, was an airy, soothing place that could seat six hundred or better, lit by gas chandeliers and an abundance of stained glass. It instantly became the center of island life.

The year after its debut, Charles Swain moved to Deal Island, on

Maryland's Eastern Shore. Tangier was sorry to lose him, for in his five years there he'd transcended the typical role of pastor. He'd been a compelling preacher, to be sure, and a balming witness to the sick. He'd marshaled the island to build a church equal to any on the Eastern Shore. He'd also embraced the town as his own, carried himself as a Tangierman, and promoted and defended the place to the greater world. His slim *A Brief History of Tangier Island* survives in libraries here and there, and when the *Washington Evening Star* published a July 1899 story asserting that many islanders "know as little about the civilized world as a child," and that "women nearly all go barefooted," and that "girls here sixteen years old will measure six feet and weigh two hundred pounds," and that Tangiermen "have signs for everything, and almost worship the moon, by which they foretell storms and all kinds of disasters," Swain was quick to do battle. "There is hardly a truth in it," he wrote to an editor, in a lengthy treatise that dismantled the story sentence by sentence. "We are not out of the world, but it would be a blessing if some newspaper correspondents were."

The incident for which he was best remembered, however—and which spoke to his devout faith and close relations with the Almighty—was that prediction about Captain William Henry Harrison Crockett's homecoming; even today, you won't stay on the island long without hearing the story. Swain left Tangier its most revered figure since Joshua Thomas, which explains the overwhelming grief the town felt when, not long at his new post, the pastor died of pneumonia. He was just forty. The congregation voted at once to name its new church for him.

Swain Memorial's bell summoned a population in the midst of an explosion—from 590 souls in 1880, to 900 by Sugar Tom's reckoning in 1890, to 1,064 ten years after that. In most respects, the crowded little town remained a throwback. Everyone walked everywhere— besides a few wheelbarrows used for delivering freight, the island was without land vehicles. The Heistin' Bridge's deck was just three feet across, and the road to Canton was the width of three wooden planks.

Children tended to stick to their own neighborhoods, rarely venturing from their home ridge to another or even from Meat Soup to King Street. Only the Methodists of Uppards ranged far afoot, and then only on Sundays: They'd walk to church on a doglegging path across the marsh and over a rickety footbridge that landed in Meat Soup at the north end of Main Street.

"The road that they made from their houses down to that bridge, they made from the ashes from the stoves they had in their homes," Jack Thorne told me. When he was a boy, he added, he heard that one Sunday, Pastor Swain's son Arthur observed a gale blowing outside the parsonage and "told his father, 'Dad, I don't think you're going to have those people from Canaan down here today.'" A short while later, they saw the folks from Uppards crawling across the bridge in their oilskins. *"Crawling,"* Jack said. "And now people won't go to church in their golf carts."

A few features of daily life were almost modern, however. The post office received mail from the mainland every day but Sunday. In 1905, a cable was laid to the island, and some of the stores installed telephones—a brief novelty, as the cable rapidly corroded in the bay's salt water. Electricity reached the stores first, too, in the form of generator-powered lights, though elsewhere the kerosene lamp ruled the night. Tangier became an official town, lost the designation, then got it again. And in the new century's first decade came the biggest advancement of all: the outboard motor.

The gasoline engine transformed crabbing. Watermen could now harvest several places in the course of a day and run multiple trotlines—the predominant technology of the time, consisting of cotton string, hundreds of feet long, held in place with buoys at each end and baited every few feet with fish, chicken, or bull lips. Crabs would grab the bait and hold on even as they were lifted from the water and shaken into a net. The new motors did much for safety on the water, too, for no longer were crabbers dependent on the winds to outrun a looming storm. Not least, they could deliver their catches to Crisfield within hours of landing them. And so Tangier continued to grow. By 1913, the island's population stood at 1,262, of whom 777 lived on the Main Ridge, 216 on the West, 107 in Oyster Creek, 104 on Uppards, and 58 in Canton.

The Army Corps of Engineers estimated that the island deployed about a thousand small boats. That underscored the pathetic state of Tangier's waterfront, which offered no protected dockage: The only deep water was in Steamboat Harbor, a long walk from anywhere and well shy of the space such a fleet demanded. The logical place for a harbor, the creek bordering Meat Soup, was only two feet deep at low water—too shallow for the typical workboat—and lacked a channel out to the sound.

During World War I, the Corps of Engineers judged the situation worthy of a fix. Its boats dredged a turning basin four hundred feet square off the north end of Main Street, then connected the basin to Tangier Sound with a navigation channel—five feet deep, fifty feet wide, and a mile long—that threaded the narrow gap between Uppards and the P'int. "This improvement has made navigation more dependable," the corps reported, "as vessels no longer have to await favorable tides to enter or leave Tangier Harbor."

The island quickly reoriented itself to the new harbor. Docks sprouted from its edge. Meat Soup, until then largely residential, became the business district. One of the new establishments there was owned by Carol Moore's great-grandfather, who'd operated a store at Canaan for years. He opened a new place just off the main drag offering both groceries and general merchandise, including shoes.

The population kept growing, and its density as well. Perhaps it was the bustle of the ridges on Tangier proper, their air of close community, that fueled a keener sense of isolation in the outer hamlets and a desire to relocate, for in the ten years following World War I, those smaller settlements dwindled. Canaan, already beset by advancing seas, emptied altogether by 1929, its people dismantling their houses and businesses and barging them to the mainland or down to the central town. A good many houses from Canaan survive today, in varying states of repair, on the Main Ridge. Ooker was born in a house floated down from the tiny Uppards outpost of Persimmon Ridge.

Then, not long after the transplants settled in, came the storm of '33.

ALTHOUGH QUICK-HITTING SQUALLS can imperil life and limb, they rarely land more than a glancing blow on the island itself. It is big weather systems, laying siege for days, that pulverize the shoreline, scatter crab pots, and push the bay into streets and homes. Every year brings northeasters, almost without fail, and the fact that they're forecasted days ahead of time does little to soften their effects. One, in April 1889, arrived as a gale that pushed water into the streets and up into buildings and kept shoving until the first floor of every home was underwater. It did not relent for forty-eight hours. Within the recollections of living Tangiermen are a few storms that nearly equaled that 1889 maelstrom. The great Ash Wednesday storm of March 1962 flooded scores of houses. A March 1984 northeaster put most of the island under a foot of water. Back-to-back storms in February 1998 created near-record tides.

But it's hurricanes that have earned the greatest fear among islanders, for the topography of their homeplace girds it little against a sustained assault from shredding winds and surging water. The nearly two hundred years since the Great September Gust have been punctuated by blows and grazings from many tropical cyclones, some of them milestones around which islanders organize their memories— weddings and births that occurred before or after, loved ones lost, homes built.

August 1879 brought a hurricane remembered as the Great Tempest, which whipped up storm surges throughout the Chesapeake. An unnamed cyclone in September 1936 put the island underwater, and tides swelled again with the Great Atlantic Hurricane of 1944. Ten years after that, Hazel brought some of the highest winds ever recorded in the bay. "Hazel, that was 105 mile," Jack Thorne told me in a conversation at his Hog Ridge home. "You talk about wind. But I was in Crisfield—it caught me in Crisfield. I was potting for crabs, and when I went in to sell my catch, the wind came up and I couldn't get back out to come home."

In August 1955, Connie dealt the island a glancing swipe, and in September 1960, Donna brought preposterous rain. Agnes, in 1972, nearly drowned the place. And in 1999, Floyd seemed poised to deal the island a fatal punch. "The tide came up higher than it had been in years and years, higher than anyone had seen it," recalled Duane Crockett, who was twenty-one at the time. "It covered the island, and it kept coming up higher and higher and we heard that it was going to keep rising for another two or three hours. And we were all thinking, 'How can it get worse than *this*? How can we go through this for two or three more hours, with the water getting higher all the time?'

"And right then, the storm changed direction, and the tide stopped rising," Duane said, his voice cracking. "The Lord tells the water how far it can go."

Actually, Floyd did not change direction: It crossed the bay near its southern end, on a straight northeast course that took it from Tangier's west side—where the island was exposed to its fiercest winds, given a hurricane's counterclockwise rotation—to its east. But a belief in heavenly intercession is common on Tangier, and has been since Joshua Thomas's day. How else to explain its survival through so many disasters that, had God not interfered, would have finished the place?

"I think of this little island often in the fact of all the storms that have come," then mayor Dewey Crockett told me during my stay in 2000. "Many you'll hear say, 'You couldn't be a waterman and not believe the Word of God.' You look at the many, many times that life has been spared because of the rough waters." We were talking at the school, where the six-foot-six Crockett served as assistant principal, and at this point he leaned back in his chair and opened his hands. "I feel," he said, "and maybe this is just being prejudiced, but I feel there has been a special anointing that has been put upon Tangier because of their strong religious stand and their strong belief in prayer. We're a blessed people."

Iris Pruitt, at eighty-eight the most senior member of the New Testament congregation, sounded a similar theme when I sat with her on her sunporch in Meat Soup. "I've often wondered about it myself," she said, her voice barely above a whisper. "Why would he protect us? He has a purpose, I guess, in protecting us." Her theory: "There's a lot of dedicated Christians here, and with the support we give to missionaries, probably the Lord thinks about that."

Be that as it may, the rising bay has steadily claimed the sandbars and marsh that once shielded Tangier from wind and tide, leaving the island wide open to weather from every point on the compass. In August 2006, tropical storm Ernesto exploited that vulnerability with high winds, heavy rains, and a storm surge that rolled through town. October 2012 brought Sandy, which pounded the island for more than two days, and swelled the bay so high that it covered the airstrip and lapped against the back doors of houses at Canton, the island's highest ridge. Inundation elsewhere was almost complete—the Heistin' Bridge was the only piece of road that wasn't covered.

Carol Moore's late father had left her his crab shanty, which stood on the north bank of the boat channel, its back to Uppards. "A few weeks before [Sandy], something told me, 'You need to go get your dad's glasses and his hat,' which he'd left in the crab house and I'd left just as they were," she said. She retrieved them. The morning after the storm, she and Lonnie walked up to the Parks Marina, at the north end of Meat Soup, to check on the shanty. "We wore boots, and in places we had to wade," she said. "I didn't want to look. I asked, 'Lonnie, is it there?' He just shook his head. It had washed away in the night." It was later the same day that she came upon the open graves at Canaan.

Damaging though Sandy was, it was minor next to Hurricane Isabel in September 2003. That storm churned twenty-foot seas and pushed a wall of water up the creek and into the harbor, shredding much of the island's industrial infrastructure. Of Tangier's eighty-five crabbing shanties, thirty-four were destroyed or heavily damaged, and because they were built over water they weren't covered by federal flood insurance. Isabel swept away shedding tanks, the big electric coolers that preserved the peeler catch, and crab pots by the thousands.

Islanders and outsiders alike wondered whether Tangier could shake off the injury. Watermen who faced the greatest rebuilding costs were unable to work during the last month of the season—and for many, if not most, losing a month's wages was unsustainable even without repair bills. Economic implosion loomed. Indeed, some crabbers gave up after Isabel and found work on tugboats and dredges.

In all the fuss over the damage to crabbing, it was easy to overlook flooding on the island itself, which damaged ninety-nine houses. All in all, Isabel was about as bad a storm as the island could survive today, in part because the economy is fragile in the best of times, but also because much of its private property is underinsured. In a marginally higher flood, Accomack County officials later reckoned, the island could see $4 million in residential losses. Only about $500,000 of that amount would be covered.

ALMOST AS SOON AS Isabel's winds had calmed, the hurricane was being compared with the benchmark storm of Tangier's long history—the worst, by general acclamation, to strike the Chesapeake region in modern times. It made landfall on August 23, 1933, in North Carolina's Outer Banks, passed directly over Norfolk—where it pushed tides nearly ten feet above normal and swamped the city's downtown under five feet of water—then trundled north over the western shore.

The track put Tangier on the cyclone's eastern side, where it was ravaged by eighty-mile-an-hour winds. At the storm's height, Meat Soup was part of the bay, with breakers rolling down Main Street and storm surge overwhelming the entire island. The flood ruined everything in the big store owned by Carol Moore's great-grandfather, forcing him out of business. "It covered every home but the parson-age," said Jack Thorne, who was about to turn nine and lived next door to the pastor. "Every home on the island. I'd say at least from the ground it would be that high"—he held his hand thirty inches off the floor, or about the height of a kitchen counter—"*in* the house."

Ginny Thorne Marshall, Jack's sister and three years younger, recalled "men going down the street in little boats." The high water tore loose a big, masted fishing vessel, sixty or seventy feet long, and carried it over the shore and into Main Street, and when the flood re-ceded it stranded the *Marian Sue* in front of what's now Daley & Son. "You talk about a strange sight," Jack said. Two other large boats, fifty feet long or better, were carried overland, too. One of them, belong-ing to Jack's uncle, "went right over Uppards almost all the way to Canaan, right over that marsh."

Folks on the mainland knew the storm was coming, and those on vulnerable ground had time to get out of the way. But Tangier failed to get the word, and most of its watermen were out in their boats. Ooker's father, Will, twenty-two at the time, was among them, and after riding out the worst of it north of the island, he came home to a scene of widespread devastation. "There was only two boats afloat when we come in the harbor," he told me in 2000. "There was a bleak-looking time."

That it was. And the outlook after such storms also tended to be bleak, for the hand they laid on the shoreline was hard. Any Tangier-man could see that his island was getting smaller, but after a hur-ricane, the changes seemed dramatic. Many islanders left after the storm of '33. From that point on, Tangier's population fell steadily.

Consider that in 1930, when Jack Thorne turned six, the census stood at 1,120. By the time he turned sixteen, he had at least 100 fewer neighbors. In his midtwenties, the count was down by 105 more. Twenty years later, another 101 were gone. In 1990, when Jack was sixty-six, the population had dwindled by an additional 155, to 659.

At the millennium it stood at 604.

More than one in five Tangiermen have died or moved away since.

Lonnie Moore prowls for hard crabs in Pocomoke Sound with Isaiah McCready and Cameron Evans, June 2016. (EARL SWIFT)

ELEVEN

HOURS BEFORE DAYBREAK I'M ABOARD LONNIE MOORE'S thirty-two-foot deadrise, *Alona Rahab*, headed out of the shanty-lined channel into Tangier Sound. A hard rain is falling, and its drops catch fire in the beam of a powerful spotlight on the cabin's roof. Beyond them, all is black: Clouds smother the moon, and at 3:50 A.M. the few lights burning onshore hide behind a heavy curtain of mist. The way ahead is an impenetrable blank. Undaunted by the heavy chop, Lonnie opens the throttle, the diesel roars, the bow rises high, and we shoot bucking and thumping into the wind and confused water beyond the P'int.

We're bound for Lonnie's hard crab pots, 425 of them strung in three-mile rows off the Eastern Shore, the nearest about ten miles from home. With us are Isaiah McCready and Cameron Evans, two rising high school juniors beginning their ninth day as Lonnie's crew. Still adjusting to the early hours, Isaiah dozes in a small berth shoehorned into the boat's forepeak. A yawning Cameron and I struggle to keep our feet on the cabin's rolling, shuddering deck.

Lonnie is hunched in his chair behind the wheel, eyes locked on radar, GPS, and depth finder, rocking with the hard thumps that accompany our collisions with water churned by wind blowing the other way. We're taking it "right in the painter holes," as Tangiermen say, but rather than slow down, he brings our speed up to nineteen knots. The boat's short nose plows into a wave, and a thick tongue of water arcs over the windows and the spotlight, filling the cabin with a weird blue-green light.

"Ride 'em, cowboy!" Lonnie shouts over the diesel's noisy throb. "We got a little wind out here. The forecast was for less than ten knots out of the south, but it's blowing harder than that." We hit a wave, and I feel my spine compress. "About twenty, it feels like to me," Lonnie says, peering through the glass. "We got flood tide and we still got whitecaps, so about twenty, I'd say." Translation: The southerly wind is blowing with the tide, rather than against it, and still the sound is foaming.

The boat dives into a wave, and another huge scoop of water sweeps over the cabin. On the GPS I see that we're rounding the southern tip of Watts Island. We rattle and thud east into Pocomoke Sound's middle, then turn due south. Up ahead, a channel marker's red light flashes weakly through the rain. Lonnie adjusts his radar to zoom in on our location and, at 4:25 A.M., cuts the engine. Isaiah crawls blinking from the berth. We're three miles west–southwest of the entrance to Onancock Creek. So say the instruments. The Eastern Shore is invisible, as is everything else beyond the cabin's fogged windows.

"Get your sea legs, boys," the captain tells his crew. "It's rockin' and rollin'." Indeed, heavy swells are moving beneath us, tilting the boat fifteen degrees one way, then the other. He flips a switch and three spotlights bathe the rear deck with stark white light. With each roll, water slaps the boat's sides and geysers up past the gunwales. "Think I should wear my coat?" Cameron asks.

Lonnie squints at the pelting rain visible in the spotlights, the water sloshing up the boat's sides. "Up to you," he says. "You'll definitely be getting some spray. That's for sure." He slips his oilskins over his shoes, hoists them up over his pants legs, untangles the shoulder straps. "I'm going to start out without mine," he says. "I'd rather be wet from the salt than wet from sweat." Cameron nods his agreement. Suited up, the three step from the humid cabin to the wind-swept, pitching deck. A sunshade over most of the boat's work area offers little protection from rain blowing sideways. Lonnie strides to the rear steering station, Cameron takes up position just astern of him, and Isaiah organizes bushel baskets in the deck's middle. Then, wordlessly—which is the way Captain Alonza J. Moore III prefers to work—they launch into a complex but fluid choreography.

Step one: Lonnie works the engine and transmission to bring the boat alongside a buoy. He drops the engine to idle, hooks the float out of the water, and loops the line through a motorized puller, which he activates with a pedal. Wheels spin, the device whines, and in two seconds, maybe three, fifty feet of line piles up next to Cameron and the pot breaks the surface. Lonnie lifts it aboard, sets it upside down on the gunwale, and unlatches the bait hatch on its bottom.

Step two: As Lonnie engages the transmission and the boat eases forward, Cameron seizes the pot, flips it over, releases its bungee closure, and shakes the catch into a galvanized metal tub on deck. He then closes the pot, flips it back upside down, grabs a couple of menhaden from a cardboard box at the stern, and stuffs the fish into the wire cylinder that forms the pot's bait box. He re-latches the hatch and tosses the pot overboard, throwing the piled line and buoy after it.

Step three: While Lonnie snags the next pot, Isaiah culls the catch, plucking crabs from the tub and tossing each into the appropri-ate bushel basket. He has to move fast, as Lonnie has refined this rou-tine to a lean efficiency, and the time between pots averages little more than a minute. With that in mind, tenacious crabs that cling to the wire mesh are left in the pots—the *Alona Rahab*'s crew will get them tomorrow—and Lonnie has custom-fitted the boat with features that streamline his movements: After threading the line into the puller, for instance, he slips his hook into two homemade brackets, where it waits for the next pot. Its placement is so ergonomically natural that he can grab it without looking.

Likewise, he uses a steering stick in place of a wheel because it's faster. Pushing it forward turns the boat right, and pulling it, left. He rarely has to give it more than a tap, an economy that saves seconds per pot. Working the stick, the transmission, and the pedal has become so intuitive, he says, that if he thought about what he was doing, he wouldn't be able to do it.

We work south along the row, the visible world defined by the spotlights: water deep green and fizzing, slapping loudly against the boat's sides; the tossing deck turning slick with mud, red moss, bits of crab, and rain, which is slanting aboard on the gusting wind; and flashes of white as gulls stalking us for handouts cross the field of light. A pot comes aboard with fifteen crabs crowded inside. "Look at that!" Lonnie hollers.

But it's an anomaly. Most of the pots bring just one or two crabs. As the sky brightens to a steely gray, Lonnie brings the hundredth pot aboard. "A quarter down," he says, and nods toward the catch so far: two bushels of clean sooks, nearly a bushel of lemons, and half bushels of number ones and twos, along with a few peelers he'll sell to Ooker. "Not too good."

"It's the moss," he tells me. "When you have a lot of moss, you don't get any crabs." Many of the pots are choked with the algae, which here seems blonder than the variety Ooker's found in the waters just off Tangier. If Lonnie were a peeler potter, a little of it might work in his favor—a crab about to molt is seeking sanctuary, after all, and the growth makes a dark cave of a wire pot. Hard crabs steer clear, however. "Yesterday there was no moss here. We caught a lot of crabs," Lonnie says. "But it was so bad a couple weeks ago I had to bring a power washer and clean the pots as I went. Couldn't even see into the pots."

We've worked our way three miles down the Eastern Shore, which in the predawn, rain-shrouded gloom is taking shape as a dark, uneven line to our east. Normally, Lonnie would head that way to the southern end of his next row, off the mouth of Pungoteague Creek, and follow it back to the north. But now, as we idle broadside to the waves, the air on deck pungent with brine, old crab, and diesel exhaust, he decides to change tactics. We'll speed up to the north end of the rows, away (he hopes) from all this moss, then work our way back down, nose to the wind. We return to the cabin. Lonnie opens the throttle.

THE ECONOMICS OF HARD POTTING leave little room for a poor day's catch. Compared with a peeler potter, Lonnie Moore has a lot of overhead to cover before he breaks even: He deploys more than twice as many pots, which are fashioned from heavier wire than those for catching peelers and more expensive. Because he plants them in deeper water farther from shore, those pots require a lot more line—which boosts their cost considerably—and he loses more of them to storms and boat traffic. His fuel costs are steeper, for while Ooker's farthest-flung pot is rarely more than five miles from his shanty, Lonnie's closest is twice that distance from home. He has to buy bait, which isn't cheap at two fish per pot, or 850 a day. He must pay his teenage crew $70 per boy per day, or a total out of pocket of $840 per six-day week.

In sum, he must catch $400 of crab on the average summer day before he actually earns his first penny. Early in the season and again in the fall, that daily break-even point rises to $500, because the boys are in school and he has to take on an adult crew, and because the razor clams favored as bait in cooler months are more expensive.

This is not, in other words, an undertaking for the weak of heart. It requires him to charge hard every day, weather be damned, and fatigue and sickness, too. It demands his complete commitment to minimizing costs and maximizing his catch. And regardless of how lean his operation or how good his luck, his success largely rides on the decisions of others, for like every crabber on Tangier he has no say in the price his crabs will bring. With the season now well under way and crabs plentiful, the crab houses are paying far less than they did six weeks ago: Number ones have dropped from $100 a bushel to $65, number twos and clean sooks from $50 to half that much, and lemons from $20 to $16.

Which makes Lonnie's choice to follow the water all the more remarkable, because for more than half of his working life, he was on the payroll of the Chesapeake Bay Foundation, first as a boat captain, then as boss over all the foundation's island programs, and later as the commander of its fleet. He chose to leave that job—and the security of a paycheck, retirement account, and predictable hours—to return to working the water after almost twenty-five years away.

Unlike most Tangier crabbers, Lonnie didn't follow his father into crabbing. He was born in November 1954 to Edna Sears— the daughter of the island's first and best-known innkeeper, Hilda Crockett—and A. J. "Junior" Moore, a World War II veteran who worked for the island's electric utility and served many years as postmaster. Lonnie also pursued land-based employment, at least at first. He worked construction for a while, then built boats with Jerry Frank Pruitt. But three years after quitting school in the eleventh grade, he was drawn to the independence and money that crabbing promised. He crewed for others for two years, then had Jerry Frank build him a forty-two-foot box-stern deadrise that he named *This'll Do*, after a beach cottage his grandfather had kept when Lonnie was a boy.

Jerry Frank charged him about $8,000 for the boat. Fully outfitted, his rig cost about $21,000, he says, "and you could go to the bank and get a loan to pay for every penny of it." It didn't seem a bargain at the time, but nowadays it would cost six times as much, easy, and because the uncertainties of working the water have made lenders skittish, most or all of it would have to be cash.

In 1986, after potting for ten years in *This'll Do*, Lonnie had a much bigger boat built—forty-seven feet long and fourteen abeam— designed to withstand the harsh conditions of winters at sea. He named it for Carol, whom he'd married in 1981, and their daughter, Loni Renee, born the following year. The *Loni Carol* proved a fortuitous upgrade four years later, when Lonnie got his captain's license and Coast Guard certification to carry passengers, and he went to work for CBF as the captain of its Port Isobel facility on the P'int.

The boat was big and seaworthy enough to carry a classroom of kids on educational outings.

At the time, the foundation wasn't winning any popularity contests on Tangier, and Lonnie's fellow crabbers were mystified by his decision. "It wasn't like I wasn't making a living on the water," he says. But by then he and Carol had had a second child, a boy named Alex, born in 1988, and here was economic security—and he also saw the job as a platform for advocacy on behalf of watermen, as well as the bay and its wildlife.

Early on, during the two years he worked at Port Isobel and the dozen he spent managing the foundation's island programs at the P'int, Smith Island, and Fox Island, it was exactly that. But the organization changed, Lonnie says: It grew slick, corporate and political, he believed, and less receptive to the bay's commercial users. "CBF used to side with the watermen occasionally," he says. "Now they always side with the fish, oysters, and crabs." Even as overt hostility between the group and Tangier eased, Lonnie's disenchantment deepened—until, halfway through his ten years as fleet senior manager, "I hated the job," he says. "I hated going to work. We had people on staff I couldn't trust. Carol kept telling me to quit." And so he did, in 2014.

He sold the *Loni Carol* to the foundation and bought the *Alona Rahab*, which is named for his granddaughters and could fit inside his old boat. Because he'd been inactive as a crabber, the state restricted him to a "little" hard crab license, authorizing him to set only eighty-five pots—a number well shy of what he'd need to earn a living, even working by himself. He traded that license and $5,000 to another crabber for a "medium" hard crab license, good for 255 pots. Early in 2016, he traded that medium license and $12,000 for the biggest license issued in Virginia, for 425 pots, and spent another $3,000 for a peeler license to keep his options open. He's thus invested $20,000 in paperwork.

Which grants him the privilege to venture onto the bay to attempt the near impossible.

BY 6:50 A.M. we're on the second row, working south from the mouth of Onancock Creek under a leaden sky. My phone reports the temperature as seventy-six degrees, but it feels far cooler in the undiminished wind sweeping over the boat from the southwest. The sound is empty save for a single deadrise a half mile off to starboard. Lonnie identifies it with a glance as the *Henrietta C.*, a fine wooden boat built by Jerry Frank and belonging to Ed Charnock.

Cameron and Isaiah have switched roles, so that Isaiah is now dumping the pots. Not that there's much to dump: Over the next twenty minutes we top off a bushel of number ones and another of clean sooks, but the red moss is worse here than it was down south. One pot after another comes up slimed and empty. Lonnie, flummoxed, decides we'll head back down to Pungoteague. "This is what I hate to do," he mutters as we speed down the coast, "run from one end to the other and find it doesn't work."

We arrive at the south end of the 120-pot row. Through the mist I can see the Pungoteague Light standing at the mouth of its namesake creek. It's a caisson-style tower built of iron and shaped like a spark plug, its bottom planted in the bay's floor and weighted in place with concrete. The bay is home to several such lights, which have stood sentry over its shoals for more than a century. This one dates to 1908.

The years have not been kind. As we draw closer I see that the light has been sheared in half, its lantern destroyed, its mangled stump no aid to navigation, but a hazard. A marker planted nearby warns sailors away. I ask Lonnie what caused such vandalism. His answer: "Ice."

Some winters freeze parts of the sounds east of Tangier, and on rare occasion the Chesapeake locks up from shore to shore. Ice halts the mailboat and the grocery store's weekly runs, and when it lingers, it infuses the island with a frenetic claustrophobia. And freezes can last far longer than the bay's temperate latitudes would suggest they might: The record, set in the winter of 1917–18, is fifty-two days. Another weeks-long siege in 1977 piled ice in immense blocks two stories high along the island's west side, and it froze so thick that sev-

eral young Tangiermen were able to walk more than two miles across
the frozen bay to an old World War II Liberty ship sunk in the open
water. Some even rode their bikes.

In the days before reliable Coast Guard icebreakers and rescue
helicopters, such maroonings could usher real hardship. A break-
bone freeze in January 1893 forced Tangier to pillage its larders and
slaughter its livestock, and still the island suffered "great destitution,"
as a news dispatch put it. A particularly harsh freeze in 1936 saw
Maryland State Police attempt to replenish the island's dwindling
food supplies by trekking with sleds from Crisfield. Along the way, a
state trooper broke through the ice and froze to death. Food eventually arrived by blimp.

Nowadays a frozen sound, and even a completely iced-in bay, is
more inconvenience than real danger: Thanks to the airstrip, islanders need not fret about starving. But worry they might when the ice
starts to break up. Pushed by wind and tide, floes become battering
rams that splinter all in their path. The Pungoteague Light's ruins
commemorate a spot ravaged not once, but twice by wind-driven ice.
An earlier screwpile lighthouse—essentially a cottage atop a spidery
tangle of steel legs screwed into the bay's bottom—was toppled here
in February 1856, in only its second winter of service.

A long roster of Chesapeake beacons has shared that fate. Floes
crushed Maryland's Hooper Strait Light in 1877, and four years later
they swept the nearby Sharps Island Light off its base and carried it
five miles down the bay with its keepers trapped inside. The Solomons Lump Light, just north of Smith Island, was heaved onto its
side by piled ice in 1893 and destroyed by another attack of floes two
years later. Ice smashed the Janes Island Light outside Crisfield in
1879, heavily damaged its replacement in 1893, and took down the
repaired light for good in 1935.

Seeing as how floes can obliterate structures built to withstand
hurricanes, they've had little trouble wreaking havoc on Tangier's
fragile shoreline. More than wind, more than waves, moving ice
grinds away sandy beach, tears great bites from upland sod, bulldozes

huge tumps of marsh into the bay. From the first days of settlement, Tangiermen walked out to their island's edges to find that freezes had left behind startling change.

At such times, the islanders couldn't help but notice that their home was shrinking, and not just at Canaan. The whole of the island, and especially its western shore, was succumbing to nature's assaults. But well into the twentieth century, they didn't have a measure of just how much was lost or at what rate. And during years when the water didn't freeze and storms didn't pound the shore, they put the matter out of their minds.

THERE WAS PLENTY MORE to think about, for change was coming to Tangier fast. In 1928, islanders rigged up a diesel generator on the Main Ridge to wires crisscrossing the island and created a primitive electrical utility. The system ran on direct current, which required special equipment if a householder were to use industry-standard lights, radios, or appliances, and its load capacity was so limited that it was fired up for only a few hours each evening, from about 5:00 to 10:30 P.M. Most Tangiermen chose to stick with their kerosene lamps. Still, the town acquired a smattering of dim streetlights, and its gathering places enjoyed safe, clean, reliable light through the long winters.

That same year, an Illinois-born waterman living in Harryhogan, a village on Virginia's western shore, received a patent for the wire-mesh crab pot. Benjamin Franklin "Frank" Lewis was pushing seventy and had spent decades trotlining crabs, when he rebelled against its taxing labor and meager yields. One hot July day in the twenties, he abandoned his trotline in the Yeocomico River and, as his son, Harvey, later recounted, sat for several hours under a tree in his yard, so deep in thought that he "didn't even hear Mom when she called him to dinner.

"Later he went to the shed and got his snips and told me to go to the store for some chicken wire," Harvey said. "Then he began to cut out pieces and fuss with 'em. That's all he did all that summer." The result was an early version of the pot: a cube of wire mesh on a heavier wire frame, with a tapering funnel built into two sides and a cylinder for bait in its bottom. Crabs entering the pot through the funnels had a hard time finding their way back out. Lewis patented the design in 1928, before he was fully satisfied with the invention, and for years fiddled with refinements.

The key improvement was based on the principle that if a crab feels itself trapped, it will usually respond by running downhill on the bottom or swimming upward. Lewis installed a wire-mesh wall dividing the pot into upper and lower rooms. He cut a mouth-shaped slit in its middle and curled the ends of the snipped wire upward, forming a one-way passage. A crab entered the trap through one of the funnels, or "throats," then—finding itself trapped in the "downstairs"—swam up through the passage into the "parlor," where escape was next to impossible.

Lewis received a patent for the refined device in 1938, and it wasn't long before Tangier watermen first encountered one. Elmer Crockett, the island crabber saved from electrocution by Half-Ass Buck, described the moment in journalist Larry Chowning's book *Barcat Skipper*: While trotlining in Mobjack Bay on the western shore, Crockett saw a pair of crabbers each bring in ten barrels of crabs— something like thirty-three bushels—on a day when he'd had rotten luck. "I could see they didn't have a trotline rig in their boats," he said, so early one morning he followed one who "went out a short distance into the Mobjack and started pulling on a rope tied to an oyster stake that was marking oyster grounds. By golly, he pulled up a wire cage that was solid full of crabs." Later, Crockett approached the potter, who cheerfully "showed me one and let me use one for a sample so I could make my own. It was the first crab pot I'd ever seen."

Crockett made thirteen pots in the space of two days, and his "partners soon saw I was making all the money. They also went and got some wire, and I showed them how to make the pots." Not long after, "the boys and I decided to take the pots home to Tangier and see how they would do over there. We were the first to bring the crab pot to Tangier."

The invention was transformative. Trotlining quickly fell out of favor in Virginia and survives today only in those shallow Maryland waters where pots have been outlawed as hazards to boat traffic. The pots Lonnie uses are little changed from Lewis's 1938 design, save for two additional throats, giving the pot one in each side, and cull rings—holes 2⅜ inches wide—cut into the pot's parlor, enabling undersized crabs to escape.

Another big change came to the island not long after Elmer Crockett and his buddies encountered that first pot, and it resulted from ice and the hardships it brought: In October 1940, the Federal Communications Commission authorized the creation of a radio-telephone link between Tangier and the mainland, citing freeze-ups through the thirties that had cut off the island from the outside world. Four telephones were installed on Tangier: one at the electric plant, two in stores, and one in the home of a prominent member of the Swain congregation. Picking up the receiver connected the caller by radio with an operator in Crisfield, who spliced the call into the mainland telephone system. Receiving a call was a little more complicated. If the recipient didn't know the call was coming, it might take a while for someone to round him up.

Over the next twenty-six years, the four telephones became fifteen coin-operated phone booths. But they still relied on the radio circuit, still worked best for outgoing calls, and were party lines— meaning an islander might not be able to make a call until a neighbor in another booth finished his.

But then, Tangiermen were inured to the limitations of technology. In 1944 their jury-rigged electric generator gave up the ghost, after sixteen years of service. The island lacked the money to replace it, so everyone went back to using kerosene.

BACK ON THE *ALONA RAHAB*, it's eight A.M. and we're nearing the north end of the row. Each pot rises from the bottom more tangled with red moss than the last. Most contain only a crab or two, and few of any size; in several, I see small crabs desperately trying to squeeze through the cull rings. "Yeah, there's more moss," Lonnie says, sighing as we motor east to his third row. "I'm hoping this south wind will carry it all above us tomorrow."

We start south. The wind comes up, and over the next half hour the seas build. A brief eruption of hard rain spatters the deck and the crew. Between pots, and over the noise of the puller, Lonnie and I attempt an oft-interrupted conversation—about the hazards that attend work on the water, the precautions one can take to improve his odds of returning to port, the sometimes harsh hand of fate. Lonnie spends a lot of money keeping the *Alona Rahab* in good condition. He's near fanatical about preventative maintenance. But sometimes things happen, he says, that you can't prepare for: chains of misfortunes that happen just so, small accidents that occur at precisely the worst moment.

He speaks from experience. On April 19, 1991, he was alone aboard the *Loni Carol* and under way at a good clip up near the red bell buoy that marks the approximate halfway point between Tangier and Crisfield, when he went aft to check something at the stern. He tripped as he walked and plunged headfirst off the end of the boat.

The water temperature was fifty-five degrees. Wind was out of the northeast at thirty miles per hour, and the seas were heavy—so much so that Tangier's watermen had stayed in that day—so the prospects for rescue were slim as Lonnie watched the *Loni Carol* motor away.

The tide was outbound. Together with the winds, it propelled him southward, and he had no choice but to go with it. Numbed and confused by the cold, losing strength by the minute, he bobbed along for two hours. "I was hallucinating," he says. "I didn't have five minutes left. I was just floating. I couldn't move my arms to swim anymore. I was gone."

Then, in an occurrence that defied almost impossible odds, two Tangiermen headed to the mainland on a beer run happened to notice something afloat in all that rough water and swung around for a closer look. He remembers being hoisted aboard, then waking up in an ambulance. Most, if not all, Tangier watermen have fallen overboard at some point. Few have been so unlucky as to face the combination of circumstances that nearly killed Lonnie. Fewer still have benefitted from the kind of blind, stupid luck that saved him.

The bad run he's experienced today is minor in comparison, but it persists. We encounter more red moss on the last row and run out of bait before we've finished it. Lonnie steers the boat for home with the day's efforts inked in red.

The wind persists as well. A few days later I reach the Situation Room ahead of everyone but Leon, who readies a pot of coffee and, while it drips, sits staring at me over his glasses. "Rough this morning," he says. "It blowed hard."

Jerry Frank Pruitt enters. "Did you go today?" Leon asks him.

"Yeah," Jerry Frank says. "It was blowing today. Blowing right good."

"Blowed hard there, this morning," Leon agrees. "Came right around to the southeast."

"Yeah, it was blowing hard, all right," Jerry Frank says. "I wasn't sure that I wouldn't have to come in, but I was able to stay out, and right after the sun come up, it settled down."

A lengthy discussion of wind ensues. Ooker bursts in the room. "Somebody let me know," he hollers. "Is it going to blow a gale *every* morning?"

So it seems. A few days later, a squall packing sixty-knot winds tears over the island, toppling stacked crab pots off their docks and into the harbor, scattering boating gear and yard furniture, swamping skiffs. A few evenings on, still another gale broadsides the island from the west. It shoves the water in Onancock Creek upstream and over the banks, flooding the town wharf. Salt water swirls two feet deep around my parked car, rendering it a total loss.

Between blows, the island broils in a wave of Burmese heat. One

Sunday morning, a big, burly tugboat crewman named Paulie Mc-Cready opens the service at New Testament Church with: "Good morning—this beautiful, hot morning." The congregation murmurs its weary assent. "It's all right," Paulie reassures his neighbors. "Ice will be here before long."

One of three crosses Ooker has erected in Tangier's marshes, this one just below the Heistin' Bridge. (EARL SWIFT)

TWELVE

HE FIRST SUNDAY IN JULY FINDS DUANE CROCKETT PREACH-
ing to a New Testament congregation augmented by former
islanders visiting for the holiday weekend. "I thank the Lord
for the United States," he tells the crowded church. "We could have
been born anywhere, but I'm thankful that we were born in America.
And more than that, I could have been born anywhere in America,
but I was born on Tangier Island, and that was a wonderful thing."

Nods all around. I'm in my usual seat in the right rear of the
church, a pew in front of Richard Pruitt of the Situation Room and
his wife, Margaretta. Leon is three pews ahead of me, seated with the
oldest of his four children, daughter Carlene. Ooker and Irene are up
in the front pew. Carol and Lonnie are on the room's far side.

Our setting is far humbler than the soaring sanctuary at Swain
Memorial. No stained glass warms this room: Here a sallow light,
more office than church, is supplied by recessed fluorescent strips. The
low acoustic tile ceiling is supported by six intrusive steel columns,
painted to more or less mimic wood. Imitation wood paneling sheaths
the walls, and burgundy shag carpet, the floor.

Up front, Duane's on a patriotic roll. "According to the Gallup
Poll, 77 percent of Americans claim to be Christians," he tells us. "We
know that isn't so. If 77 percent of Americans were, in fact, Christians,

schools would still begin each morning with prayer and Bible read-ing." He looks up from his notes and scans the room. "Godless and heathen organizations such as the ACLU would not have any influ-ence on the politics of the United States."

I look around. The remark causes no stir in his audience, at least none that I detect. I'm reminded of another distinction between the island's churches. New Testament has no pastor—it's led by a hand-ful of male elders who rotate preaching duties. All are island natives. None is formally schooled in theology, and all hew more or less to the style of worship Joshua Thomas introduced to Tangier. It is plain-spoken, rife with over-the-left phrasing, and studded with lessons taken from island life. It can be self-effacing and humorous, swing from bold to contrite to teary. Evidently, it can also trash the First Amendment.

His preamble complete, Duane gets around to the real subject of his message: "When we make up our own rules and regulations out-side of what's in the Bible," he says, "then we put ourselves in what's called a yoke of bondage. I have done that in my own life so many times."

He cites, as an example, his relationship with a combination gas station and wine shop on U.S. Route 13, the main highway running up the Eastern Shore. "I've thought many a time, 'I would not get gas to a place called the Wine Rack. I will not do it. Lord knows how many people have gotten drunk from wine they've bought there. I will not buy gas to the Wine Rack.' It was something I started in my own mind, that I couldn't gas my car up there—and then everybody who gassed their car up there became a heathen because they did that.

"I was coming back down to Crisfield one day. The gas light started flickering." A knowing chuckle rises from the congregation. "I had to gas up. I said, 'Now, I can take a good chance and wait till I get down to Princess Anne, or I can pull over here and gas my car up.' What did I do?" He pauses for two beats. "I did the commonsense thing. I pulled my car over and I gassed my car up at the Wine Rack."

"I'm not going to be judged on what kind of gas I put in my vehicle," he says. "But you see, I had put myself under a yoke of bondage. And to redeem myself to the congregation I have to say I haven't gotten gas at the Wine Rack since the day the light went off in my car."

THE YOKE OF BONDAGE: Tangier knows a thing or two about that. After Charles P. Swain's death, and through much of the twentieth century, the island's Methodists held fast to practices that struck visitors as quaint, even backward. For starters, the Sabbath was observed to a degree long abandoned on the Virginia mainland. Many an island family did no cooking on Sundays, as it might be considered work; devout Tangiermen instead prepared their Sunday meals on Saturday and strove to preserve them until time came to eat.

Some islanders wouldn't wash their hair or wash dishes, either, out of fear that it constituted work. They avoided any activity involving the use of machines, which they defined to include such simple devices as scissors. A believer did not mow his lawn, ride a bike, or work on his boat on a Sunday, and he didn't snip coupons from the newspaper, either.

The faithful might forgo eating dessert with a Sunday meal as an unnecessary pleasure and thus displeasing to God. The Sabbath was thought to nix child's play, too. "I remember when I was a little boy," Duane says later in his sermon, "I used to like to carry around with me a bushel basket lid, and everywhere I'd go I'd pretend I was steering a boat."

"Well, one Sunday morning I went to Sunday school. I had went out the yard, steering my boat. I left [the lid] at the edge of the yard, and when I come back from Sunday school I came back through the yard and picked it up, and Pop-Pop Wes was out on the porch, waiting for me. He always called me Wayne. He said, 'Wayne, come here a minute.' I said, 'What, Pop-Pop?' He started singing this little chorus to me. These were the words:

> *You must not play on Sunday, Sunday, Sunday.*
> *You must not play on Sunday because it is a sin.*

You can play on Monday, Tuesday, Wednesday, Thursday,
Friday, Saturday, until Sunday comes again.

"He said, 'You ought not to be toting that lid through the yard on a Sunday.' And I never did anymore," Duane tells us. "And let me say this: to the point that I did *nothing* anymore on a Sunday at all, because I felt like I was hurting God by doing that."

Enforcing the Sabbath wasn't just the province of Swain's minister but the town government, for in most respects Tangier behaved as a full-on theocracy. No municipal decision was made without the church's assent. No new idea could take root without its backing. And the town's ordinance book included a blue law that went far beyond those elsewhere: It outlawed "loafing on store porches and streets" on Sundays and decreed that all residents be "in church during the hours of service, or in their homes."

The island has never been populated solely with the faithful; from the start it was home to scalawags and backsliders, to drinkers and brawlers. But even the resolutely unchurched respected the place and power of Swain Memorial in Tangier life, and recognized the conformity that peaceable coexistence with the Methodist majority required. Alcohol was surely present, but it was imported discreetly and consumed on the sly. A drunk islander would wade the marsh before he'd be seen weaving down the Main Ridge. Even the saltiest Tangierman strove to keep his language clean around his neighbors. And with rare exceptions, all observed the blue law.

Such an exception sparked one of the most notorious episodes in island history. On a Sunday in April 1920, seventeen-year-old Roland Parks—grandfather of Annette Charnock, whom I sat with at the wedding—went to his family's closed grocery store on the Main Ridge to fetch ice cream for his ailing mother. The town sergeant, C. C. "Bud" Connorton, saw him at the store and, citing the blue law, tried to arrest him. When Parks insisted on completing his errand, Connorton followed him home and shot him on his front porch.

The boy survived the wound, and Connorton was sentenced to a year in prison. He served a fraction of the stretch before receiving a gubernatorial pardon, and he returned to Tangier and his job as the island's cop. Five years later, as he sat beside an open window in an island café, Connorton was shot and killed by an unidentified gunman.

A great many islanders were said to know who was responsible, but none talked. The tight-lipped included God-fearing Swain regulars.

On Tangier, piety and lawlessness can keep close company.

WHICH BRINGS US TO A SAGA that cut Tangier to the bone and continues to reverberate among its people today. At its center was James C. Richardson, who arrived on Tangier in November 1943 as the new pastor of Swain Memorial. He was a native of Hampton, Virginia, at the bay's south end, and had started preaching in neighboring Newport News at seventeen years old. When he moved into Swain's parsonage with his wife, Elva Curl Wilson, and their fifteen-month-old daughter, Grace, the dynamic, blond-haired preacher was a rising star in the Methodist Church's Virginia Conference. He wasn't yet thirty.

Richardson quickly won the hearts and minds of his new congregation, which was storied throughout the conference as an especially active, Bible-smart, and conservative bunch. Then, in 1945, he welcomed into his ministry an island native named Stella Thomas, who was returning to Tangier after years of overseas missionary work. She brought some newfangled ideas with her—ideas that today's evangelicals might not find radical, but which came as a surprise to Swain's old-fashioned and biblically literalist worshipers.

I won't detail the ideas here, because they really don't matter to the story. Suffice to say that they excited Richardson, along with some in the congregation, but struck many more as deviations from Scripture. And it probably didn't help that the majority sensed a bit of the holier-than-thou in Richardson's close supporters, an attitude that the Methodism that had so long piloted the island was not good or strong enough to suit them.

The situation came to a head when, in the summer of 1946, Richardson began leading daily morning services in addition to the regular Sunday worship, to accommodate his flock-within-a-flock. Church trustees petitioned the Virginia Conference to have the pastor removed. Richardson asked for another year on the island. The conference's bishop turned him down.

So it came to pass that after pondering his situation, Richardson decided to quit the Methodist ministry and start a second church on Tangier. He rented a house on the West Ridge and moved his family—which now included a second daughter—out of the parsonage. "My earliest memory over here is the day we were leaving the church and moving into the house on West Ridge," his older daughter, Grace Kimpel, told me when she visited the island in the summer of 2016. "My grandmother was pushing my sister in the baby carriage— she was four months old or so. She was pushing the baby carriage with one hand and carrying an upright floor lamp with the other.

"It felt like we were going on an adventure, but I had a sense that it was important."

Richardson didn't leave alone. Fifty-five of Swain's women and four of its men followed him—only a modest share of the Methodist membership, but an exodus that split families, separated husbands from wives, and sowed hurt and distrust in scores of Tangier homes. Those who left found themselves banished from their social circles and branded as deluded and disloyal by those who stayed put.

Stella Thomas's sister rented Richardson an empty one-room store she owned in Meat Soup. In November 1946, what was then called the House of Prayer, and would later become the New Testament Church, held its first service there. The next morning, Richardson discovered that someone had sunk his boat in Tangier's harbor.

So the lawlessness began. The island's wheezy electrical plant had failed two years before, and under cover of darkness, persons unknown vandalized the storefront sanctuary, tearing up hymnals, hauling out chairs and tossing them onto the roof, and several times pitching the small organ into the harbor. Prowlers regularly lurked outside the Richardsons' house on the West Ridge. The harassment gradually increased in brazenness over the winter and spring of 1947.

Then, with summer, it got completely out of hand. During five nights of prayer meetings in August 1947, the House of Prayer came under siege from noisy mobs hurling bricks, cans, and shells at the converted store. An islander confronted Richardson as he crossed a bridge over the Big Gut with an armload of groceries and did his best to knock him into the water. His followers found copper paint poured into their boat engines, their windows broken, their presence ignored by kin and lifelong friends. Richardson's boat was again sunk. Rumors flew that the former pastor and his family would be kidnapped or forcibly run off the island. The mayor and a member of the town council advised that they could not guarantee his safety.

One Saturday night in late August, while Richardson attended a conference in Indiana, vandals once again broke into the makeshift church, dumping slop jars full of excrement on the floor and chairs. The next day the congregation met in the home of a member, who soon lost his fishing nets and a storage building to arson. For another six weeks the attacks continued unabated: Someone broke the pillars supporting the church's front porch, and again dumped slop jars in the sanctuary, and broke out the windows, and painted graffiti on the walls, and ruined a second organ.

Richardson met these trials with "surprise or sadness," Grace Kimpel says, but she does not recall seeing him angry: "My dad had the idea of outliving and outloving his critics and adversaries." Still, by early October he was clearly frustrated, which was manifest in letters he wrote to the mayor and Swain's lay leader. "It has been suggested to me that these things continue because the young men who are responsible for them feel that they have the approval and moral support of the Mayor, the Town Council, and the leaders of the community," one letter read. "I do not find it in me to believe that the people of Tangier approve such actions. Yet, I have waited in vain to hear any protest from any official of the town, or to see any printed notice to discourage such actions and attitudes."

Incredibly, the town had taken no official position on depredations that had been going on for nearly a year and seemed destined to bring serious injury or worse—and Richardson's letter failed to shame it into doing so now. A week later, he wrote an open letter to all of Tangier, the gist of which was contained in a single sentence: "It is a time for sinners to think on their ways."

Perhaps it struck a chord among the pious. Perhaps it even prompted the lawless to self-examination. We'll never know, because just two days after that, on October 13, *Newsweek* published a story titled "Trouble in Tangier" that put the crisis before all of America. "As in many isolated communities, religion is taken seriously on Tangier, and for more than a century Methodism has been its only denomination," the story read. "Last week reports reached the mainland that the island was in turmoil over religious differences. Old friends were snarling at each other. Families were split. Prowlers hurled rocks through church windows. Boats were sunk at their docks. And a minister, afraid for his life, huddled with his wife and two children in a little house with shattered window panes."

It went on to report that Swain's members denied condoning the attacks. "They blame them on young rowdies who have overheard family discussions of the dispute," the magazine reported. "With amusements nonexistent, they say, some youthful residents might start drinking and commit acts of vandalism."

Embarrassed by the story, Virginia governor William Tuck enlisted Accomack County Circuit Court judge Jefferson F. Walter to bring a halt to the unrest, and at month's end the judge convened a grand jury to look into the matter. Its members summoned the mayor, town council, and a procession of church and community leaders, along with several islanders recently returned from the war—the presumed "rowdies" mentioned by *Newsweek*. The grand jury returned no indictments, but the judge made clear to all that any further harassment would be answered swiftly and harshly. And with that, the trouble stopped.

Richardson moved his fledgling church out of the store and into his own home, knocking out a hallway wall to enlarge the living room so that fifty-one chairs could be shoehorned within. Children squeezed three abreast on the stairs, and an adjoining room seated more adults. The pastor stood at the front door, often using a blackboard to illustrate his messages. "Not everybody could necessarily see my dad," Grace Kimpel recalled, "but everybody could hear him."

There the New Testament congregation met for the next eight years, until it raised the money to build its spartan two-story church down at the seam between King Street and Black Dye. The first service there was held on Easter Sunday in 1957. Though dwarfed by Swain, it thrived for years, offering a brand of evangelism as unassuming and stripped of frill as its building. Richardson shared its leadership with a handful of church elders, who took turns at the pulpit. Most worked the water, and all were men.

Children grew up in the church—Jerry Frank Pruitt and Ed Charnock among them—and younger islanders joined the congregation of their own accord. But eventually New Testament's original faithful aged and started to die faster than they were replaced.

PIETY AND LAWLESSNESS—examples abound of these competing facets of the Tangier character. Take the shirt factory down near the south end of the West Ridge: A Baltimore clothing company opened the plant in 1919, employing sixty island women in the daily creation of "dress and work shirts as well as rompers of different qualities and styles," as an Eastern Shore newspaper reported in 1927. "That the employees are contented is proven by the fact that some have been there since the factory was opened."

The factory burned. Contented though its workers might have been, island tradition holds that their men weren't pleased with the independence that jobs afforded their wives and daughters; the plant's destruction was no accident. Fanned by a strong southerly wind, the fire consumed the building, then raced across the marsh and devoured the Heistin' Bridge, too, and only the frantic efforts of bucket brigades kept it from pushing on to Black Dye, where it might have leaped from house to house up the Main Ridge. No one was ever made to answer for the crime.

I could offer further examples from history, but let's instead con-sider a happening three weeks after Duane Crockett's sermon on the yoke of bondage: One night late in July, someone steals into Swain Memorial's small office, which occupies an addition on the church's west side, and makes off with about $3,000.

The news reaches the Situation Room two days later, when Hoot Pruitt, Swain's current lay leader, informs the regulars that only $600 of the loot was in cash, with the remainder in personal checks and useless to the thieves, assuming they aren't so stupid as to try to cash them. The money was not stored in the church safe, he says, but it wasn't lying in plain view, either: It had been cached in piles of pa-perwork in three different parts of the room.

"Somebody knew where it was at," Hoot says. "They went right to it."

"Somebody very familiar with what was going on," Jerry Frank muses. "And they'll probably get caught."

"I hope so," Hoot says.

"When did this happen?" Leon asks.

"Sunday night after church to Monday morning," Hoot says. "A lot of hanky-panky people hanging around there." He adds, after a pause, "And they might not have done it."

"Inside job," Ooker theorizes.

"The way it looks," Hoot says, visibly stricken, "you have to think so."

A few hours later, an air of apprehensive curiosity hangs over Swain as the Wednesday evening church service begins. A deflated-looking John Flood paces the front of the sanctuary. "By now you've probably guessed that it's going to be a little different this evening from a normal Wednesday evening service," he says, his breathing labored and transmitted over the PA by his headset microphone. "You know me. I don't like secrets. I try to provide any information I have.

"I'm sure by now everybody knows on the island that we were robbed Sunday night or early Monday morning. The tithes and col-lections were taken, and some of the money from the men's prayer breakfast was taken." He pauses for a long moment, during which he studies his congregation, and we listen as he catches his breath. "The thing I really want to ask is that you not let this harden your heart. I have to tell you that I feel so violated by this. I have trouble putting into words how I feel about it. I will be honest with you: I felt anger, along with the hurt. They didn't take *my* money. They didn't take *your* money. The moment it touched that plate, it became *God's* money. They took *God's* money."

His tone softens. "We haven't had to sit in the dark, in the heat, or in the cold yet. The Lord will provide," he says. "I want to ask you to do an unusual thing, and that is to pray for the person who did this, because that person clearly needs our prayers." He then offers up a prayer. "We pray for the person or persons who did this," he says. "We pray for the investigators who are looking into this."

A few minutes later, when the pastor solicits prayer requests, a waterman suggests that perhaps Swain's best course of action is to put the theft behind it, to forgive what's happened. Marlene Mc-Cready, Jackie Haskins's wife, voices her agreement. "Yeah, we want to be Christian-like in this," she says, though she adds, "This is a hard blow on our church. This is what we need to get us through the winter, even."

Ginny Marshall, Carol Moore's aunt, is sitting in the pew behind me. "It's the worst thing that's ever happened on Tangier," she whispers. "There's been other things, but this is the worst, in my opinion." It strikes me as an overstatement, what with the cholera, the deaths of islanders in boats and in combat, and myriad other tragedies that have befallen the place over the past two centuries. But then, I reflect, the church has always been the rock on which the island depended in hard times. The church, whether Swain or New Testament, has been a refuge against all the uncertainties of life on a tiny island isolated and buffeted by big water. Church has been central to the essence of Tangier, and Tangier central to church.

Perhaps erosion threatens both.

PART THREE

EYEING
THE END TIMES

Above: The wave-ravaged western shore of Uppards, April 2017. (EARL SWIFT)
Below: Uppards, as seen from the northeast—an angle that clearly shows the islet to be nearly as much water as marsh. The breach is visible as a gap in the far shoreline. (EARL SWIFT)

THIRTEEN

———

ASBURY PRUITT WAS THAT RARE MAN ABOUT WHOM NO ONE, it seems, can muster any comment short of praise. He's rarer still for another reason: He was the first Tangierman who thought to track and record his island's disappearance. When he took up this task in 1964, his neighbors viewed it as a mild curiosity, an eccentric hobby. Ultimately, it proved indispensable to understanding the forces at work on his home and the relentlessness of the encroaching bay.

A quick portrait of the man at the time he launched his research, the consensus of those who knew him: of average height, with a bit of a belly, but physically strong and graced with preternaturally youthful skin that masked his age. Humble and sweet tempered. So slow to anger that few recall seeing it happen—when displeased, he'd merely hum to himself. So opposed to gossip that if he sensed its approach, he'd turn off his hearing aids. So deeply sentimental that he often wept in conversation. A teetotaler and a serious student of the Bible. An elder and preacher at New Testament who illustrated his messages with examples drawn from island life.

He was forty-five when he took up his measurements. By then he'd married an eighth-generation islander, served on a Coast Guard buoy tender during World War II, and had three children, among

them Jerry Frank. Before and after the war, he'd worked the water.
The family lived in a house built for his grandfather William Asbury
Crockett, the lighthouse keeper who'd drowned on his run to get the
mail.

In 1958, Asbury had left the crab and oyster business to work
for the U.S. Navy, which had used the Chesapeake as a gunnery
range for nearly fifty years. In 1911, it had anchored a retired battleship a few miles southwest of Tangier and shelled, torpedoed, and
strafed it for decades. It later parked other derelict warships nearby
and pounded them, too. Now the navy positioned a pair of stripped
Liberty ships just over two miles from Tangier's western shore and
dispatched waves of jets to assault them with smoke bombs.

Asbury was one of four Tangier civilians hired to man two navy
spotting stations off the West Ridge, and to offer feedback to the pi-
lots on their accuracy and the angles of their attack. He was the boss,
overseeing Short Ed Parks and Short Ed's brother-in-law, Charles
Pruitt, and Pat Parks, all veterans of the service. Before long two
other islanders joined the crew, and for twelve years these six did their
duty as the air over peaceful Tangier was shredded by navy attack jets,
as many as eighty a day, ripping low over the water and the islanders'
crab pots.

At some early point in his tenure, Asbury took note of the stations' commanding view of his island's western edge and the effects
of the waves there. It was a rare perspective. Tangiermen knew their
home was shrinking; Canaan was going fast, and they'd seen Oyster
Creek consumed and the P'int whittled, too. But Asbury now saw it
occurring before his eyes. "The ground would seem to wave up and
down," he said, "and finally drop into the water like chunks of a melting iceberg."

The sight alarmed him. And so, after nearly two hundred years of
settlement on Tangier, Asbury Pruitt became the first of his people
to seek an answer to the question: Exactly how much are we losing?

On January 8, 1964, Asbury hammered an iron pipe into the
marsh, not far from the main navy spotting station, and measured

the distance from the pipe to the water's edge. To my mind, that date and that act rank as turning points in Tangier history. With those measurements, Asbury engaged the island in a quest to understand an old and increasingly critical challenge. Until that day, Jerry Frank told me, erosion was a subject of "some concern" among Tangiermen. "But they didn't keep up with it much. They didn't pay much attention."

Asbury repeated the exercise every January 8 for decades, and his measurements provided the first year-by-year chronicle of nature's wear on the place. "He was able to keep pretty good track of what was happening," Jerry Frank recalled. "He was very concerned about it, he was."

Rightly so, because the island's losses were far greater than anyone had imagined. During each of the first seven years of Asbury's record keeping, he found that the Chesapeake tore away, on average, twelve feet of shoreline. After that, the damage increased year by year. In January 1973, he found that eighteen feet had disappeared. In 1974, his measurements showed that thirty-seven feet of island had vanished in the previous twelve months.

By January 1975, his records covered an entire decade, during which Tangier's shoreline had retreated by 159 feet up at the island's north end, near the navy's main spotting station. Down south, where a second spotting shack stood below Hog Ridge, a terrifying 181 feet had disappeared, 44 in the last year alone. Pond and marsh that had once separated that shack from the Chesapeake were gone. "You can stand right on the platform of that station," Asbury said, "and throw something right into the bay." He said this to reporters alerted to his annual ritual. His measurements thus earned mention in the *Virginian-Pilot* and other regional newspapers, which prompted the government to step in. Virginia convened a task force of federal and state officials to study the situation.

Its most immediate worry was the island's new airport. In December 1965, the Corps of Engineers had answered a long-standing desire of Tangier watermen by excavating a channel from the town's harbor westward, through marsh, to the Chesapeake. Until this

"North Channel" was dug, a crabber looking to set pots on the island's west side had to head east into Tangier Sound, circle wide
around the spit, then traverse more than a mile of bay. A boat channel
slicing through the island's waist, separating Uppards from Tangier
proper, seemed an easy fix to this inconvenience and a no-fuss project
for the corps. It proved so easy, in fact, that dredgers dug the passage
in less than a month.

The project was bundled with another that promised to make
island life easier: The corps pumped the dredge spoils into a long
rectangle of soggy marsh behind the houses on the West Ridge. Over
the next three years, the state graded and paved the reclaimed land
into an airport with a 3,050-foot runway and a large airplane parking
area. Governor Mills E. Godwin Jr. dedicated the facility in March
1970, even as Asbury Pruitt's measurements made clear that it wasn't
long for the world.

Five years later, with the bay clawing at the asphalt runway and
surf breaking less than one hundred yards from the houses in Hog
Ridge, the Virginia Airport Authority moved to armor the airport's
crumbling south end with a curving riprap seawall. Crews placed
the thirteen thousand tons of rock in record time, and still the waves
snapped off huge pieces of the airstrip and stole the lights at its
south end.

Meanwhile, the state task force issued its report and recommendations. Studying maps, the panel found that between 1942 and
1960, the spit had shifted radically eastward. The west side's land loss
was even worse than Asbury Pruitt's records suggested: an average
of eighteen feet a year from 1850 to 1942, twenty feet a year from
1942 to 1967, and twenty-five feet a year since. And while the riprap
guarding the airport might halt erosion there, it would not stop catastrophic loss of marsh and upland elsewhere. The task force focused
on two possible solutions: extending the riprap seawall along the entire western shore of Tangier proper, all the way up to the new boat
channel's mouth—far and away the best answer, but expensive—or
attaching mats made of old tires to the same stretch of shoreline,

which the group admitted was an ugly and untested alternative, but cheap.

Whatever remedy the state chose would only slow, not stop, the island's continued destruction. Because the largest waves tended to approach Tangier from the northwest, the future looked bleak for Uppards and the new boat channel's western entrance. "One would anticipate continued retreat of the island north of the channel," the report warned. "This would result in an offset configuration at the entrance of the channel resulting in eventual shoaling and exposure to wave action on the south bank of the channel. Thus, future corrective action for Tangier Channel, maintained by the Corps of Engineers, is foreseen."

Remember that. It will be important later. And remember that this was in 1976.

It so happened that Virginia did not produce the money to build a seawall that year, and didn't even find the $100,000 necessary to experiment with a mat of tires. The following winter, when the bay froze solid, ice tore away eighty-five feet of the western shore. Asbury Pruitt had to move his pipe inland, then had to do it again, and still the water advanced. In January 1978, he found that it had drawn another thirty-one feet closer.

Down below Hog Ridge, the navy was forced to abandon the southern spotting shack. Jack Chandler, who would later marry Leon's daughter Carolyn, went to work for Asbury around that time. "The very first day I was there, he took me down to the old spotting station on the beach," he told me. "The sand had washed from underneath it. It was sitting in the bay." Asbury pointed past the ruin, to the deep water offshore. "He said, 'That out there used to be forest.'"

Little more than two years after riprap had saved the airport's south end, a second state study predicted that the rest of the runway would be underwater within a decade. "An emergency situation exists on Tangier Island which must be remedied immediately," it concluded. "The erosion of the island is so severe that it will wash into the

Chesapeake Bay in the very near future." Tangier had a host of other ills, the document said, but if the erosion went unchecked, "the other problems will not need to be addressed."

That grim forecast resonated in April 1979, when Asbury was shocked to see that the western shore had retreated by seventeen feet in only three months. He also discovered telephone lines in the surf, six feet from the shore. The phone company told him it had installed the lines forty-five feet inland just fourteen months before.

That same month, a delegation of high-ranking officials came calling, led by U.S. senator John Warner of Virginia. He walked the airport runway. He eyed the water drawing close. He spoke to Asbury Pruitt and other Tangiermen. "If we don't stop the erosion soon, a piece of American history will disappear," the senator told reporters who'd followed him there.

His visit seemed to light a fire under the state and federal governments. That summer, Virginia signed an agreement authorizing the Army Corps of Engineers to start designing a seawall. The millions it would cost to build it wasn't on hand, but islanders were encouraged that *something* seemed to be happening. In due course, the House of Representatives folded the seawall project into the Water Resources Development Act of 1980. President Jimmy Carter's administration complained, however, that corps engineers had not adequately tested the effectiveness of a wall, rendering the project premature. Further study was needed.

The House passed the bill anyway, but it bogged down in the Senate. When more than a year had passed since Warner's visit to the island, the exasperated senator appealed to his colleagues. "Is there any process through which we can expedite—on an emergency basis— the authorization of a project to save historic Tangier Island . . . from further erosion damage that threatens its very existence?" he asked in a letter to the committee handling the matter. "Time and tide wait for no man." While senators dithered, Warner wrote, "Tangier Island has been sinking—and continues to sink—not so slowly into the Chesapeake Bay."

The bill failed. Warner promised he wouldn't give up and, nearly three years later, sponsored a new measure authorizing the corps to study the situation and estimate the cost of a seawall. That bill took more than a year to make it through the congressional sausage works. In the meantime, the water threatened another key public investment.

In the early 1980s, the federal and state governments elected to end the island's reliance on septic tanks and sewer ditches with a modern waste treatment system, and combined the project with a new waterworks. Contractors tore up the roads to lay pipe, built cinder block pumping stations on all the ridges, and erected the water tower. The work dragged on for years and made a muddy eyesore of the place. Then mayor Dewey Crockett complained that islanders had to "wear hip boots to get into the church on Sunday."

The centerpiece of the effort was a lavish sewage treatment plant and garbage incinerator west of the airport runway. Although situated on one of the island's highest tracts, it was separated from the fast-advancing bay by only two hundred-odd feet of marsh. The $3.5 million investment was still under construction when, in January 1983, Asbury Pruitt found that the gap between building and water was closing fast. "I mean to tell you," he told a reporter, "she's a-going."

Congress wrote and passed another bill financing a seawall, only to see it vetoed by President Reagan. The bay kept coming. Every outside expert agreed that Tangier's circumstances were dire. Every visitor seemed to consider the place worth saving. But twenty-one years after Asbury Pruitt made his first measurements, the federal government continued to sit on the problem. Islanders were not alone in their frustration. "Delay and deferment have characterized the federal effort," the *Virginian-Pilot* editorialized. "Tangier Island may end up as Tangier Reef if Congress doesn't get off its collective duff and approve the seawall the island desperately needs."

TANGIER HAD BEEN in difficult straits before and had always managed to somehow navigate them. "We are here until he [God] says otherwise," Duane Crockett told me one afternoon at the island firehouse, where he works part-time as a bookkeeper. "I know the Lord works in some ways where things have to get to the point where we don't know what to do—and right then, we find out that the Lord has everything under control."

The island had reached such a point. In October 1985, Junior Moore, Lonnie's father, worried to a reporter that "with the rate it's going, we won't be here for another generation," a comment that seemed downright optimistic when Asbury's next measurement revealed that another thirty-seven feet of island had crumbled away. Then fortune shifted. It might have been the high cost of doing nothing that finally brought the reprieve—a realization in Washington that taking no action would surely claim the sewage treatment plant and squander a $3.5 million public investment, *plus* cost taxpayers additional millions to replace it. That by comparison a wall was a smart and economical use of the public's money.

Whatever the case, over the next two years Congress got serious about doing something, and by April 1988 all the pieces of a deal were in place: The federal government would contribute $2.68 million toward a 5,700-foot riprap wall along the western shore, stretching from the boat channel south to the barrier built to safeguard the airport. In turn, the state pledged $1.47 million, and Accomack County, $200,000. At a town meeting, Tangiermen voted for a fivefold increase in their real estate taxes to drum up their $200,000 share of the cost. The vote, by show of hands, was unanimous.

And so barges laden with granite boulders appeared off Tangier's western shore in 1989, and a crane positioned them, piece by piece, until the coastline was sheathed in rock that towered higher than the ground it protected. The armoring stopped the erosion fast. Not an inch of ground has slipped into the bay there since.

But it did not halt the Chesapeake, which took aim at other parts of the island with new ferocity. Among its favorite targets was the North Channel's western entrance, just as the state panel had predicted so many years before. The effects are obvious with a trip to the slab. The cut through the marsh, originally seventy-five feet wide, is at least four times as broad now. Every storm reams it wider, and you can gauge the damage by a power pole planted over on Uppards: In the first five months of my stay on Tangier, I watched the ground around the pole's guy wires retreat eight to ten feet, until the anchor for one wire was underwater.

Just as vexing, the widening channel opened the harbor to westerlies, so that any weather beyond a light breeze turned the haven rough. "We used to have a good harbor before they opened up that channel," Leon lamented one afternoon at the Situation Room. And on another day: "Over there where the crab shanties are at, we used to have fifty crab floats overboard, and not once did any of them break loose. Not once did the tide run high enough to move those floats. What happened? That channel is what happened."

The channel's unhappy side effects didn't end with rough water. Waves carried sand and silt ripped from Uppards into the passage, too. "It didn't take long before the bars on either side of the channel started to get shallow," Short Ed Parks told me when I visited with him in his home on the Ponderosa. "There was one man on the island who was against opening that channel. His name was John Parks, and just about everything he said would happen has happened. He said, 'You build that channel, it'll ruin the creek.'"

That would be "Junk John" Parks, so called because he salvaged and recycled for his keep, on top of crabbing. He was Leon's uncle. "He was right," Ed said. "I didn't think so at the time. When you were discussing things around the dock, as men do, he'd voice his opinion, and some would get hot about him. He was sure about it, though. He said it would make things shallow where you didn't want it that way, and deeper where you don't want it deep."

The Corps of Engineers recognized the channel's vulnerability early on. Six years after the seawall's completion, it had already devised a solution: a rock jetty extending from the southwest corner of Uppards and shielding the channel's mouth. At the time, the corps reckoned that such a project would cost about $1.2 million, with the federal government ponying up $900,000 of that amount and the remainder split among state, county, and local governments.

But like the seawall, the jetty project moved from conception to approval to financing at a glacial pace. Construction finally appeared imminent eight years later, in the spring of 2004. It didn't happen. As islanders were to learn, whenever money seemed in hand for the jetty, the corps thought additional study was necessary, and whenever the corps' penchant for study seemed sated, money was elusive. Years passed. The jetty's price tag climbed to $3 million by 2007 and eventually to $4.2 million. At the same time, the shoreline at Uppards steadily eroded, changing the jetty's design parameters, prompting more study.

And along the way, a curious phenomenon unfolded. The jetty's size and form shifted in the minds of many Tangiermen. No longer was it seen for what it was—a simple and rather limited ploy to stave off shoaling and protect the harbor from wind-driven waves. No, now it was conflated with other, grander schemes that had come and gone over the years, proposals that called for armoring much of Uppards and Tangier's exposed southern underbelly. The jetty was reshaped in the public imagination into a seawall not unlike that guarding the airport—a seawall that promised the island's survival.

It is this mythical seawall to which Tangiermen seem to refer when they offer up prayer requests for "the seawall" at church, or when older islanders say they hope to live long enough to see "the seawall" built: not a jetty, but some amorphous product of wishful thinking, a mass confabulation of what has been considered, planned, hoped for, and promised for years.

That might explain the euphoria that gripped the island when then-governor Robert F. McDonnell and corps officials pledged in November 2012, three weeks after Carol Moore stumbled on the sprung-open graves at Canaan, that the jetty would be built, and soon. The promise came at a ceremony on the Tangier waterfront covered by the *Washington Post*. A corps official told the paper he was "cautiously optimistic that construction will be finished in 2017."

Islanders were convinced that this meant construction would start in 2016, Ooker among them. They learned they were mistaken when the corps and representatives of the Chesapeake Bay Foundation met with townsfolk in the school cafeteria in December 2015. Yes, the officials told them, the initial phase of the project was funded and was ready to start—but that initial phase was more study.

Those in attendance were shocked and angry, and a few of them were vocal about it. "We usually sit there politely and listen," Principal Denny Crockett told me. "But when they said it wouldn't start in 2016, that ruffled our feathers. Each year the channel gets wider and there's less and less you can do with the money, and people were upset."

The stormy meeting ended with the corps assuring the islanders that they'd see physical work on the jetty begin in 2017 and could expect it to finish in 2018. "They said 2017," Ooker told me. "They didn't give an actual month, but they said in 2017 it would begin."

As it happened, the corps did not start the work in any month of 2017.

"They do studies, then they study the studies," the mayor complained as we fished up pots. "I know that's their procedure, but it gets frustrating. We're at the point now that it's like me coming across a family in a boat that's sinking, and I say, 'I'm going to rescue you, but I have to study it first.'

"They know we're losing land and running out of time. Sometimes you have to just go ahead and *do* it."

ONE LATE AFTERNOON, I climb into Carol Moore's skiff and we speed away to Canaan. Carol noses the boat against the shore—a foot-high bluff of peat and oyster shell undermined by tide and currents—and I scramble out with a small anchor and drive its blade into sod that squishes and quakes underfoot.

From here, our routine holds that Carol, seeking solitude, strikes off to the east, while I prog the scant remains of Canaan or wander down the island's west side. So off she goes, eyes cast down, on the lookout for arrowheads, bottles, pieces of the past. I pick my way west along the water's edge, swatting at flies, tracing a meandering route around mudholes and tidal pools as the coastline bends southward. Bright-billed oystercatchers alarmed by my intrusion circle, shrilling, just overhead.

The shore consists mostly of sod bristling with broken stalks of spartina grass or smoothed slick into mudflat, and all of it carved into deep scallops. As I walk, waves roll into these concavities, leap their sides with loud slurps, and tear away chunks of soil. Now and then I encounter sod torn from the shore and thrown back in great lifeless blocks, some the size of ottomans.

Later, as we motor back down Uppards's west side, Carol points out a break in the shore that islanders call "the breach." It looks to be about thirty feet wide and filled with water that runs well back into the marsh. I can make out an old boat abandoned on a muddy bank fifty yards in.

It doesn't much look it, but the breach signals a new emergency. Uppards has had inlets cutting into its marshes from the east for as long as the place has been mapped. The largest of these is a club-shaped body of water called Tom's Gut. On maps of old, it traversed about halfway across the island's width. But with the erosion of Uppards from the west, the marsh between gut and bay narrowed—and at the same time, Tom's Gut ballooned in size—until the waters met.

The gut's expansion is a textbook by-product of sea-level rise: Upland turns to wetland, while marsh drowns, giving way to mudflats and, in time, open water. A 2006 article in the journal *Global Environmental Change* showed that what's happening to Tom's Gut also contributed to Holland Island's demise. Between 1849 and 1989, some eighty-eight acres of Holland's high ground was lost, the study found. Most of it—a little more than fifty-three acres—was obliterated by "edge erosion due to waves around the island perimeter." But the remaining thirty-four acres of high ground turned to marsh.

But back to this breach and what it portends: Uppards has been cut in two, and its interior is now exposed to the erosive action of the bay. It stands to be scoured away inside and out, and the damage will accelerate as the opening widens. As we eye the breach from the boat, Carol says it's only been six or seven years since "it was still so narrow that we could barely get a skiff through it, and it was so shallow you'd get hung up every time. Now two workboats can go through there, side by side, it's so deep and so wide."

The Tangier Town Council has been asking the corps since at least 2011 to plug the breach with dredge fill. To date, it hasn't happened. Seeing how long it's taken the agency's other Tangier projects to become reality, it could be a while yet. Meanwhile, every day, Tom's Gut grows and Uppards shrinks.

A well-stocked aisle at Daley & Son. (Earl Swift)

FOURTEEN

—

HIS SKIFF'S OUTBOARD "AIN'T GOT A REVERSE OR A NEU-tral," so Cameron Evans has to cut the engine thirty feet from Ooker's crab shanty and let the boat coast the rest of the way in. He's earning pocket money by bustering up the mayor's peelers this dark Saturday morning, and he'll deserve every penny: Tangier is gripped by a blustery chill. Its water tower is all but invisible behind a caul of misty rain.

The skiff glides alongside Ooker's dock and thumps against a piling, which Cameron hugs to bring us to a stop. Shielded from the elements in a slicker and waterproof pants, he scrambles onto the deck and strides past the cats to the shedding tanks. In one, Ooker's placed a flounder, olive gray and a foot long, that blundered into a pot. It spreads pancake-flat and motionless on the bottom, laying low among a gang of crouching peelers. Cameron plucks several crabs from the water and moves them to other tanks, then eyes the busters in the tank closest to the shanty. He scoops out two soft crabs with a hand net and carries them to the cooler, but opts to leave a third for his next visit; it just seconds ago pulled free from its shell and is too soft, too frail, to pick up and move.

He makes this judgment without touching the crab. At sixteen, Cameron has already developed a waterman's almost extrasensory ability to appraise crabs with a fleeting glance. Few other Tangier teenagers can boast his level of skill. But then, few others share his interest in the island's chief industry—or, for that matter, his passion for much of the life that his father and grandfathers enjoyed as boys.

While some other island youngsters sit indoors, entranced by TV and video games, Cameron is almost always outside, using a compound bow to hunt stingrays down off the spit with his classmate Isaiah McCready, or roaming the island with his camera, or casting for rockfish off the docks. Hunkering down in duck blinds in the cold and dark of winter. Digging for clams in the mudflats. This kid's a true Tangierman. Has mud between his toes, as islanders say. Lives for the place.

Two busters have hung up and died while shedding. He separates them from their old shells. A quick examination determines they're recently deceased, so he carries them into the shanty, wraps them in plastic, and places them in an enormous chest freezer almost brimming with similarly wrapped softshells—hundreds on hundreds of crabs that Ooker will provide to his buyers over the winter, when live animals can't be had.

I ask Cameron what he plans to do once he graduates with the class of 2018. "I really don't know anything as of yet," he says. He pours dry cat food into small mounds on the floor. Sam Alito, John Roberts, and Ann Coulter are blurs crossing the room. "I'm keeping my mind open."

"Do you think you'll go to college?"

"Yeah, I think I want to go," he replies. "But I really don't know what I want to do. I'd like to have a job that's outside and all—I wouldn't want to work in an office."

Around us is the closest thing to an office he'd likely occupy on the island. Coils of rope and piles of rusty netting. A strong bouquet of brine. Walls decorated with a horseshoe crab a foot across, fading Eskridge family snapshots leached of their reds, and a bumper sticker that gives away its age: NEW CENTURY. SAME GOD.

Back outside, Cameron scoops six dead peelers from a tank and piles the corpses on its wooden edge. They'll make good fishing bait. The breeze stiffens. A gull on the shanty's roof struggles, feathers splaying, to keep its perch. Cameron scans the tanks a second time. "Not much else to do," he says. "There'll be more later." We head back to the skiff.

"So, would you like to stay on Tangier?" I ask once we're aboard.

Cameron shoves us away from the pilings and starts the outboard. "I guess if I could, I probably would," he answers. "I'd like to stay in the state, I know that much. But I really don't know much more than that."

We putter from the shanty and into the harbor. Of all the kids I've met on the island, here is the one I can most imagine choosing to take up the work of his forebears—the one I can picture as a boat captain, chasing down pots in a deadrise. I am not alone in this assessment. Ask old-timers who among the island's boys reminds them most of themselves coming up, and they'd tell you it's Cameron. Ask them whom of those nearing graduation they could imagine forgoing the mainland and still doing well for himself, and they'd tell you: Cameron Evans.

Yet there'd be wishful thinking to their answers, because everyone on Tangier knows that Cameron has the potential to succeed wherever he goes and in whatever he chooses to do. He's gregarious, funny, and kindhearted, a deeply decent kid who's active in Sunday school and the youth programs at Swain Memorial. His lean, square-jawed features should season well. And he's sharp: He posts good grades while shouldering an ambitious load of classes—physics, dramatic lit, Advanced Placement psychology, third-year Spanish.

In fact, if Cameron were to announce that he'd decided to remain on Tangier, I believe that many island adults would urge him to reconsider. It would hurt them to do it, but as much as they appreciate a teenager who shares their love for the place, and as much as they'd relish seeing their work and traditions passed along to so capable a successor, they'd know he was making a mistake.

They face an uncomfortable reality: Their island's future depends on young people remaining here, working the water, raising families. But few islanders would wish the hardships and uncertainties of their own lives on the children they love—especially those children, like Cameron, with almost unlimited options.

THE LAST TANGIER BOYS to follow their fathers onto the water as a group are nearing forty today. Islanders blame a state clampdown on crabbing licenses nearly twenty years ago for the dearth of young captains since. That seems a reasonable view, until you consider that Tangiermen were later exempted from the license freeze's most bothersome particulars and that young people were fleeing for the mainland years before the new regulations came along.

What really seems to be at work is far more basic: The island's boys want no part of crabbing's long hours, physical demands, and financial instability. As for Tangier girls, what waits for them if they stay? Marriage, perhaps, though that is not the sure thing it was in their parents' time, for available mates are few. Raising children, if they do marry. Keeping house. A job at one of the restaurants, maybe. Driving a tour buggy. Cutting grass.

Both boys and girls have grown up watching their fathers grow old before their time, their parents fret over disappointing harvests, their mothers' lives stunted by the absence of meaningful opportunity—and they've grown up, too, witnessing the wider, faster, more glamorous world on satellite TV and the internet. Which is to say, the currents tugging at Tangier's young people are much the same as those that have emptied rural towns across America, from farm burgs in the South to Native villages in bush Alaska. A kid doesn't have to see much of the world beyond the town limits—or the water—before concluding that he'll miss out by staying put. Island kids have an additional impetus in their home's ongoing destruction: Even if they wanted to stay, could they realistically expect to live out their lives here? And if not, wouldn't they waste valuable time by investing, even for a few years, in a place where they were doomed to have little to show for their efforts?

Though older Tangiermen lament the diaspora, they have quietly encouraged it. "Go to school as long as they'll let you go"—that's the advice that my landlady, Cindy Parks, and her late husband, Charles, gave their two sons. The decades that Charles spent crabbing were never easy. "Charles used to tell the boys growing up, 'If you can do anything other than what I'm doing, do it,'" Cindy told me. "He knew how difficult it can be to keep a roof over your head. You come across two or three days of good crabs, and something goes wrong with your engine. You try to put aside a little bit, and a storm takes your crab pots." A lot of money passes through a crabber's hands, but even in a good year, he doesn't get to keep much of it.

Principal Nina Pruitt, married to a waterman-turned-tugboater, recognizes that a life on Tangier might not have the appeal it did when she graduated from high school in 1980. "I get asked a lot whether I encourage my students to stay on the island," she said. "And the answer is no. Just because I've chosen this life for myself and decided to stay doesn't mean I think they should do the same, to keep this island afloat.

"I encourage them to do what they want to do."

IN THE SPRING OF 1914, journalist J. W. Church and a photographer stepped ashore at Tangier half expecting to find "a striking case of inbreeding" in the population. The two were on assignment for *Harper's Magazine* and knew little of the place except that hundreds of islanders shared a handful of last names. The mainland oystermen and villagers Church consulted before their trip described Tangiermen as "mighty cur'us folk" and spoke of "queer goin's-on over yonder."

The reporter's fears were only bolstered when the islander who put them up, Captain Ed Crockett, told them, "Once in a while one of our boys goes over to the Eastern Shore for a wife, but most generally we Tangier folk kinda like to flock to ourselves."

"My thoughts," Church wrote, "went back to the succession of graves we had passed, and now I wondered if we would not find a densely populated asylum for defectives tucked away somewhere on Tangier."

The journalists found no such thing. Close cousins surely married in the early days of settlement, but by 1914—when the head count was nearing its peak—a great many islanders were the children or grandchildren of come-heres. The gene pool has broadened further since: Though virtually everyone born on Tangier can trace his or her lineage back to the original Joseph Crockett, he or she can also trace it back to a large and varied cast of mainlanders.

But not because the island attracts many outsiders who choose to stick here. Rare are those come-heres who put down stakes without an existing family connection, and those who do rarely stay for long. For some, Tangier's unbuffered weather and austere landscape prove depressing, especially in the winter. The island's gossip can wear. Some simply grow bored. "Folks will come here and have their island experience. They'll be attracted by this idea that it's an idyllic place," Jean Crockett explained to me. "And the truth is that it is not an idyllic place, and they won't stay. People have to have a reason to live *here*. Otherwise, you wouldn't be willing to put up with the stupid mess. Because there's quite a bit of stupid mess."

With its youngsters departing and its numbers unreplenished by a steady supply of newcomers, Tangier faces a danger unique to island communities. Because it's marooned from other population centers and the services they offer, its viability hinges on the health of a few key enterprises, without which daily life would become taxing, if not impossible. Each operation relies on a critical mass of patronage to remain solvent—in other words, each has a tipping point. And as the population falls, their tipping points draw nearer.

The first such outfit is Daley & Son. In 1986, JoAnne and Terry Daley Sr. became the fourth owners of the store, the island's sole source of packaged food, fresh produce and meats, and over-the-counter medicines. The grocery's three long, narrow aisles are heavy on canned and boxed goods, as you might expect. Medicines dominate one wall. Lining the other are coolers and freezers offering everything from frozen pizza to hummus. At the store's rear, produce is displayed in bushel baskets and meats behind a glass counter that dates to the mid-twentieth century.

The shelves are replenished every Thursday, when the Daleys haul eight to eleven thousand pounds of consumables from the mainland aboard their forty-five-foot deadrise, the *Working Dog*. "We try to stock all the basics," JoAnne told me one afternoon at her home in King Street. Twenty-eight or more cases of milk per week. Ninety dozen eggs. A mountain of bread, mostly white. Drinking water in half liters, liters, and gallons, and distilled water for golf cart batteries.

Her son, Terry Jr., worked at the store in his teens, then followed the water until his father's death in 2013 brought him back. JoAnne has since ceded the store's day-to-day operation over to Terry Jr. and his son, Lance. Not all of their top-selling fare is endorsed by the American Medical Association. The store sells a year-round average of fifty to sixty cartons of cigarettes per week (Marlboro Reds are the most popular) and up to eighty cartons a week in midsummer. The store also goes through twenty to fifty cases of potato chips a week, twelve bags to the case. Cookies, crackers, and candy by the bag and bar. A lot of cake mix. Enough chocolate syrup to float a skiff. And so much soda in the summer that the *Working Dog* makes special runs a couple of times a month. On each, it carries five to six hundred cases, which supplies the store, the island's restaurants, and six outdoor soda machines that Terry Jr. owns. On a dry island, sugar rules.

All of this could be had on the mainland. Walmart Supercenters stand a few minutes' drive from both Crisfield and Onancock, and the Maryland port has a well-appointed Food Lion. From a distance, it might appear Tangier could survive if Daley & Son were to fail. But island life would be a logistical nightmare. Trips to the mainland are expensive and time-consuming and beyond the ability of many elderly islanders.

And those mainland stores don't allow their patrons to run a tab, or "tick." Three out of four Daley customers do, and given the hit-

and-miss nature of a waterman's pay, it's more than a convenience. "I have some customers pay every two weeks and some by the month," JoAnne said. "I try to give them a break in the winter, and they'll be able to pay their bills over the summer. Most of them do it."

When I asked JoAnne whether she could envision a time when the population could no longer support the store, she told me that yes, sadly, she could. "It will probably come down to it," she said. "But it'll probably be on my son and grandson, on them two, more than it will be me."

I raised the subject with the men. "We might not be doing as good now as we did a few years ago," Lance allowed. "We've lost a lot of good customers, people who've passed away."

I asked if he worried about it. "Not really," he said. He paused, then changed his mind: "Well, I guess we do."

ANOTHER MUST-HAVE ENTERPRISE is the mailboat, which has operated so reliably, for so long, that it could be mistaken for a government service. Actually, it's a family business—Joshua Thomas's descendants have carried the mail, passengers, and freight between Tangier and Crisfield since the turn of the twentieth century. Captain John W. Thomas, Joshua's great-grandson, started the service and ran it until his death in 1934. His son, Captain Eulice H. Thomas, spent sixty-some years as a skipper. He eventually turned over the helm to his son, Rudy A. Thomas Sr., who expanded the family business to include tour boats operating out of Crisfield and Reedville.

Rudy Jr., born in 1956, started making the Crisfield run for his father before he was old enough to drive. When he married Beth Parks in 1978, the company charged its passengers ten dollars each way or twelve dollars per same-day round-trip. Rudy was always reluctant to raise it, even as the population—and thus the company's pool of customers—dropped by more than one hundred during the 1980s and kept falling. "Once [the fare] hit twenty dollars," Beth said, "and the cost of fuel continued to go up, I'd say, 'We need to raise it,' and he always said, 'No. Twenty dollars is plenty enough to go to Crisfield.'"

Rudy Jr. was able to hold firm because the shrinkage in his customer base was at least partly offset by an uptick in trips per islander. Thirty years ago, a run to Crisfield was a special occasion. Today, scores of Tangiermen keep cars there, and commutes to college, tugboating jobs, and medical specialists are routine. It helped, too, that passenger fares composed only part of the boat's income. The company contracts with the U.S. Postal Service to carry the mail and with FedEx and UPS to deliver packages. The *Courtney Thomas* also carries the bulk of the island's general freight, from boxes of soft-shell crabs to building supplies. "We carry pretty much everything," Beth said. "All the Sysco products for the restaurants. Stuff from the hardware stores. We carry dead bodies—and we don't charge anything for dead bodies. That's an order from generations back. Rudy said he heard that from his grandfather: 'We don't charge anybody for their last journey home.'"

But a diversified income stream only slows the inevitable. Trips to the doctor stop when the patient makes that last free journey. Fewer islanders produce less freight. And overhead is high. The *Courtney Thomas*, built in Louisiana in 1989 to carry crews to offshore drilling rigs in the Gulf of Mexico, is a confident craft that can shoulder a heavy payload in all but the fiercest weather, but its diesels drink a lot of fuel.

And there are no painless means to boost income or tighten spending. Raising fares would likely suppress ridership, and switching to a smaller boat would save money but slash capacity, especially for freight. In 2011, three years before his death at age fifty-seven, Rudy Jr. told an interviewer, "Sometimes I wonder if this is where it's all going to end. I don't know how much longer we're going to be able to sustain things around here."

Were the boat to stop running, Mark Haynie, who already makes round-trips to Crisfield seven days a week throughout the year, could assume some of the burden in the *Sharon Kay III*. But that vessel is far smaller than the mailboat, with limited room for cargo—and many older islanders are uneasy about winter crossings in anything smaller than the *Courtney Thomas*. Without the vital link it provides to the greater world, the island would become an even tinier place.

"I guess if we sat and thought about it too much, we'd all get out of heart," Beth said. "But the proof's in the pudding. There are a lot of senior citizens, and we don't have a lot of young families having children."

THEN THERE'S THE BIG ONE. Of all the disasters that might befall the island, nothing but a killer hurricane would bring the end as surely and quickly as the demise of Tangier Combined School. Not one of the island parents I've talked with would allow their children to boat across unpredictable Tangier and Pocomoke sounds twice a day to attend classes on the mainland. "You'd see people leaving like rats off a sinking ship," Cindy Parks told me. "They'd say, 'I'm not putting my kids on *that*.'"

Islanders have watched with growing angst as the school's enrollment has fallen. When Nina Pruitt graduated in 1980, the student body numbered about 120. As of the start of the 2016–17 academic year, it stood at 67. By the 2018–19 school year, it's projected to drop to 54 and to 53 the year after. It's expected to rebound slightly in 2020–21, to 58, but that promises to be only a temporary reprieve. Babies are in short supply.

Rhonda Hall, Accomack's assistant superintendent for instruction, told me that the numbers aren't an immediate cause for worry. "I don't think that the board has gotten to that point where they say, 'Okay, if the student body falls below fifty, or whatever, we're going to close the school,'" she said. "I've been here as an assistant superintendent since 2005. It's never been suggested that Tangier is becoming too expensive to run."

Instead, the district has addressed the decline by combining six elementary grades into three classrooms and relying on computer streaming for foreign-language courses and other specialized instruction. It's also decided that some capital improvements aren't likely to happen: "What will come up is something like, for instance, a playground," Hall said. "We'll talk about how much sense it would make to build a full-blown playground there for so few children.

"It's a struggle. But we've gotten it down to a science now."

Even if enrollment slips considerably, former principal Denny Crockett figures that Tangier Combined will remain safe. "When you look at what you'd have to do to replace the school, it would be at least as expensive," he said. "You'd have to have a boat just for that, with a licensed captain and a licensed mate." The cost would be a minor hurdle next to the journey itself. Students on Maryland's Smith Island are boated to mainland schools from eighth grade up, but their commute is half the distance of that from Tangier to Onancock and across comparatively protected water. The sixteen-mile crossing for Tangier youngsters would take an hour or more, assuming the waters were smooth—which is hardly a safe assumption. "If you have an exam," Denny said, "and you have to go over to take that exam and it's blowing twenty-five miles per hour and you get seasick on the way over—well, educationally, it wouldn't be very conducive. So I think Accomack County is stuck with us."

But islanders do have one cause for concern, and it isn't the number of students in the school. It's their teachers. "The average age is older than dirt," as Nina put it. Some are in poor physical shape. It's likely that a couple will retire in the near future and conceivable that two or three others might, too. "[And] with the dwindling population," Nina told me, "teachers who retire won't be replaced."

Hall pointed out that Nina could retire herself. "She could retire at any time, and she has five or six teachers who could retire right now. They could go at any time.

"It isn't really when the number of students starts declining that causes a problem," she said. "But if we had a lot of teachers retire at one time, that *would* be a problem."

For Tangier, it could be existential. Close the school, and the households with children would relocate—and considering that all but a few children live in two-parent homes, we're talking a third of the population. They'd waste no time about it. The other essential enterprises would implode. And as it did at Holland Island a century ago, wholesale abandonment would loom large.

ON A FRIDAY IN MID-AUGUST, word reaches the Situation Room that Cook Cannon has just fallen out of a tree and broken both of his feet.

"That didn't hurt none," Ernest Ed Parks observes.

"All he can do is crawl?" Leon asks.

Lonnie arrives with details. He understands that Cook broke a toe on one foot and at least three bones in the other. And he didn't fall from a tree; he lost his balance while standing on a pile of brush at the P'int and fell with his feet locked in a tangle of branches. We all wince. I wonder who will pull the air conditioner from the window once the weather cools.

The conversation turns, as it has in several sessions over the past month, to the presidential campaign. Most members of the group have been die-hard supporters of Donald Trump since he announced his candidacy, and the reasons for that go beyond party affiliation or his stands on abortion, same-sex marriage, and immigration: Tangiermen reckon that he'd cut through the government red tape that has kept the island's salvation on hold. That he'd put a stop to incessant corps studies and spur the agency to action. That he'd force Congress to find the money. That he'd recognize Tangier as a town imbued with patriotism, reverence, a strong work ethic, old-fashioned values—traits they deem central to his "Make America Great Again" campaign slogan—and thus worth preserving. That he'd use his experience in business and as a builder to *get things done*.

But today there is a naysayer in their midst. Long before Trump became the preemptive Republican nominee, Lonnie was forecasting his defeat, famously predicting: "A couple of months from now, people will be asking, 'Trump *who*?'" Ooker memorialized the remark by writing "Trump who?" in ballpoint on the wall near the coffee maker.

Now Lonnie announces that voting for Trump or Hillary Clinton is equally wasteful. "They're the same person," he tells us. "They're exactly the same. You vote for one, it's the same as voting for the other."

"I'm like you—I don't think either of them is worth nothin'," Hoot Pruitt replies, rubbing his hemispherical belly. "I ain't of very good heart, but I'll vote for Trump before I'll vote for her."

"You're wasting your vote," Lonnie shoots back.

As a group, the others in the room tell him he's wrong. Lonnie attempts to keep the back-and-forth going, but no one bites. When it's obvious debate has ended, Bruce Gordy leans forward in his chair. "Do you think there will be another generation of watermen from Tangier?" he asks. A long silence commences.

"I guess I want to be optimistic," Lonnie finally says, choosing his words slowly, "and say you're going to have a few. It might not be an actual generation, but there'd be a few."

Bruce: "How many, you think?"

"I don't know," Lonnie says. "Three, four. Maybe five." After several wordless seconds he continues: "You hear that if they lifted the restrictions, there'd be more boys going on the water, but I don't think that's true." He looks around the room. He is broaching a subject far closer to Tangier hearts than the upcoming election. "There are lots of licenses for sale," he says. "If boys wanted to go on the water, they could do it right now. But you don't see anybody buying them."

The group sits in silence for another long moment. He gets no argument.

Homebound Tangiermen study their cell phones aboard Mark Haynie's *Sharon Kay III*, November 2016. (EARL SWIFT)

FIFTEEN

WHEN I FIRST SAT THROUGH SUNDAY MORNING SERVICES at Swain Memorial, in 2000, the pastor was a big-bellied, suspender-clad fellow named L. Wade Creedle Jr., born and raised in the tobacco country below Richmond. He was in his ninth year on the island, and his prayers and sermons were simple, straightforward, and fine-tuned to the concerns of his congregation. He talked about how crabbers were faring and about storms, rough seas, and money troubles. More important, he had the good sense to share his pulpit with the Tangier-born Dewey Crockett, who stood a head taller than the pastor and who had a soothing baritone and folksy delivery reminiscent of Garrison Keillor, had Keillor grown up talking backward. Dewey's presence was welcoming, reassuring; sharp of mind but gentle in spirit, he led with a kindly chuckle, a hand on the shoulder, an "Amen, brother."

The church around me was packed, and these Methodists, many of them obviously unschooled, rose from the pews to deliver prayers deeply informed by scriptural study. The pastor's wife, Nancy, banged out foot-stomping nineteenth-century hymns on the keyboards, and everyone sang so that the high-ceilinged sanctuary filled with sound. And I, no follower of organized religion, found myself thinking, "So this is why people go to church."

But nowhere is Tangier's decline in population more apparent than in its churches. By early in the new century, Swain Memorial's big Sunday morning service rarely drew more than two hundred to a sanctuary built to seat three times that number. And Swain's situation was happy compared with the scene down the road at New Testament. The congregation there had seen its old originals fall away, and while their children had largely stayed true to the breakaway flock, they were getting up in years themselves. The third generation of "Holy Rollers," as everyone called them, wasn't nearly as large. "We had got to the point that on a Sunday morning we'd maybe have thirty," said John Wesley Charnock, who grew up attending New Testament. "On a Sunday night we'd have only fifteen or twenty." Some prayer meetings drew only six or seven. The day was coming, it seemed, when Tangier might again host just one congregation.

Then two once-unthinkable events occurred. First, in the summer of 2009, the Methodist Church got a female preacher.

Patricia Stover was a native of Virginia's Shenandoah Valley and kin to farmers and preachers—both of her grandfathers led churches, and Stover felt herself drawn to the pulpit from childhood. "I was always talking to Jesus," she told me. "He was kind of my invisible playmate." She resisted the call. After earning degrees in biology and history at James Madison University, she married, gave birth to a son and a daughter, divorced, and worked a succession of jobs—as a biochemist, grants manager, state functionary—while "wrestling with the Lord something awful," as she put it. Finally, nearing fifty, "I surrendered and said, 'Okay, I'll do what you want me to do.'"

She entered the ministry in 1994 and pastored rural churches around the state. She was beginning her fourth year at a church in the Blue Ridge southeast of Roanoke when her district superintendent told her he'd found her a "wonderful appointment opportunity" that suited her evangelical bent and love for close-knit communities. You're going to Tangier Island, he announced.

So Pastor Stover landed at Swain and met with the church's board, led by Principal Denny Crockett. The introduction went well.

"I really did feel welcome," she said. "I had wondered about that, because I knew they were very conservative, and I was told that if ten years earlier someone had said they'd have a woman preacher, they would have said, 'That won't happen.'"

The parsonage, more than a century old by then, had fallen into disrepair. During its lengthy renovation, the pastor bunked in a small King Street cottage. Otherwise, her inaugural two years seemed to go smoothly. "Everything looked great on the surface," she said. "Money coming in. They had tithers. And they knew their Bible—they'd get up in class meeting and just amaze me." Beneath that healthy veneer, however, Pastor Stover "could feel an undercurrent in the church. I didn't know what it was. I couldn't quite put my finger on it." Was it because some of the congregation, both men and women, couldn't accept her in the pulpit? She didn't want to think so, but she wondered. Whatever its cause, she was "convinced that there was a spirit of schism there," she told me, "and a spirit of lawlessness."

Some of the faithful attending Swain at the time acknowledge that there was, in fact, a brooding unrest in the congregation. They cite a common source: The wider Methodist Church was becoming too liberal, especially in its views on homosexuality and its internal debates over whether to ordain gay and lesbian ministers. "I ain't got nothing against them," Cook Cannon said, summarizing the thoughts of many. "I ain't a basher or anything like that. But it ain't right. I don't feel like being in a place that supports that."

The United Methodist Church did not permit homosexuals in its leadership at the time and does not do so today. The church does not permit its ministers to perform same-sex marriages and doesn't allow such ceremonies in its sanctuaries. But that's not to say that these issues haven't been argued for years. In fact, they've split the national membership down the middle, and some fear that a Methodist summit on LGBT roles in the denomination, scheduled for 2019, will trigger a permanent breakup. That they were discussed at all was enough, it seems, to upset many on Tangier.

Duane Crockett, who was deeply involved in Swain as a Sunday school teacher, youth leader, and organist, identified a more immediate, local problem. "There was just no participation anymore—it was the same few people doing everything. Everybody would say, 'I can't do it. I'm too busy,'" he said. "And a lot of the older, devout people had passed away."

Pastor Stover agreed that a very few were, in fact, running things. But, she recalled, they were "loud and kind of controlling" and "had a stranglehold on the church." Nancy Creedle backed her up. "You had a few—quite a few, I might ought to say—who wanted to run the church," she said, adding, "That's a common situation."

Against this backdrop, Swain Memorial dispatched its delegates to the 2011 Virginia United Methodist Conference.

EVERY JUNE, the United Methodist Church gathers thousands of its Virginia ministers and church members to tackle the denomination's business for the coming year. The 2011 conference was held in Roanoke, in the state's mountainous west. As is the custom, Swain sent two delegates to the meeting: its pastor and Eugenia Pruitt, a member of the laity who had represented the congregation at the conference for more than ten years.

Eugenia is the daughter of Edwin "Eddie Boy" Parks, Leon's first cousin. She's the granddaughter of Miss Annie Parks, whom islanders regard as the closest thing to a Tangier saint since the days of Charles P. Swain. Stocky, with a frizz of dark hair, and graced with a booming voice, Eugenia was sixty-nine years old in 2011 and had been saved for fifty years.

"When you're elected to be a delegate, I take that very serious," she said one afternoon, as we sat in her home on the West Ridge. "So before conference you get this book. It's a handbook, like, and you're asked to read this whole book so you know what's going on. And in it are all the resolutions that have been proposed, that the conference is going to vote on." The document in question is the annual *Book of Reports*, a dry read if ever there was one, and deep in the 2011 edition were the eighteen resolutions for that year. One of them, Resolution 13, captured Eugenia's attention.

Titled "Effective and Constructive Peacemaking Between Palestinians and Israelis," it opened by noting that the church sought "to act as an advocate for peace in the Israeli-Palestinian conflict." The preferred route, it declared, was "the creation of two independent sovereign nations, Israel and Palestine, living side by side in peace and in economic justice and cooperation." It noted that "a viable Palestinian state must have a sustainable financial foundation" and argued: "To this end, positive financial investment for the Palestinians must be encouraged." And it committed the church to "study and make recommendations for concrete measures" that could be taken by its boards, agencies, and members "to encourage, aid and assist the Palestinian people in their efforts in nation building." That was the meat of the resolution, but that is not what Eugenia took away from her reading. She understood it to call on "the Methodist Conference to donate money to Palestine for them to become a state," as she put it to me.

"Well, when I read it I stayed very troubled. I prayed about it until I let Ooker read it. I saw him up to Fisherman's Corner. He was up having dinner, and as he read it he was doing this." She shook her head. She showed it to Dewey Crockett, who did the same thing, and to Duane Crockett, too. Eugenia drove with her husband, Fred, to Roanoke. She spoke against the resolution. She voted against it. It passed nonetheless. She and Fred drove home. The following Sunday morning, she followed church custom by giving a conference report to Swain's congregation.

"I don't think anybody thought she'd say anything at all—that it would be some boring report," recalled Jean Crockett, who served on the church board. "This was *not* a boring report."

NO MATTER THAT the resolution did not call on the Methodist Church to give money to Palestine, as Eugenia interpreted it to say, and as many Tangiermen believe it said to this day. Some at Swain objected to it on other, more fundamental grounds: It encouraged the creation of a Palestinian state, which they viewed as a de facto attack on Israel and which, in turn, ran counter to their reading of Scripture. "Israel is God's chosen people," Duane informed me. "If there's no Israel, there's no kingdom. If there's no kingdom, there's no king. And if there's no king, the whole Bible is a lie."

One might argue that the Israel of the Bible and the modern nation-state of the same name are different entities entirely—that the former refers to a people and an ancient, vaguely defined territory, while the latter is a creation of secular law and politics dating only to 1948. But that isn't accepted thinking among many of the faithful on Tangier. As they see it, Israel is Israel, and any support for Palestine constitutes aid to Israel's enemies and is sufficient to provoke God's wrath. "[Palestinians] got one aim in their life, one aim, and that's to kill everybody in Israel," Cook Cannon explained to me. "I support the Jews. The Bible says all the land belongs to the Jews, and they've never gotten their land back."

"I think that's something that will make God mad."

Duane, who knew beforehand that Eugenia planned to denounce the resolution, followed with his own comments to the congregation: "I said, 'If you stand with our decision to stand against this, will you please show that by standing?'" Some in the congregation were confused by all the fuss and weren't sure what such a "stand" involved. Nevertheless, most of them stood.

The service ended, but the disquiet did not. Pastor Stover said she "tried to explain to them that our resolutions are a statement, that there's no teeth to it, that there's no action. I said, 'There's no money involved in the resolution whatsoever.' I said, 'It's just a statement, and some people support it, and it's just one of those things where we agree to disagree.'" But many in Swain, convinced that the Virginia Conference had turned on Israel, remained up in arms.

The subject came up again at a Swain board meeting. "I asked the board if I could write to the district superintendent and the bishop," Duane said. "The board voted unanimously to let me do that." He also asked whether he and Carlene Shores, Leon's oldest daughter, could write to every newspaper in Virginia, "to let the people know what was going on." The board agreed to that as well.

The letters appeared in dozens of newspapers late that summer. I found one preserved on the website of the *Herald Courier* of Bristol, Virginia, in the state's far southwest. "The members of Swain United Methodist on Tangier cannot support the building of a nation, who is Israel's sworn enemy," it declared, adding, "We call on members of the Virginia United Methodist Conference to take a stand against this unbiblical resolution by contacting your pastor, District Superintendent, and Bishop, and stand on the Word of God."

Soon after, with no groundswell of outrage evident in the state's other churches, the congregation found itself at an impasse. "It began to consume our board meetings and our committee meetings: What are we going to do next?" Duane said. Their answer was to stop sending a share of Swain's weekly collections and tithes to the larger Methodist Church—to support only the Tangier church and its missions. Some members also thought it wise to confiscate all the Methodist hymnals from the pews.

So things stood when, early that fall, Tangier earned a visit from Tammy Estep, the district superintendent for the Eastern Shore. Her message was firm: You can't remove the hymnals—they're the property of the United Methodist Church. And if you want to leave the Virginia Conference, that's up to you—but understand that Swain Memorial is UMC property, too. The building stays Methodist.

The protesters hadn't counted on that. Their ancestors had built the church and they'd assumed it theirs. But sure enough, the Methodist Church held the deed to the property. What had seemed an easy break with the ever more liberal outside world, they now saw, carried a steep price. The Swain sanctuary was the repository of more than a century's collective memories—christenings, graduations, marriages, and funerals; prayers as storms bore down, babies were born, and neighbors ailed; rejoicing and grief across generations. This was no simple building. This was a spiritual and emotional home.

Building or no, Carlene was the first to go: She quit her post as Swain's musical director and left the church that December. Shortly after Christmas, five Tangier-born ministers who preached on the mainland came to Swain, advocating cool heads. "We didn't dispute anything they had to say," Jean Crockett recalled. "But they mostly talked about our heritage, and our heritage was ending." The following night, the congregation met to decide what to do. Some argued for staying in the Virginia Conference and promoting Israel from within it. Others suggested that they should all quit the conference, every last one of them, leaving the Methodists with a church no one used.

It came to a vote, which was well short of unanimous. And so the second once-inconceivable event occurred. Swain's congregation split, though no one knew how many would actually leave until a Sunday in January 2012, when Pastor Stover found her congregation smaller by about eighty people.

"I was hurt to have to leave the church," Eugenia said. "I don't want to say it was easy for me to leave. It weren't. But I left with no doubts about leaving. That Thursday night, when they had all those preachers come talk to us, I knew that night as I left that I was never coming back. I got into the golf cart with Fred and said, 'Honey, I'm not going back.'

"Those that stayed, I would say a lot of them stayed because of the church—the building," she said. "We knew that if everybody stood together and left the conference, they'd have come and shut the doors. They'd have boarded it up. But I believe in a year or two we could have bought the church back. What were they going to do with that big building? They'd have sold it back to us."

THE SPLIT, like that in 1946, divided families. Carol Moore left for New Testament, while her mother, aunts, and her uncle Jack stayed at Swain. Ed and Annette Charnock remained at Swain. Ed's daughter Danielle left. Allen Ray Crockett stayed put; his brother Dewey, a man who for decades was synonymous with Swain, chose to leave. Bobby and Lisa Crockett left. Lisa's mother remains a Methodist.

Like the earlier schism, this one bred hard feelings. "It was like a death," Nancy Creedle told me. "It was so hurtful to be going down the street and to see people I'd seen for so many years in church, and they turned their heads like I wasn't there. I kept praying that the Lord would help me act like I should act."

But in time, those wounds began to scab. New Testament, having teetered on the brink of extinction, enjoyed a new vitality, with Sunday attendance equal to the church up the road. And among Swain's depleted number, Pastor Stover found a cause for celebration. "As soon as the split occurred, the undercurrent wasn't there anymore, and this sweet spirit emerged," she told me. "It was just beautiful to see."

In 2013, the time had come for Pastor Stover's replacement. Island-born Robbie Parks, a Methodist preacher supervising a central Virginia district, realized that one of the pastors he oversaw might be a good fit for the job. He spent a Saturday with John Flood, who was pastoring three churches in rural Prince Edward County, and his wife, Delores. "I told him all the good things about Tangier," he recalled. "I told him all the negatives about Tangier."

John Flood remembers the conversation well. "He said, 'There's been problems there. The church has split and there's been a lot of turmoil. Healing needs to take place.'" Not long after, he and Delores made their first visit. Both were charmed by Tangier and the islanders they met. The feeling was mutual. "As soon as they stepped in the door and said hello, I knew that John Flood was the right man," Pastor Stover recalled. "I just knew it. He just seemed like he was warm and fuzzy and loving, and just perfect."

In the years since, Pastor Flood has managed to bridge much of the chasm. He's done it with a quiet, folksy manner born of his upbringing on a rural tobacco farm and a pragmatism honed in his first career running a trucking company. Years working the fields equipped him with insight into the uncertainties of weather and harvests, the inevitability and variability of seasons—and, by extension, the travails of life on the water.

He has made the entire island his ministry, regardless of church affiliation. If the Maryland State Police helicopter touches down to medevac a Tangierman to the mainland, Pastor Flood is at the airport, offering encouragement to patient and family. If a Tangierman's in the hospital, the preacher is sure to turn up at his or her bedside.

"The Lord sent Tangier a great gift with Pastor Flood," Eugenia said. "For a while there were hard feelings, and most of them blamed me. Down through history, I'll be the one who broke up the church. But the Lord can work in spite of us. There was a lot of forgiveness, once Pastor Flood came."

One evening I sat in on the congregation's annual meeting with its district superintendent, Alexander B. Joyner. It opened with testimony from Hoot Pruitt. "We're blessed to have our pastor, because he knows how to get people in line—not in a rough way, but in a gentle spirit," he said.

Annette Charnock took the floor. "They all love him, whether they're a member of this church or not," she told the Reverend Joyner. "He's like a gentle giant."

Marlene McCready: "He's the shepherd of the whole island."

Joyner, who was an Eastern Shore pastor during the 2012 drama and who followed it closely, stepped up to a lectern. "It's a real joy, as your district superintendent, to hear such support for your pastor," he told the gathering. "It's especially gratifying to hear it at *this* church.

"Not because it's an unloving place," he added quickly. "But as you know, Tangier can be a difficult place."

Ooker buys peelers from hard-potter Tabby Crockett. (EARL SWIFT)

SIXTEEN

BACK IN THE DAYS BEFORE CRABBERS PUT THEIR PEELERS into pump-supplied tanks to molt, many chose not to bother with the shedding process at all. Instead, they sold their peelers to a shedding house that bustered them up and sent them to market.

Nowadays all the island's peeler crabbers run their own tanks. But one of them still maintains a vestige of the old shedding operations: Ooker buys the peelers that hard-crab potters catch. He pays fifty cents for each. Some days he buys hundreds; I've seen him buy seventy at a time from a single crabber. He built a pier extending from his shanty to the channel's edge, to make it easier for his hard-crabbing neighbors to drop off their baskets, and pays what he owes them once a week or so.

On paper, it's good business. He more than doubles the number of peelers in his tanks, and the crabs he buys for fifty cents will fetch six or seven times as much once they're bustered. But shedding has its risks. A July storm sweeps over Tangier with winds so fierce they topple poles carrying electrical lines out to the shanties, knocking out power to the pumps that circulate water over the peelers. Crabs die by the hundreds in Ooker's tanks and those of his neighbors. When it happens, he's out not only the crabs and the money they'd bring, but the money he owes the hard-potters for crabs he'll never sell.

And then comes the mysterious die-off that strikes the tanks of peeler crabbers every year, a phenomenon they say did not occur in the days when peelers were kept in overboard floats. One day in the Situation Room, Ooker announces he's running three tanks of his own crabs and three of peelers he's bought, and the die-off is in full effect, especially among the crabs he got from hard-potters. Of the purchased crabs, last night he had thirty busters and thirty-three dead. This morning, he found another forty peelers had died over-night. Of the crabs he caught himself, last night he had fifty-five busters and five dead; this morning, seven more had died.

Lonnie Moore says he's finding two or three dead crabs in every pot. "It's happening everywhere," he says.

"A lot of times," Ooker says, "those crabs'll have a different look to them, when they're dyin'."

"A goin'-away look," Lonnie suggests.

"Yeah. They don't have that healthy color to them."

Eleven days later, the die-off has worsened. In pots and tanks alike, crabs are expiring by the bushel. "All dyin'," Richard Pruitt says with a sigh. "It does this every year."

"It ain't the right water," Leon says. "It ain't like it is overboard." That's as much an explanation as any Tangierman can offer. They can't say why the die-off ends after two or three weeks every summer, either.

The carnage is still under way when I join Ooker on the *Sreedevi* for an abbreviated day on the water. It's his birthday—mine, too—and he and Irene plan to head to the mainland on the evening boat, then drive down to Virginia Beach for a night at an oceanfront hotel. Just after sunup, a stultifying heat has already descended on the harbor. As we motor out toward the sound, we pass Short Ed Parks in the *Princess Sky*. He's dressed for the weather in shorts, knee socks, and white shrimp boots.

The first pots come up sparkling. The water has been crowded lately with gelatinous comb jellies, which ooze through the mesh, reflecting and bending the light of the low sun like prisms. Trapped with them is an abundance of small hard crabs. Ooker pronounces them "shit-eaters," which marks the first time I've heard him curse. "This time of year, you catch a lot of crabs, but they're no good," he explains. "They're peeler size"—that is, far smaller than the five-inch minimum for hard crabs—"and they look like peelers. They look like they're going to turn."

But instead of molting within a few days, Ooker says, they just hang in that in-between state for weeks. "They call them junk crabs, shit-eaters. If they do turn eventually, it seems like they take the whole month of August before they do." He tosses them back over the side, one by one. "Not much you can do with them." For now, anyway.

Some of the pots contain a peeler or two—Ooker's first row yields twenty-six of them—but they're almost lost among the junk crabs. One pot comes up with eleven crabs. He has to throw back every one. "You can see why a peeler crab wouldn't want to go in there," he says, "with all that commotion."

He has the radio on, and we listen to a Virginia Lottery ad touting a $478 million jackpot. He recalls a past lottery offering a similar payout; Ed Charnock told him that if he won, he could buy a lot of crabbing supplies with that kind of money. Not me, Ooker chuckles. "If I won that, I think I'd take up my peeler pots."

Another row. "A lot of crabs. A lot of junk crabs. A few jimmies in 'em, but very few peelers." He sighs. "I'm losing my drive." After four lackluster hours on the water, Ooker takes a break to eat a ham and mayo on white. We bob in gentle swells off the spit, close enough to hear an atonal chorus from the pelicans and black-backed gulls congregated there. "Right there, at Whale Point," he says, nodding toward Canton's friable southern shore, "is where we would have put some of those barges."

The barges are a sore spot. In the spring of 2011, with salvation from the Corps of Engineers years away, a would-be Tangier rescuer appeared on the southern horizon: A Hampton Roads salvage company was willing to donate several empty barges for use as makeshift breakwaters around the island's battered shoreline. U.S. representative Scott Rigell, the island's Republican congressman, got behind the idea, calling it an imaginative way to stave off erosion until a more permanent solution came along. And because the company would clean, tow, and sink the barges at its own expense, it would cost taxpayers nothing.

That June, Rigell visited Tangier to meet with Ooker and one of the salvage company's officers. All were excited by the idea. Unorthodox as it seemed, it had a precedent: When a ferry terminal opened near the southern tip of the Eastern Shore after World War II, it was shielded from the bay's winds and waves by nine concrete-hulled cargo ships sunk end to end offshore. The ferry stopped running when the Chesapeake Bay Bridge-Tunnel opened in 1964, but the ships remain on guard, though profoundly decayed and sprouting with grasses, shrubs, even a few trees. They now protect Virginia's Kiptopeke State Park and serve as crowded roosts for pelicans, gulls, and migrating waterfowl.

At Tangier, the three men envisioned barges used in similar fashion. They'd be sunk singly or in groups off fast-eroding points: at both ends of the boat channel, alongside Uppards, and across fast-eroding Whale Point. Rigell wanted to get started right away—in six months, he said, "but ideally, half of that." He did not "want to see this held up," he announced. "I really don't want to see any pushback from any government agency."

But pushback he got. The corps balked, and the state, too. The Chesapeake Bay Foundation piled on. "One of those corps guys asked me, 'How would that look, a bunch of rusty barges off your shoreline?'" Ooker recalls. "I said, 'They'd look real good when a storm's approaching.'"

The chief objection, the mayor explains, was that the barges would have been planted on top of underwater grasses. He was frustrated by that, because if the erosion continues, the grasses are doomed anyway. "It's the same principle as fighting wildfires—you give up some to save more," he tells me. He shakes his head. "Common sense is not so common anymore."

ONE MORNING I DRIVE TO NORFOLK, where the Army Corps of Engineers occupies a large, wedge-shaped office building on the Elizabeth River waterfront, to meet Dave Schulte, the lead author of the 2015 *Scientific Reports* article that hastened my return to the island. He turns out to be a thickset fellow with a shaved head, whose well-considered speech and three-piece suit give him something of a professorial air.

In his office, and later over lunch, Schulte outlines the years he's spent studying the island and mulling its future. He arrived at the Norfolk district in 2001, after years handling forestry, game management, and wetlands restoration projects in Virginia and Georgia for the Defense Department. One of his early corps projects took him to Tangier, where he participated in a comprehensive study of the island's erosion dilemma, aimed at preserving its usefulness as habitat to a wide array of bird, turtle, and insect species, and to marine life in the seagrass beds offshore.

Uppards figured prominently in the work, and he can recall his first reactions on stepping ashore there. "It really struck me how low everything was," he says. "The cemetery was still there. There was a mobile home, too, a fair distance from the water. I remember a few fig trees. There were pine trees there, too—it wasn't a big stand, fifteen or twenty trees, but there were real uplands there."

That was in 2002. The study recommended a string of breakwaters along Uppards's western shore, as well as a jetty at the mouth of the channel—a more elaborate and far more protective version of the jetty contemplated today. Corps projects require a positive cost-benefit assessment; if the agency's estimate of what the public stands to gain from a project doesn't meet or exceed that project's costs, said project is discarded. In this case, the corps reckoned the island's importance as habitat was a worthy investment.

But corps projects also require shared spending from federal, state, and local governments. "We were very frustrated," Schulte tells me, "because Tangier couldn't come up with its share of the project. That's not unusual. They asked for the state to pay their share, but the state couldn't come up with the money, either."

The study offered a prediction of what might happen to Uppards were the project not undertaken: the scenario that has been unfolding since, which Schulte describes as "pretty grim." That has been "sorely frustrating, because we sit here at the corps with this old study that could have addressed a lot of the problems the island's experiencing," he says. "It's 2016, fourteen years later, and we're just finishing the internal steps necessary to address one small piece of that proposal. It seems like it takes a long time to go from saying, 'We've got to fix this problem,' to actually fixing it."

SCHULTE STAYED ALERT to news about Tangier in the years that followed, stayed interested in the fate of the island and its people, and stayed up on the literature about the changing bay. Even before he arrived in Norfolk, there was a lot of such information, for the Chesapeake had emerged as a favored setting for scientists studying the effects of climate change. In a 1991 paper, for example, two University of Maryland scholars studied four vanishing Maryland islands and marshes on the nearby mainland, searching for evidence of shifts in the rate of sea-level rise. They found that the islands began losing land at an accelerated pace in the mid-nineteenth century. Even more profound acceleration followed in the twentieth, when tide-gauge records showed that the rate of sea-level rise in the bay was "more than double the long-term trend of the last several thousand years."

In 1995, another paper found that more than forty-four thousand acres of Chesapeake Bay coastline had washed away over the preceding one hundred years. The historical record unambiguously "shows land loss occurring since at least the mid-nineteenth century," the authors wrote. "It appears that land loss is not a new phenomenon, but the rate of loss has likely accelerated in the past century."

The following year, an article in the *Journal of Coastal Research* reported much the same thing—that both sea-level rise and shore erosion were slow "from the middle seventeenth century until about 1850," at which point charting them required "the insertion of a sharp inflection point."

Clearly, the middle of the nineteenth century marked a turning point for the bay and its islands. Studies published after Schulte became involved with Tangier built on that notion. For example, a 2006 article in the journal *Global Environmental Change* used Holland Island as its focus, reporting that land loss there averaged less than a quarter acre per year from 1668 to 1849 but underwent a fivefold jump between 1849 and 1989.

And the literature suggested that sea-level rise and land loss to this point, dramatic though they've been, pale next to projections for the coming decades. A prognosis offered in 2010 by scientists writing in the journal *Estuarine, Coastal and Shelf Science* promised that the Chesapeake's water level will rise ever faster, achieving "increases of approximately 700 to 1600mm by 2100, a projection we consider to be very likely." That's a jump of twenty-seven to sixty-three inches.

What would that look like on Tangier? Another *Journal of Coastal Research* article found that in all of 1998, tides rose to a meter or more over then-mean sea levels for only three hours on the island. But if sea levels rose as expected, by 2100 the Main and West ridges could be underwater for as many as 4,400 hours per year, or about half the time. Long before that the place would be uninhabitable, because Uppards would dissolve away altogether, rendering the harbor useless.

How Tangiermen might react to such a scenario was forecast in a 2013 paper in the journal *Nature Climate Change*. The intensity of people's attachment to a climate-threatened community signals how they'll respond to government strategies for dealing with the threat, the paper said. "Consequently, adaptation strategies that directly affect attachment to place may not be supported, and different strategies that allow people to remain in their current place are more likely to be successful."

Indeed, I've heard islanders say as much. Ooker summed up the feeling of many when we talked about mainlanders who think saving Tangier would cost too much, and that the islanders should pick up and move. "We've been here for a couple hundred years or better," he said, "and it's our home." Or as Carol Moore put it to me: "I'm staying until I got one foot on the side and one foot in the water."

SCHULTE UNDERTOOK his own 2015 study with two coworkers: Karin Dridge, a geographer, and Mark Hudgins, chief of the Corps of Engineers Norfolk District's Hydraulics and Hydrology Section. The corps financed the paper, but because it did so through a scientific grant, Schulte "could say things that the corps can't," he tells me. "I could talk about the value of the resource. We [the corps] can't put a dollar value on it, but this time I did."

It wasn't the value of Tangier's human presence that the paper focused on. "Losing the town is, of course, a huge loss, but I wanted people to know that there's more to it than that," Schulte says. "Island habitat is extremely valuable, because it's not in abundant supply. I wanted to show what would be lost, just in terms of habitat—and in dollars—if the island were allowed to wash away."

The paper pegged wetlands and underwater grasses with ecological values, which are not what you could buy them for, but what they're worth *per year*, based on their status as wildlife habitat. Using a conservative projection of future land loss, the authors estimated that by 2063, Tangier, Uppards, and their outlying islets would lose more than 431 acres of wetlands, with an estimated annual value of $1.75 million in 2015 dollars. By 2113, the islands would lose 629 acres of marsh—in other words, pretty much all of it—and habitat worth $2.54 million per year. "Bird nesting will decline," the paper warned, "accelerating the decline of these species."

Just losing Uppards would usher the destruction of subaquatic vegetation off the island's leeward east side, covering more than 370 acres of the bay's floor. Schulte and company reckoned that was worth $3 million per year. Over decades, the cumulative value of lost habitat would run into hundreds of millions of dollars. "We concluded," he tells me, "that saving the island is well justified."

The authors did not put a dollar value on losses to the island's human inhabitants, but they made clear that they'd be steep. They expected Tangier "to lose land at an exponential rate," they wrote. "South of the seawall, Tangier will experience significant land loss. The projection indicates it is very likely that the sand spit, which provides some protection against incoming wave energy from the south, will be lost. Tidal creeks winding through the islands will widen significantly, encroaching into the upland ridges. Most of Uppards is predicted to be inundated by 2063, reducing the protection provided to the Town of Tangier."

Actually, losing what remains of Uppards would eliminate *any* protection to the town during the winter, when winds blow in from points north. "Without Uppards, you can't really have a Tangier," Schulte acknowledges. "You lose your protected harbor. You lose the entire seafood industry."

Tangier would be uninhabitable in less than a hundred years, and likely by 2063, Schulte and his partners predicted. Its ridges, with the possible exception of a few isolated knobs of high ground, "will be converted to a mix of intertidal and high estuarine marsh." This was not fully accepted on Tangier, especially the study's memorable last line: "The Tangier Islands and the Town are running out of time, and if no action is taken, the citizens of Tangier may become among the first climate change refugees in the continental USA."

The mayor and much of his constituency maintain their skepticism about a human role in global warming, and the town itself has since sold T-shirts with the silk-screened legend: I REFUSE TO BE A CLIMATE CHANGE REFUGEE. Schulte nods patiently when I ask him about that. "The subtleties of sea-level rise are something they're just not able to comprehend," he says. "They'll see a spot that had grass and is now marsh. They'll know that trees grew in a spot, and now the trees have died and they can't grow new ones. But they don't ascribe that to sea-level rise. That's flooding."

Schulte's paper prescribed a chain of rock breakwaters around the west, north, and east edges of Uppards, positioned a short distance offshore. The space between the breakwaters and the fast-eroding shoreline would be filled with a combination of beach and sand dunes, with the deepest such buffer on the island's exposed western flank. It also advocated another breakwater armoring the beach below the airport—essentially a further extension of the existing seawall. Finally, it suggested using dredged sand to rebuild five uplands once home to human settlement on Uppards and Tangier proper, and planting them with loblolly pines.

Schulte and his coauthors were thinking big, and such thinking carried a price tag to match—about $20 million to $30 million, they estimated, which they tried to offset by restating what was at stake: "If no action is taken, significant wildlife habitat will be lost, as well as the culturally-unique Town of Tangier, the last offshore fishing community in Virginia waters of Chesapeake Bay."

I tell Schulte that I find the article compelling but can't help but feel it's overly optimistic—that based on the changes I've seen since 2000, and the retreat of shoreline at Canaan that I've witnessed in just the past few months, I figure you'll be able to drive a workboat over most of Tangier by 2063. No chance the town can last anywhere near that long.

Schulte nods. Corps studies are required to use computer-modeled sea-level projections, or "curves," he tells me. The source for the figures used in Tangier's curves is the tidal gauge records for Sewells Point, on the Norfolk Naval Base. "The problem is with the Sewells Point data set itself," he says, which is "one of our longest sets on record, and we use the entire data set to establish averages." Because the data starts at a point when sea levels were rising at a much slower rate than they are today, averaging its entirety understates the severity of the current problem; it "gives a falsely rosy picture of what's happening."

"You have a choice," he says. "You can chop off those early years and use the more accurate, more recent data, or you can recognize the low-scenario curves"—the most conservative projections of future sea-level rise—"as unlikely." He and his coauthors chose the latter path, using a curve that split the middle between best- and worst-case scenarios—though, as the paper admitted, the worst case was looking more realistic all the time.

"It *is* a conservative estimate," he tells me. "I don't think they have fifty years. When newspapers interview me, I always say twenty-five to fifty—and I think it's probably closer to twenty-five, because the rate is continuing to go up."

I ask him about the proposed jetty at the west end of the boat channel. "That'll help them a little bit with that particular problem," he says, "but it doesn't do anything about the bigger problems." Build the jetty without somehow armoring the rest of the island, and eventually Tangier will consist of that jetty and the existing seawall to its south. The island itself will be gone.

"Anything we're going to do is going to have to get started in the next couple of years," he says. "Let's say the interest level was high enough. You could do a mini Poplar Island over there. That's an option."

IF THERE ARE TWO WORDS guaranteed to provoke an exasperated shake of the head from Tangiermen, they are "Poplar Island." Located about sixty miles north of Tangier off Maryland's Eastern Shore, Poplar was one of the bay's many inhabited islands in the mid-nineteenth century. Its principal settlement, Valliant, boasted a hundred citizens and cattle farms, a school, a general store, a post office, and a sawmill.

That sawmill may have hastened its destruction. In felling Poplar's dense woods, its inhabitants robbed their home of the tree roots binding its soil. From about 1,100 acres that stretched in a four-mile-long crescent before the Civil War, Poplar broke into pieces, and each dwindled rapidly. Islanders quit the place by 1920, and seventy years later, the island had been reduced to a five-acre islet and three tiny outlying knobs of land. All seemed destined to become shoal.

But unlike so many other islands that have been allowed to disappear, Poplar was saved at the proverbial last minute: The U.S. Fish and Wildlife Service, decrying the bay's shrinking inventory of remote island bird habitat, partnered with the State of Maryland and the Corps of Engineers' Baltimore District to rebuild the island to its original size.

The project used silt that the corps dredged from the shipping approaches to Baltimore harbor. Dredging the channels was nothing new—the corps has performed this duty for generations to keep the nation's major harbors open to deep-water traffic. But it had typically disposed of the silt by dumping it at sea. Restoring an island in the Chesapeake made more sense, and Poplar was only thirty miles away—a short run for scows lugging the dredge spoils.

So, beginning in 1998, the corps built a riprap dike approximating the island's 1847 shape and size and started dumping dredged material three years later. By early in 2016, Poplar measured 1,140 acres, bisected by a high spine. On the eastern, more protected side of that ridgeline, the corps built varied wetlands: high and low marsh, ponds, tidal flats, and islands for nesting birds. To the west, it mounded two large upland cells with silt that will eventually reach an elevation of twenty-five feet and will be forested with pines.

In 2014, Congress authorized expanding the new island by 575 acres, combining open-water habitat with additional high ground and marsh. When the entire project is completed around 2040, it will have used about 68 million cubic yards of dredge spoils, or enough to fill more than 2 million twenty-foot shipping containers. The projected price tag: about $1.4 billion.

Here's the part that deeply aggrieves Tangiermen: This lavish, decades-long effort and heroic outlay will create a habitat for birds by the thousands and a wide range of other wildlife besides. But the human population will be zero. For that kind of money, Tangier could be turned into an impregnable fortress. As Jerry Frank Pruitt put it in the Situation Room: "They can build islands but they can't save an island? I don't go for that much."

One of the challenges Tangier faces in drawing that kind of

money is its location. Had Poplar not been convenient to Baltimore, and had the corps not needed to keep Baltimore's shipping lanes clear, that island would have gone under. Tangier is a long way from any such dredging operation, which worries Schulte. "We've already lost so many islands in the Chesapeake Bay," he says. "We only have a couple left. Are we really going to let them disappear, too?

"Right now, it looks like maybe we will."

Standing water occupies the footprint of a demolished house in Black Dye. (Earl Swift)

SEVENTEEN

L ATE ON THE MORNING OF MONDAY, AUGUST 15, A CONTIN-
gent of mainland visitors arrived at Tangier's firehouse for a
meeting with members of the town council. Most of the island
was unaware of the gathering, and even the council wasn't sure what
it was about—the visitors had sketched out the agenda in a few vague
sentences. The Tangiermen were excited, just the same.

The business at hand dated to the late spring, when Gregory C.
Steele of the Corps of Engineers' Norfolk District had had lunch with
John Bull, commissioner of the Virginia Marine Resources Commission. They were discussing the disposal of dredge spoils in the open
bay near the mouth of the York River—a corps practice since the
1980s, but one that had made the state increasingly uncomfortable,
due to its possible effects on the blue crab.

"We were saying, 'Yeah, we really shouldn't be dumping that
dredge spoil on crab spawning grounds,'" Bull recalled. "And one of
us said, 'We ought to take it all and dump it on Tangier.' And we were
laughing about it, until we both said, 'Well, hold on a minute.'"

Now, more than two months later, the men led a party eight
strong. Steele, who heads the Norfolk District's Water Resources
Division, was accompanied by five corps specialists—a civil engineer,
planners, policy people—and Bull, by a second VMRC higher-up.

Not present: Dave Schulte, who told me he was unaware of the meeting.

Representing Tangier were town manager Renee Tyler and three members of the council: Norwood Evans, who works for the electrical co-op; James Parks, Jerry Frank Pruitt's brother-in-law; and Anna Pruitt-Parks, Jerry Frank's daughter. Not present: Ooker, who told me he didn't know of the gathering, either, because he didn't check his email. He never checks his email. "I don't do email," he said. Everyone on the council knows he's one of the few elected public officials in America who doesn't find it useful. But instead of phoning him, the folks arranging the meeting emailed him.

In any event, he wasn't present to hear the visitors lay out a proposal for Tangier's salvation that would make another Poplar Island of the place: The corps would build cells over and around the island, then fill them to create high ground or sculpted wetlands. The primary aim was not to protect people—as Dave Schulte's *Scientific Reports* paper suggested, the island's wildlife habitat holds more value than the town's infrastructure. But if the concept were to win approval, Tangiermen would be saved along with the birds.

"I was excited," Anna told me later. "I'm not saying there's any such thing as climate change, but if they're willing to spend the money and try different things to fight climate change, so they can see whether they'd work in other places, I'm okay with that."

When I asked about how the meeting happened without the mayor's participation, Anna was visibly irritated. "Ooker knew," she said. "I guess I shouldn't be saying it, but Ooker's a good mayor for about six months of the year. When it's crab-shedding season, though, he ain't worth shit. Love him to pieces, but when those soft crabs are shedding, you can't get him away from them.

"And I mean, there's been so many reporters, so many articles, so much attention, since that [*Scientific Reports*] article came out last December, that I think he's just been exhausted and totally forgot about it."

Whatever the case, Ooker is present two afternoons later, when the council convenes an official town meeting to share what went on at the sit-down. By then, a few details have trickled out, enough that fifty-eight expectant islanders assemble at New Testament for the session. It opens with a prayer from Duane Crockett. "We thank you, Lord, for your concern for our island," he says. "And, Lord, there's a lot of hearts that need to be changed in the government to do this."

James Parks outlines the corps' proposal. He points to a rather crude corps rendering of what the new Tangier might look like: a squarish blob reminiscent of the island's size a century ago, but only vaguely recognizable as Tangier. Inlets and concavities in the shoreline have been smoothed flat. The tidal creeks wriggling through the interior are gone, along with a lot of the creek between Canton and the P'int. In place of a spit, the south end of the island ends in a wedge.

"It's going to take a three-year study," James says. "They explained that. Gonna take two years to get to the three-year study, possibly. So that's a five-year waiting period to see if this project will take off."

Anna explains that the corps can't justify saving Tangier for the sake of its human inhabitants—it costs too much for too few people—but that officials seemed to think that "what they done at Poplar Island" might work. "They haven't looked real close at what that will do to the living area," she says, referring to the town. "Once they get to expanding the island, will they have to elevate the living area to keep it from flooding? Will they have to blow dredge material into this area? Will they have to elevate the homes to accommodate that? All of that will be looked at in that three-year study."

Ooker notes that the project will require adjustments from the island's crabbers. "Some of the places, like where maybe some scrapers work, will be covered up," he says. "But it's a grand plan." He looks around. "Any questions?"

"Yeah, I got one." It's Eugenia Pruitt. "Are they aware that our marshland, every day, is quickly leaving this island?" she asks. "Whatever's in this marshland is doing a whole lot more damage than the erosion." Large, ravenous rodents devouring the wetlands are a continuing worry of Eugenia's. She's told me that "these marsh creatures" are "tame as can be, and they're bigger than a muskrat. I've seen one on my front walk. I seen him come in my front gate, and walk right up the walk. He'd sit in the yard and look at you." What these animals might be, she can't say. Nutria have been a torment to the Eastern Shore, but no one has seen any on Tangier.

"In another year the whole sound's going to be on West Ridge because of them—whatever's in there," she tells the gathering. "I don't care if it's a mouse eating it up; something's eating it up, and it's quickly going away. Anybody can see that."

Town meetings on Tangier, like those anywhere, frequently veer off topic. "Well, that's all part of the project, too," James says, eager to change the subject.

Eugenia: "If we got to wait three more years, there'll be none of it left in the middle."

James: "That's exactly what we told them."

Eugenia: "There'll be nothing left."

"Everybody knows how vulnerable our harbor is," Ooker interrupts, trying to steer the discussion back to the corps' proposal. "If you got a crab house or anything out in the creek here, you get a wind." If the proposal becomes reality, he says, "there won't be any more rough water in the harbor area, because it'll be closed off. The only open water'll be coming in from the channel, east and west side."

"You'll be able to go to your crab house on a moped," James kids.

Duane Crockett stands. "I think it's very sad to human nature that they would do all of this for birds and wildlife," he says, "but I'm not going to strain out a gnat and swallow a camel, either. I don't care why they save our island, as long as they save it. And I look at this plan as a divine thing."

"Like Duane said, it's sad [that birds count more than people]," Ooker says. "But whatever it takes, I'm fine with that."

An islander asks how long the project will take. James and Ooker say they don't know for sure, but years—more than ten. Perhaps many more than ten.

Eva Marie Pruitt, who lives in Black Dye, speaks up. She shares Eugenia's concern about rodents in the marsh. (Unsolicited, she's told me, "I've heard that if one of 'em gets ahold of you, they hold on until you're *dead*—and I believe it.") "So this has nothing to do with those things that are eating our marshlands?" she asks. "I don't know what they're called . . ."

"What was that?" James asks.

"Muskrats," Anna says.

"Oh, no," James says.

"As far as the muskrat problem," says Inez Pruitt, sister to James and mother to Anna, "I say whoever wants to get permits and start trapping them, go right ahead."

A convoluted exchange on muskrats ensues. It ends when waterman Gary Parks voices a thought that no doubt lurks in many minds: "You know, we've been told this a lot."

"Exactly," James agrees.

"I'm probably one of the skeptical ones about it," Gary says. "I've got my doubts."

"I've got mine, too, Gary," James says. "I do. But it's like they said, we got to start somewhere."

Anna announces she forgot to mention something earlier. "After our meeting Monday with the corps, yesterday Senator Tim Kaine sent his regional manager and two of his legislative aides out to talk about the plan," she says. "The week before he was called up to run for vice president he'd called us and was supposed to come out here to Tangier to sit down to a meeting with us. . . .

"I guess, hopefully, if he doesn't get vice president, then we'll see him out here and he'll be helping us," she says. "I'm going to be selfish and want him helping us for our project instead."

Her mother takes this thought a step further. "Vote that he don't get in," Inez says. "Vote the right way."

"Well," the mayor says, "we can only hope that if he does become vice president, he'll still remember us, maybe."

THE FOLLOWING DAY, the mood in the Situation Room is ever so slightly jaundiced. "It's as if they just thought of it: 'Hey! We can blow fill!' We've been asking for that for years," Bruce Gordy complains. "Fifty-eight people were there out of 450, so you know how much interest there is. One reason is that they've heard this all before. They've heard it thousands of times."

"In 2016 we were supposed to be putting stones down," Leon says. "Do we have stone? Three years to study it. *Three years!*"

"We'll all be dead and buried and forgotten about," Bruce says, blowing e-smoke.

"I don't even believe 'em," Leon mutters. "They've lied so many times in the past. They've lied more than Hillary. Why didn't we get the jetty in 2016 like we were supposed to?"

"They used the money for something else," Ernest Ed Parks suggests.

"And they'll use it for something else again," Leon snarls. "We got more history here than any place in Virginia. Virginia don't care about history. Maryland—now, Maryland cares about history. But not Virginia."

Richard Pruitt enters. The room falls silent as he fixes a cup of coffee, then settles into a chair next to the trash can. The fruit flies over that way are worse than ever, and a swirling cloud forms near his left shoulder. He doesn't seem to notice.

"You ever seen the tide make so low?" Leon asks the room. I'm curious myself, because while out on the water yesterday I noticed that the harbor was drained dry: The creek between Canton and the P'int was an exposed mudflat, and the bottom under most of the shanties was exposed, too. The breeze bore the complex scent of brine, methane, and rotting crab exoskeletons, which lay heaped beneath the shedding tanks.

"The whole bay's gonna come in the creek at flood tide," Allen Ray says.

"The tide's a funny thing," Leon muses. "It's an awful funny thing."

The sound of the ambulance's siren reaches the room. Conversation halts. "Somebody's sick," Leon says quietly.

Bruce, peering out the window toward the road, announces, "Going up Meat Soup way, and quick."

"Somebody's sick, or somebody fell," Leon says. They catalog all the likely prospects up at that end of the Main Ridge: Jean Autry and Strickland Crockett, both elderly and failing. Ginny Marshall, who is eighty-nine. Iris Pruitt, eighty-eight. The list goes on.

Ooker strides into the room. "Milton fell on his dock," he says. That would be eighty-five-year-old Milton Parks, owner of the Parks Marina at the top of Meat Soup and father-in-law to Jerry Frank, first cousin to Leon, close cousin to Allen Ray and Richard. It seems a boat was coming in to dock and tossed a line to Milton. He gave it a pull, and it turned out the rope wasn't tied to the boat. He lost his balance and went down hard, breaking a hip.

Jerry Frank takes his leave as the conversation returns to yesterday's meeting. Ooker complains that there was too much talk about muskrats. "You don't think they're the problem?" Bruce asks, suppressing a smile.

Ooker tilts his hat back, looking weary. "I wanted to say a few things, but I thought, 'No.' I just let it go. I thought, 'Lord, we got one of the greatest proposals presented to us, and we're getting hung up on muskrats.' She said, 'This jetty won't do any good if we don't take care of these muskrats.' I thought, 'Yeah, that's good. I *wish* our biggest problem was muskrats.'"

The subject shifts yet again, this time to the good luck dipnetters have been having. Only a few Tangier crabbers use hand nets, and only for a short spell in late summer. "The netting's good up where Hearn Island used to be," Ooker says, noting that his son Woodpecker was there yesterday "and caught fifty-five in an hour."

"I know it's been good," Leon says, nodding.

"He said if the tide hadn't been so low, he could have caught a hundred or more."

"Where exactly was Hearn Island?" I ask.

"It was up above Fishbone Island, on the way to Smith," Ooker says. He pauses.

"The muskrats got it."

A MONTH AFTER THE MEETING at the firehouse, I drive down to Norfolk to visit with the corps and to hear the details of the proposal for myself. I don't necessarily distrust the town council's interpretation of what was put on the table, but seeing as how the proposed jetty has ballooned in island minds into something bigger than it is, it seems prudent to cut out the middleman.

I'm ushered into a conference room and seated at a table with Gregory Steele and a senior member of the party he led to the island—Susan L. Conner, chief of the Norfolk District's Planning and Policy Branch. "It's been confounding, how we as an agency can best address what's happening at Tangier," Steele opens. "It's a very difficult proposition. You have to look at where is the taxpayer best served, in terms of the limited dollars we have."

He walks me through the vexations facing the corps. First: Saving the town itself as a flood-management project is a no-go. It would never meet the agency's cost-benefit analysis. Second: Even if the corps were able to somehow find the money to preserve the settlement, the most efficient means "would likely turn out to be, okay, we'll relocate the town, move 'em to the Eastern Shore." That is not a viable solution to the island's dilemma, however. "Tangier has a cultural and economic aspect to it that makes that kind of move really unattractive to the people there," he says. "And they *are* the engine and the hub of the blue crab fishery in the bay."

So the corps found itself at an impasse, until Steele's lunch with John Bull. Together, they had the first glimmer of how they might solve several problems at once: The corps could use the by-product of

one of its key missions—improving navigation—to perform another
of its primary directives—preserving and creating ecosystems. Almost
as an afterthought, it could save the town, flood management being
another of its missions. And the state would no longer fret about
mountains of silt getting tossed into the Chesapeake.

But a glimmer is all that it is, Steele emphasizes. "This is not a
slam dunk," he tells me. "This is a Hail Mary. If we were to come up
with a project to save Tangier, a whole lot of pieces would have to fall
into place. We have to have authorities, and those will be difficult to
get. We have to have appropriations."

Conner explains that before the corps can begin its three-year general investment study—the inquiry the town council mentioned—the
agency must receive authority to conduct it from an assistant secretary
of the army. The corps itself cannot seek that authority; the request
has to come from stakeholders in the project, those being Tangier
and the state. They have to make a strong case. The army reviews all
such submissions and it nitpicks. Getting through the process is a
long shot.

Even assuming it clears that hurdle, the study would be no sure
thing, because only ten are green-lighted per year nationwide. If the
Tangier study were one of them, it would cost $3 million, and the fed-
eral government would pay only half that. And even if the research
concluded that the project was a superb use of the public treasury and
of unqualified benefit to the nation, there would still be the matter of
financing, the biggest challenge of all.

Steele says he doesn't know how much the whole grand scheme
would cost, but it could conceivably run upward of $800 million—
"the equivalent of giving everyone on Tangier $2 million." That could
prompt resistance in Congress, which might prefer to spend a tenth
as much: "Instead of spending $2 million per person, you could spend
$200,000 per person and get them a really nice house somewhere else."

And if it gets through Congress intact, the project could still just
sit, Steele tells me. "The corps has I don't even know how many bil-
lions of dollars in projects that have been authorized but never con-

structed," he says. His summary of the proposal: "This is a heavy lift for all of us."

At this point I'm both glad that I made the trip and depressed at what I've learned, because it is clear that the town council put a rather optimistic spin on that meeting at the firehouse. Tangier does not have, as James Parks told his fellow islanders, a "five-year waiting period to see if this project will take off." The corps doesn't yet know whether the concept is even worthy of study.

And that study, if it were to happen, might well conclude the idea is a bad one. A danger of such projects, Steele says, is "inducing risk—creating protection that works until it doesn't, and the results are catastrophic." Build the island up, and you might prompt a false sense of security in its occupants, when in reality they remain on a tiny piece of hurricane bait in the center of a great and unpredictable body of water. "The community is going to think, 'We're good,'" he says, "when really, no, you're not—the right combination of factors can bring about a disaster."

Still, he and Conner stress, there is reason for hope. The state seems enthusiastic: "The fact that VMRC sees a lot in this is a big win," Steele says. The army might be intrigued by a project that marries the corps' three primary missions. The island represents a "resource that has become much scarcer and therefore more significant." And there's an intangible allure to Tangier: "The cultural component of an isolated population seems to resonate with people," Conner points out.

"At the least," Steele says, "we think the idea would pass the red-face test."

THE CONVERSATION TURNS, in time, to what a reconstructed Tangier Island might look like, not only in its footprint but on the ground, from its streets, and out the windows of its homes. The blocky, indistinct rendering displayed at the town meeting made clear that if this concept becomes reality, most Tangiermen would live farther from the water. Reaching either the Chesapeake or Tangier Sound would require a longer morning commute. The beach might no longer exist. The Heistin' Bridge could well be superfluous.

Much of this might be welcome; most islanders, I'm sure, would trade a pretty view for the security of a sturdy shoreline. Still, these are people who, time and again over the past century, have demonstrated a staunch resistance to change. That yen for preserving the status quo has colored most every controversy and collective worry. Tangiermen aren't known for their eagerness to seek new and better ways to do things.

"If we were to identify a plan and go forward with it, the island would look very different from the way it looks now," Steele acknowledges. "We've learned things at Poplar, so whatever we did at Tangier, if we did something, might not look anything like Poplar, either." For instance, Conner says, instead of riprapping the island's entire circumference, the corps might opt for "natural approaches to reduce flood risk," such as rebuilt marshland around the town and tree plantings.

Steele raises another possibility. "It could be that we build cells like at Poplar, too," he says, "and we move the town to one of the completed cells—move the whole town, lock, stock, and barrel, and leave the current townsite [to revert to marsh]." This is not a prospect that came up at the town meeting, and it provides me with the grist for much later thought. Among Tangiermen's proudest attributes is their connection with their little piece of the planet, with particular houses and lanes changed little over time. While the world beyond its shores has been refigured in concrete and steel, Tangier today is much as it was generations back. To a population raised on chart and compass, geography matters.

What happens if all that is scrambled? If the town were, as Steele suggests, picked up and moved, it would not likely take the same form elsewhere; it wouldn't make sense to separate its homes into three far-flung ridges, when a snug grid of streets would be cheaper and far more efficient. It might be a nicer town in which to live, in most respects—neater, more neighborly, with shade trees and a park, perhaps. Shorter distances between homes, the grocery, and church might encourage walking. Tourists might find it more the postcard-ready village they seek and expect, in place of the wind-scoured, cluttered fishing town that it is. But would it be *Tangier*?

Conner cautions that even if all goes as hoped, this ambitious rescue will move slowly. "From where we sit now, it's at least a few years before we can get a study started," she tells me. "I don't want to throw a number out, but we're most likely talking about five to eight years before the study would be finished."

Steele: "At the midpoint of that study, we'll have a pretty good idea of whether this'll work—whether we'll have traction."

"We try to manage expectations," Conner says. "When you tie all of these timetables together, it's unlikely you're moving dirt in less than ten years."

We shake hands not long after. I ride the elevator down to the building's ground floor, turn in my badge, and step outside into the summer heat. The Elizabeth River, a three-pronged estuary that curls through the heart of Hampton Roads, is swollen with a flood tide and laps high along the bulkhead that rings the corps' property. If sea-level rise and subsidence combine here to the drastic degree forecast by scientists, the corps will be among the first of the region's land-owners to experience them.

By that time, Tangier's fate will be sealed, unless the corps pulls off this Poplar Island concept. I walk to my car with the conviction that it will take a miracle—in getting it through the army's bureaucracy, in getting it financed by Congress, and in getting it finished before a storm muscles up the bay and renders the whole thing moot, at least for the island's human inhabitants.

A minimum of ten years before earth is moved?

Nature has time on its side.

PART FOUR

A PEOPLE ANOINTED

Pastor John Flood, left, and former principal Denny Crockett shoot the breeze in their golf carts, Memorial Day 2016. (EARL SWIFT)

EIGHTEEN

WHAT WITH TANGIER'S LONG HISTORY OF BIBLE-THUMPING, its physical isolation from the mainland's sharp-elbowed bustle, and its detachment from its countrymen's obsessions with social media and reality TV, it's often cast as a throwback to more innocent, more godly times—straight out of a Norman Rockwell painting, with morals intact, air fresh, and entertainments wholesome, a place where family comes first, where the men are God-fearing, self-reliant, and strong, and the women, faithful, loving, and good cooks all. Tourists come in search of this fabled place: a vestige of America untouched by road rage, fast food, and snark; where anonymity is impossible, accountability unavoidable. And when they come for an afternoon, or spend just a night or two, that is what they find.

Most of what you read about Tangier bolsters this image. The only reputable history of the place, published by a Methodist minister in 1999, is titled *God's Island*. The 1973 *National Geographic* package on Tangier, written by the late Harold "Spike" Wheatley, the school's principal at the time, showcased worship at Swain as much as any other facet of island life: Besides the picture of Annette Charnock at her wedding, there were photographs of the preacher and of Miss Annie Parks during services, and a lovely, ethereal image of Bruce Gordy's wife, Peggy, in the sanctuary with her two small daughters.

Across the years, hundreds of other newspaper and magazine stories have made hay of religion's influence in all things, beginning with that piece by J. W. Church in the May 1914 issue of *Harper's Magazine*, which reported that "the minister is a benevolent despot whose word is law" and that he oversaw a church whose "members and adherents embrace virtually every adult in the settlement."

"As fishing is the sole industry of the island, so is religion, of the sternest and most uncompromising sort, the only intellectual stimulus or recreation," Church wrote. "No alcoholic drinks, playing-cards, dancing, or frivolous amusements are tolerated or apparently desired by the fisher-folk of Tangier. Life is too serious a matter for such things."

The town had always been "singularly free from crime or misdemeanor." So why, Church wondered, did it need a town constable? Sugar Tom's son, Captain Ed Crockett, explained that the cop's job was "to look after what strangers come ashore." Church pressed him to reveal more:

> "A few years ago," he said, "things come to an awful pass here. There got to be a regular spell o' swearin', an' it wa'n't only on the boats, but right on the street within hearin' of the childer. So thirty of us met right here in this room an' formed a Law an' Order League, an' we pledged our sacred lives an' property to put a stop to this wickedness. I told them that a man could be fined five dollars for swearin' in any state in the Union, an' it ought to be the same here. So that's what we decided to do, an' we told Bud Connerton [*sic*], the deputy-sheriff, to give every man who swore a fair warning, and the next time to fine him five dollars."
>
> "Did he make any arrests?" encouraged Ellis [Church's photographer].
>
> "Forty-three the fust week," said the Captain, "an' none since. The boys soon decided that swearin' was too expensive to be careless about."
>
> "How about strangers?" I asked.

Captain Ed eyed me suspiciously. "We warn them *twice*," he said. "We've only had to fine one. Bud, he's great on doin' his duty."

That would be the same C. C. "Bud" Connorton who shot Annette Charnock's grandfather a few years later and came to a sad end himself.

That *Harper's* story set the stage for a great many that followed over the next eighty years. The island's unsullied reputation seemed affirmed for good when, early in 1998, Warner Bros. sent location scouts ashore, on the hunt for a fishing village in which to set a planned Kevin Costner and Paul Newman movie called *Message in a Bottle*. They eyed the town by land, boat, and helicopter and decided it was ideal for some outdoor scenes.

The studio called the film a "poignant romantic drama." A fisherman, played by Costner, writes a love letter to his dead wife, slips it in a bottle, and chucks it out to sea. It washes up on a Cape Cod beach and into the hands of a vacationing Chicago newspaper researcher, played by Robin Wright. She feels compelled to track down the loving soul who wrote such a beautiful letter, and does—and so on, in the manner one expects of a film based on a Nicholas Sparks novel.

Warner Bros. proposed to pay the town $5,000 for its brief use of public property during the shoot and to hire locals to build sets. Tangier's mailboat, restaurants, grocery, and inns stood to turn a little business. Excitement ran high. But the six-member town council had a peek at the script and saw that in one scene, Costner's character undressed Wright's, and a violation of Christian mores followed. The characters also drank beer and wine, and Paul Newman, playing the part of Costner's father, threw around some PG-13 language. One council member complained of characters using the Lord's name in vain, too.

So in a unanimous vote, the panel denied the studio permission to film on the island unless it cleaned up the story line. "Our Town

Council is made up of Christian people," then-mayor Dewey Crock-ett explained to the *Washington Post*. "We just couldn't accept it." Warner Bros. opted to take its shoot elsewhere.

The vote attracted a flurry of press attention, much of which marveled at the backbone Tangier had displayed in standing up to Hollywood. "Look up *chutzpah* in the dictionary, and you'll undoubt-edly find a thumbnail sketch of this Eastern Shore island town in the middle of the Chesapeake," the *Baltimore Jewish Times* editorial-ized later in the month. "Whether or not you agree with this plucky town's perspective, you have to applaud its refusal to succumb to the Almighty Green."

A good many Tangiermen were not cheered by the decision, however. The night after the council's vote, they crowded into the school auditorium for a boisterous town meeting. "Every soul on the island turned out for that one, to see the fight that would ensue," said Anna Pruitt-Parks, whose father, Jerry Frank, was on the council. The twenty-seven who spoke were split pretty evenly for and against. Passions were roused. A petition asking the council to reconsider its decision drew two hundred signatures. It changed nothing.

The following week, Beth and the late Rudy Thomas Jr., who ran the mailboat and had circulated the petition, were visited by a family friend, Keith Ward, who had a charter boat service on the mainland. Come down to my boat, he told them. I need to show you something.

In the vessel's cabin was Paul Newman. "He said it looked like a place he really would have enjoyed," Beth recalled.

I CAME TO TANGIER in 2000 having read the *Harper's* story and scores written since. I'd followed the movie saga. And in my six weeks on the island I noticed nothing that caused me to question its image as a devout community striving to live by the Scripture. To the contrary, I became convinced that practically every Tangierman was in church on Sunday morning and that worship was no weekly chore—it was the island's radar, its compass, its guiding star.

Both churches were thoroughly integrated into every aspect of life. The absence of alcohol sales was the most visible manifestation of religious influence, but others abounded. I heard no one swear. The mailboat didn't run on Sunday. The school library circulated no dangerous Harry Potter novels. No municipal decision ran counter to the faith, and no wonder—Dewey Crockett was not only mayor, but he also served as Swain's music director and de facto associate pastor, the school's assistant principal, and town undertaker. Denny Crockett was both school principal and a Swain lay leader. I spent a day with Cook Cannon, who was running the sewage treatment plant at the time, and he likened the outside world to the waves tearing at the shore—erosive and sneaky, worthy of constant vigilance.

In their homes, islanders told me of the many times Tangier had been spared from sure destruction by the Almighty's benevolent hand, and just as often they spoke of a recent high-water mark of Tangier faith. Revivals are typically hosted by each congregation in spring and fall and are led by visiting preachers for five or six nights straight. In March 1995, however, the two congregations combined forces into a single run of nighttime meetings, lay preachers from each alternating at the pulpit. At first, the results were unremarkable—as Jack Thorne told me, "There were just a few, one or two, getting saved after several nights." But on the sixth night, a Sunday, the revival's last scheduled service, a member of Swain's congregation came to the altar to pray. Marshall Pruitt, a New Testament elder who was up at the pulpit, climbed over the rail to kneel beside him. With that, Jack said, "the love come in."

"That's when people started coming up," recalled Marshall's widow, Iris Pruitt. "And they kept coming up. People were just under conviction."

Dozens of people came forward to be saved, so the leaders of both churches elected to extend the revival by a day. The same thing happened, so they extended it again. The same thing happened, and the same again the following night. "Sometimes we wouldn't have even started, and somebody would go forward," said Jean Crockett, Dewey's widow. "It was not orchestrated by men, you could tell."

In all, more than two hundred people were saved, in a town of fewer than 700 inhabitants. The islanders talk about it still—as I heard Duane Crockett do at New Testament one Sunday morning. "Anyone who was present will well remember," he said from the pulpit. "People we've known our whole lives stood up, preached simple sermons, and at the conclusion of the sermon an invitation would be given, and thirty and forty people a night would respond to the Gospel. It was a sight that I will never forget for my whole life.

"It was preached by ordinary men who believed the Word, and it is the Holy Spirit who takes that Word and convicts the hearts and brings people to salvation. I've seen grown men sob to the Gospel before. I have seen people break out in sweats during invitations, and clench on to benches, and shaking, and won't let go of that, like there's a battle that's going on within them. That's the power of the Gospel."

All of which is to say that I left Tangier in the spring of 2000 and returned sixteen years later, confident that most all of the island was, as they say, right with the Lord.

THE CLUES that such is not the case are nuanced and likely missed by tourists, but they pile up. For instance: One late-summer evening, I witness what appears to be a drug transaction in King Street, not far from the museum. I'm on my bike when I come upon a golf cart stopped in the middle of Main Ridge Road. It's occupied by a thin man in his late thirties or early forties, whom I recognize but don't know. Another islander, of roughly the same age, is standing beside the vehicle, and they perform a quick, furtive exchange: folded cash for a small packet.

Another clue: One afternoon I'm riding up Hog Ridge, having spent an hour down at the spit, when I encounter Ernest Ed Parks working on his pickup truck outside the closed Sunset Inn—once the biggest and most comfortable of Tangier's lodges, now abandoned and gone to seed. The stripped remains of a golf cart molder in waist-high weeds out front. Mildew stains the siding. While talking with

Ern, I remark that it's a shame that such a fine building has come on hard times. "Used to be a right nice place," he tells me. "But they just let it go. Walked away from it." After a pause he asks, "You ever been inside?"

He leads the way up a broad set of wooden stairs to a deck that hugs the building's south side. We push our way through the drooping branches of a magnolia that, untrimmed, has grown to block the deck. We try the front door, find it locked, and continue to the inn's west side, which overlooks the south end of the runway and the bay beyond. A sliding glass door there has been kicked or blown inward, and we step over it into ruins: The large room's wall-to-wall carpet is filthy, waterlogged, and wrinkled, and the Sheetrock walls bloom with mold. The smell of cat is overpowering. Trash is strewn across the floor, including an empty bottle of cinnamon schnapps.

We venture from room to darkened room, each more ravaged than the last. Empty bottles of bottom-shelf gin and whiskey figure prominently. In two rooms lie soiled mattresses, sheetless but equipped with makeshift pillows, and both looking recently used. I'm eager to leave.

A few nights later, my visiting fiancée and I ride our bikes past the Sunset Inn, and through a window we see the light of a smartphone's screen moving in one of the rooms. Once I know to look for it, I detect evidence of transgression all about: skinny, ravaged-looking islanders meeting at the slab out past the dump; flotillas of empty beer cans lifted from the marsh by storm-driven tides; mournful shakes of the head among churchgoers at the mention of a neighbor's name; references to a bend on West Ridge Road clustered with hardscrabble trailers as the Devil's Elbow. It's hard to miss one tiny home, measuring no more than eight hundred square feet, in which ten people live: Often when I pass, its yard is filled with children in almost medical need of a scrubbing, and when I ask about the place others call it "a sad situation."

Duane Crockett hinted that all islanders do not adhere to the straight and narrow in his sermon about the '95 revival. "Search your hearts, church," he told us. "Think of a place on Tangier that you wouldn't be caught coming out of. Or a person on Tangier it would tarnish your reputation for being seen with. I can guarantee you: That may be the very person or place that the Lord will have you to go to, to share the Gospel with them.

"Iris has been in [many such] homes on Tangier to witness to people about their soul's salvation," he said, referring to New Testament's senior member. "And I mean some homes where, if I were to be going in the door, I might turn around and look to see who's going by and would see me going in. That's the kind of home I mean."

IN THE CLUTTERED OFFICE of Swain Memorial I talked to John Flood about the island's unsaved. Next to the mainland, Tangier is a godly place, he told me. But before he came to the island, he thought it was godlier than it is.

"That was the big shocker for me," he said, "because my impression of Tangier was that on Sunday morning everybody was in church, and if you weren't in church you stayed in your house with the door closed. That may have been the case eighty to one hundred years ago, but it's certainly not the case today."

By Pastor Flood's reckoning, about half the island attends one church or the other. Of the other half, the great majority are living solidly moral lives.

But not all. For evidence, we need not look beyond the room in which we sat—the site of the July theft, which remains high in the island's collective mind two months after the fact. At Swain's August board meeting, members worried that the church's custom of publishing the income generated by each week's offerings might have invited the crime. The numbers, often impressive, appeared on the third page of the bulletin handed out on Sunday mornings, sandwiched between a list of approaching island birthdays and a schedule of Swain's upcoming prayer and choir sessions. That week's bulletin had noted that the previous Sunday, offerings had totaled $4,273. The board decided to suspend publication of the collections for "two or three months," to see whether anyone objected or even noticed.

At the September board meeting, it's obvious that they have. Denny Crockett, the board chairman, reports that he's been approached by one of the church's biggest givers. "He wanted to know why we'd done that, and he wanted to know whether something was being hidden," Denny says. Several in the room express shock. Why would anyone think that? What would they be hiding? Denny holds up a hand. "There's paranoia both ways," he says. "I've always believed, in any organization I've ever been in, in transparency. We have all the best intentions in the world—but this is a good person who asked this."

"People need to feel comfortable that what we do is open," a board member suggests. "That what we do is led by the Holy Spirit." It doesn't help that no one has been arrested for the theft. Accomack County sent an investigator over who is said to be planning another visit. But if there's a strong suspect in the case, word hasn't leaked out. On rumor-rich Tangier, this has bred talk that Swain's leadership has a good idea of who's responsible and is sitting on the information.

"Surely nobody thinks it was an inside job," a board member says.

"That they do," says another.

Pastor Flood stands. "Gossip is something you're going to have," he counsels. On a recent trip to Crisfield, he was approached by someone who thought the money had been returned. A mainland cop told the pastor he'd heard a similar story—that the thief had turned himself in to church leaders and apologized, and all had been forgiven.

Hoot Pruitt is troubled that a decision aimed at protecting the congregation is, instead, creating doubt and division. "If it's going to upset people," he says, "we need to go back." With that, he moves that the congregation resume publishing the collections. Grace Pruitt, Carol Moore's mother, seconds the motion. It passes in a unanimous voice vote.

TWENTY-FOUR HOURS LATER, any surviving notion I might harbor that Tangier has escaped the evils of mainland life is dashed for good.

The monthly Tangier Town Council meeting is gaveled to order in the town hall, an asbestos-clad hut that once housed Asbury Pruitt's navy spotters. Principal item of business: Sylvia Bonniwell, who drives a tour buggy and lives at the southern end of the West Ridge, appears before the panel to complain of widespread lawlessness. Speeders on the island are brazen and "need a little course on how to drive a golf cart," she says, adding, "They've come close to killing five people this summer." Drug transactions have become flagrant. "They don't even try to hide it," she says. Also, littering's pretty bad.

"On the drug situation," Councilman "Colonel" Ed Parks says, "you are empowered to make a citizen's arrest."

Sylvia blinks at him. "I'm afraid they'd blow my head off."

"Well," says Colonel Ed, "you can't have your cake and eat it, too."

Looking on is Rob Baechtel, a Metro D.C. reserve cop who's moved into a big house on the West Ridge with his wife, Barb. "There's a meth problem on this island," he announces. "I've seen meth being smoked on this island. One of the problems we have is that John [Charnock, the town cop] doesn't have the experience to be able to testify in court." Rob suggests an undercover officer with a track record in drug arrests be assigned to the island and partnered with John. Said officer, on witnessing a drug transaction, would have the probable cause to intercede. "John doesn't have the probable cause," he says, "because he doesn't have the experience."

Ooker points out that an undercover cop won't stay undercover for long. "If we have a guy over here you know how he stands out. It's not like it is on the mainland."

"Having another cop work with John will give John the experience to do such work on his own," Rob says.

Ooker turns to Sylvia. "Your concern is ours as well," he tells her. "Those who've been caught have just got a slap on the wrist and get back before the police boat."

The meeting reveals a host of other challenges the town faces: leaking water pipes, disintegrating roads, and a sewage treatment plant frequently overtaxed and always expensive. Failing wells, too—

the island has ten tapped into an aquifer a thousand feet down, but only three are functioning; of those, one was taken offline in 2016 because it was fouled with bacteria, and the water from the others is so loaded with barium, sodium, and fluoride that a past doctor and dentist both advised their patients not to drink it. Bottled water sells well at Daley & Son.

But no difficulty is equal to the island's struggle with drugs, and it's waged this fight for a while: An entry in the town council's minutes of October 4, 2001, reported that "a large number of pills were sold this past Sunday by one individual. This person has a history of this." Another entry, from a meeting in December 2006, read: "Several people have asked if the Town would be willing to put a light by Bill's dock because there is a lot of dealing going on, and a light might be a deterrent."

When I raised the subject with Inez Pruitt, the physician's assistant who runs the health center, she told me that most Tangier drug users "are in their forties. We had a guy who got hooked on heroin. That's not something that just pops up on Tangier Island." One first-rate, well-regarded waterman died of an apparent overdose.

"At least once a month I say to myself, 'Inez, you ought to go off and get training for drug addictions,'" she said. "I don't think it's as prevalent as it seems, but we know everybody, and this has affected just about every family on the island."

I also broached the topic with Nina Pruitt during a visit to the school. "We have a big prescription drug problem," the principal said. "A big alcohol problem. I think it's gotten worse.

"Tangier and alcohol has always been a strange combination, because it's always taken one of two forms: you either drink to get stone-blind drunk or you don't drink at all. There's never been any social drinking."

THE MAN WITH THE MOST DIFFICULT, thankless job on Tangier is rolling at a glacial pace through King Street in a subcompact Chevy Aveo hatchback. John Wesley Charnock calmly scans the houses and yards on both sides of Main Ridge Road, offering a wave or nod to the playing kids he passes. He has been town sergeant since 2009. It's up to him to keep the peace among his neighbors. It's an almost impossible balancing act.

Not because Tangier is any sort of Dodge City—its drug and alcohol issues usually play out behind closed doors and only rarely spill into the streets or involve violence. Rather, what makes his task so demanding is that his every act of law enforcement puts him at odds with a relative or an in-law. "Every time that phone rings, I know it's going to be a family member or a close friend," he says. "It's not like I'm never going to see these people again. I have to live here with them."

Today, as is his habit, John presents a smile to those he passes in his four-cylinder squad car, which has power enough to outrun a golf cart but not the four-wheelers and motorcycles some islanders gun around at night. I'm riding shotgun, having climbed in outside the parsonage for the sort of pairing that big-city reporters routinely make with big-city policemen. The car's cockpit is snug. The back seat is far too small to carry any but the most petite and docile prisoners.

Now on our second loop of the island's road circuit, we creep past the new health center and bend around Swain's churchyard, headed into Meat Soup. "I've made probably forty arrests the entire time, and a lot of them has been the same people," he says. "That's physical arrests—I've made a lot more with summons, of course. Out of that forty arrests, I think only eight or ten people have been involved."

John keeps a small office in the old health center, just up the darkened hallway from the Situation Room. On its wall is a video camera and a flat-screen monitor that link him to a magistrate in Accomack County, who decides whether an arrestee requires jailing on the mainland. The technology has simplified the job immensely from the days when every arrest required a boat trip.

"I have very little trouble with the younger people. The problem is with people in their thirties and forties," John says as we patrol. "It all

stems from drugs or alcohol, all the crime around here, pretty much."
Any mischief the island's teenagers get into is likely to be vehicular—
underage golf-cart driving, running without lights, dragging down
the airport runway—and the product of boredom.

"When I was a teenager coming up, there were five or six places
we could go to hang out," he says. "There were places we could shoot
pool. We had places with jukeboxes, and we could go dancing. The
rec was open for us to play basketball two or three nights a week. We
played volleyball. We played baseball."

We pass a golf cart headed the other way, two teen girls looping
the circuit. Night after night, you'll see kids circling the island's road
system—up one ridge, over the marsh, down another ridge, over the
marsh—over and over. Sometimes they do it for hours straight. "As
far as things for the teenagers to do," John says, "the island has really
gone backward."

We slow even further outside Daley & Son and swing left past
Fisherman's Corner, then through the narrows between Carol's house
and Jerry Frank's place. John lives a few doors beyond. Across from
his bungalow, Kim "Socks" Parks is flying the Israeli flag from a pole
in his yard. Ahead, a navy Seahawk helicopter is sitting on the airport's apron, rotor spinning. "Probably ordered lunch," John guesses.
"They do that right often. Lorraine's will run it over to them." We
cross the Long Bridge and turn down the West Ridge. Halfway to
the Heistin', we pass another Israeli flag flying outside Ooker's place.

Most of the crime he encounters is against property, John tells
me, and really, there's not much of that. "It's like I've told the town
council time and time again: We're fortunate around here, because
the guys who have their addictions, they all make good money," he
says. Many work on the water. "If that weren't the case, you'd have a
lot more break-ins."

John handles a domestic assault now and then, almost all involving alcohol or drugs. "By the time you leave, everybody—the victim
and the suspect—is mad at you," he says. "The worst thing about the

job, the thing that hurts, is that there's people I've known all my life, that I grew up with, and they'll turn their head away when I come near."

Wearied by the job's pressures, his predecessors have quit well short of John Wesley's seven-plus years in the post. Ooker served as Tangier's part-time cop in the early oughts, and he told me he didn't enjoy it. "Somebody would call you about somebody else, and both of them your friends," he said. "That wasn't a good job, knowing everybody." A later officer complained to the town council in 2006 that he was "swamped with phone calls" about "people drinking and causing trouble," and from "people wanting to press charges against someone else," then failing to follow through with their complaints. The island needed a second officer, he said, "because one person can't do this job."

We rumble across the Heistin' and turn north on Main Ridge Road. Like most of the islanders who've held the job, John took up police work after working the water. For fifteen years he worked with his older brother, Ed "Eddie Jacks" Charnock, then got his own boat, a thirty-eight-foot box-stern deadrise that he named the *Valerie Faith*, after his wife. "She's helped me through some tough times," he tells me. "If it weren't for her, I don't know how I'd have made it. It weren't problems of habit or anything like that, but financial, mostly, trying to make a living on the water.

"Working on the water, you need to have somebody who can handle what you bring home. You can be making good money one month and the next month wondering what you're going to do." He kept the *Valerie Faith* for fifteen-plus years, then went to work for his uncle, Charles "Puge" Charnock, selling bait to the island's crabbers. After a half-dozen years there, the town sent him to a police academy in Hampton Roads.

We pass a small child standing at the roadside in Black Dye, studying us through folding opera glasses. Another piece of the balancing act: Just about everyone in town has an opinion of how you might better do your duty. John's as quiet as his brother Ed and bears such talk with a bemused stoicism. He shrugs as we pass again through Meat Soup. "They allow me to use discretion," he says of the council, "to do things the way I think it should be done."

We come to a stop. My bike, which I'd left leaning against the chain-link fence ringing the parsonage, isn't there. "My bike's gone," I tell John.

"You sure you left it here?" he asks. I reply that I'm quite sure. "Okay, let's go look for it," he says.

We begin another slow circuit of the town, he peering off to the left, I to the right. We crawl down the West Ridge, John pointing out bikes parked in yards along the way. "That it? How about that one?" Then, as we negotiate the jog onto Hog Ridge, there it is, leaning against a fence near the Sunset Inn.

Tangier's cross-adorned water tower looms over the school on a misty December morning. (EARL SWIFT)

NINETEEN

I N MID-SEPTEMBER OOKER HAS RIGGED THE *SREEDEVI* FOR scraping, as he does for two or three weeks every year. One morning, after we motor past Uppards's eastern flank to the few remaining dots of marsh in the shallows below Smith Island, he cuts the engine. We float beside Long Tump, a grassy shelf that no longer lives up to its name; fifty yards long at most, it barely clears the tide. It is quiet here, empty of traffic save for Leon and Short Ed Parks scraping a quarter mile to our north, their boats gleaming bright against the low, ragged green of Smith's coastline.

The weather has cooled over the past few days on the back of an easterly breeze. Gone is the drenching humidity and brutal heat of the Labor Day weekend, when Ooker was moved to erect a giant rainbow-striped beach umbrella over his steering console. The morning could not be more pleasant. But when Ooker hauls the scrape up and dumps its contents, he finds just three doublers in a tangle of eelgrass. "Before the cold snap, I was getting fourteen, fifteen doublers a lick," he tells me. "That's the trouble with the snaps. They feel nice, but they're no good for the crabs. A shedder wanting a cold snap is like a farmer wanting a drought."

These are the closing days of the scraping season. As a rule, the shallower the water, the quicker its temperature will shift with changes in the weather, and the sooner crabs will respond. The animals like heat, so scrapers working the shallows are always the first to feel the economic effects of autumn's chill. Ooker will return to potting in a few days, and at month's end Leon will lay up his barcat until next May. Even for potters, the season brings changes. Peelers in the shedding tanks will take longer to molt. "In the middle of the summer it might take three days," Ooker says. "But when the days get short, it can take a week."

Another lick catches a huge porcupine fish, which islanders call a thorn toad. Like its cousin the northern puffer, it's self-inflating, but boxier and covered with spikes. "That'll bite," Ooker says, grasping the fish with one hand while searching for something to stick in its inch-wide mouth. He comes up with a crab shell. The thorn toad chomps down on the thick piece of exoskeleton and splinters it with a loud crunch. "Don't want to get a finger stuck in there," Ooker advises. He lobs the fish overboard.

A few yards to our west, a stained and gouged concrete cylinder rises from the water, the stump of a wooden pole jutting from its top. Fifty yards to the north is another, and beyond that a long chain of them crossing the open water. They're the remains of an old power line from Tangier to Smith—and the work of a come-here who is still talked about on the island seventy years later.

Henry Jander was a Connecticut building contractor who visited Tangier with his wife, Anne, while on vacation. They were so charmed by the place that they sold their New England home, bought an old house in the marsh below Hog Ridge, and moved here with their children in 1943. At first, islanders were uneasy about the new arrivals. Both were college-educated sophisticates, which were in short supply on Hog Ridge. They shared a yen for art and classical music. And their surname sounded suspiciously German. A war was on.

But Henry Jander turned out to be a handy man to have around, and soon found himself elected to the town council. When the old

electrical plant failed, he took the lead in trying to get service re-
stored, making the island's case to the Rural Electrification Administration. He was rebuffed—the REA judged Tangier too small to
warrant its help—so Jander and other town leaders sold shares in a
town-owned system, promising to repay their investors with interest.
They raised enough money to build a new Meat Soup powerhouse
equipped with two war-surplus generators. When Jander couldn't
track down transformers for the system, he went back to the REA.
While the agency didn't produce the equipment he needed, it took
another look at Tangier's situation. Smith Island, it noted, was also
in the hunt for help, and the REA saw that by combining the two
populations, it could create a customer base that qualified for federal
assistance.

With that, the Chesapeake Islands Electric Cooperative was
born. By 1947, these poles, girded with concrete, carried power lines
across Uppards and the crabbing grounds to Smith. Tangier's lights
returned that winter.

THAT IT TOOK an outsider to lead the drive for electricity illuminates
a curious aspect of the Tangier character, one that natives readily acknowledge: As individuals, the islanders are fiercely independent and
self-sufficient—modern-day cowboys, or so they like to think. As a
group, however, they show precious little initiative. "Nobody wants
to jump in and volunteer," Anna Pruitt-Parks told me one day at the
firehouse. "People seem to think that, 'Okay, you got elected to the
town council, *you* take care of it,' especially if getting involved gets in
the way of being on the water."

I heard much the same from Nina Pruitt, the school principal.
"When people come from the county or state and we have a meeting,
maybe thirty or forty people might come to it, and twenty of them
will be aged sixty-five or over. And you have to imagine that all those
officials go home thinking, 'Why should we spend all that money on
so few people, and all of them so old?'"

Denny Crockett, the former principal, figures that the island's faith might play a role in its lack of gumption. "We're a very religious community, and I think that sometimes we put it in God's hands," he said. "We do believe that God takes care of things." But, he acknowledged, "there's also the view that sometimes God expects us to take care of ourselves."

The collective lethargy, or apathy, or whatever you choose to call it, showed itself early in the summer, when the grapevine carried word that homecoming had been canceled. Much lamentation followed, for homecoming is the island's biggest party, a days-long affair that draws those who've moved away back to their birthplace. This time, the one Tangierman who put it together every year couldn't manage the job. No one offered to step in.

I saw it again one summer afternoon in the Situation Room, when those assembled discussed a rumor that the state might cancel the 2016–17 oyster-dredging season, an important source of wintertime income. "There ain't no question that they can do it," Leon growled. "They can do anything they want, any time they want. But I thought there'd be more talk about it than they're doing."

"There ought to be," Allen Ray said.

"Somebody better call somebody," Ernest Ed Parks said, "and find out."

Bruce Gordy: "That'd be a good way to start."

Allen Ray: "We need one man to go down there who can talk."

"Need a man with education," Leon said.

Jerry Frank Pruitt entered the room. Leon asked whether he'd heard the rumor. "Yeah," Jerry Frank replied. "I know that Billy Boy"—William Ayers Pruitt, a Tangier-born former head of the Virginia Marine Resources Commission—"put on the website that we'd all better get down to the meeting. I think it's Wednesday. It's very important for people to go. They're talking about stopping oystering in Virginia."

"What are they thinking?" Leon wondered.

"Don't know," Jerry Frank replied.

I spoke up. "So, who will go?"

"Somebody should," Allen Ray said.

"Somebody *needs* to go," Jerry Frank agreed.

"But who will do it?" I asked.

Allen Ray: "Somebody should."

Me: "Will *you* go?"

"Well," Allen Ray said, "*somebody* should."

At its meeting two days later, the VMRC didn't suspend the season after all, but not because Tangiermen spoke against the idea. Not one islander showed up.

OOKER PULLS UP his scrape and we motor north, into a wide expanse of shallow water that crabbers call the Knoll. "Like the one in Dallas, the grassy knoll," he tells me. "Only here it's seagrass, of course." The line of stumped power poles glides by to our west, forming a de facto dividing line between bay and sound. Just to our south, where the poles crossed Goose Island, is where Elmer Crockett brushed a drooping wire with his head while hunting geese, and Half-Ass Buck saved his life. Used to be that the wire ran north to the southern shore of Smith Island, then across its marshy underside to a powerhouse in Ewell, the biggest of its three villages. In the intervening decades, however, Smith's southern coast has retreated to the north and east. Now the poles run a straight line into a patch of open water and stop.

One has to wonder how long Tangier would have endured life without electricity, had not Henry Jander intervened. One must wonder, too, how many islanders would have succumbed to illness or accident without the health center in which the Situation Room meets. It would not have been built without Oscar J. Rishel, one of Swain Memorial's come-here pastors, who spearheaded the project after suffering a heart attack in the parsonage. His personal advocacy with Virginia officials brought the well-equipped building to Tangier in 1957.

Rishel also played a principal role in finding a doctor to staff the building. Dr. Charles Gladstone, a general practitioner who arrived to treat the sick during the worldwide flu pandemic of 1918, had stuck around for thirty-six years, earning his keep by charging a small weekly subscription fee from every household. When Gladstone announced his retirement, the pastor led the scramble to find his replacement. The task occupied Rishel and state medical officials for four years before Dr. Mikio Kato—a thirty-three-year-old bachelor from Kobe, Japan—arrived in April 1957. It was Kato who delivered Ooker at his parents' home in King Street in July of the following year.

Jander and Rishel were among a parade of come-heres who accomplished what Tangier was unable or unwilling to do for itself. The most beloved, hands down, was Dr. David B. Nichols, who started visiting Tangier on his days off in the late 1970s, when the island hadn't had a resident doctor for more than a decade. Nichols, who ran a family practice on the western shore, was the next best thing. He came every week for the next thirty-one years.

Nichols grew to so love the island that he learned to fly helicopters and had a helipad built beside his mainland practice, the better to shorten the trip. In addition to treating virtually every Tangierman, he hired Inez Pruitt and mentored her transformation from high school dropout to physician's assistant, encouraging her to obtain first her GED, then a degree from the University of Maryland, and, finally, to navigate the labyrinth of state licensing. Nichols also helped lay the groundwork for the new $1.4 million David B. Nichols Health Center, which opened in September 2010. He died of cancer four months later, at age sixty-two, but his modern clinic is still staffed weekly by a fly-in doctor and is otherwise in Inez's capable hands.

Another part-timer saved the P'int. In 1959, George Randolph "Randy" Klinefelter, a Pennsylvania insurance executive and avid sailor, bought the islet and renamed it Port Isobel for his wife. Over the ensuing years, Klinefelter filled the marshy island's center with shipped-in soil and planted a thick forest of pines on its now-solid ground. Had he not done so, much of the P'int would have vanished by now, leaving Tangier's east side unprotected. Klinefelter bought Watts Island, too, which was fast crumbling into Tangier Sound. When he found that its cemetery was washing away, he rescued the gravestones and created a monument to the Parker family, who farmed the island in the nineteenth century, in the woods at Port Isobel.

Finally, in 1988, Klinefelter donated the 250-acre P'int, and a lovely retreat he'd built on its shore, to the Chesapeake Bay Foundation. In doing so, CBF told its members, he "opened up a treasured and historic piece of the Chesapeake to thousands of students, teachers, and citizens." Klinefelter served on the CBF's board for eleven years. Upon his death in 2007, the organization mourned the loss of "a friend, trustee, and one-of-a-kind donor." That he was.

Perhaps no come-here had as lasting an impact on Tangier, from so short a stay, as Susan Drake Emmerich. She moved to the island in June 1997, when the relationship between Tangiermen and the CBF was at its worst—and no surprise, because objectively speaking, islanders were poor stewards of their island and its waters. The marshes were studded with their discarded kitchen appliances, bicycles, and outboard motors. Litter made eyesores of the ridges. Watermen routinely threw trash, including motor oil, overboard; the harbor's shallows had acquired a sharp-smelling and colorful sheen. And Tangiermen had nothing but enmity for environmentalists, who warned that the bay's blue crab population was overfished, teetering on collapse, and would rebound only with tighter regulation of the commercial harvest.

Emmerich, a doctoral student at the University of Wisconsin, came armed with powerful ideas for improving the fishery, the island's long-term economy, and communications between the watermen and CBF. First idea: that the bay was God's creation and its stewardship a Christian duty—and that Tangier's ingrained disregard for the en-

vironment thus conflicted with its beliefs. Second: that the CBF and
government officials overlooked the importance of faith in their dealings with the island, much to everyone's sorrow.

Over a few short months, Emmerich challenged the islanders
to examine their relationship with the bay and its bounty, and to
bring their behavior on the water into compliance with the scriptural
teachings they otherwise strove to follow. The effort culminated in
fifty-eight Tangier watermen—Ooker among them—signing a Watermen's Stewardship Covenant, in which they pledged to obey the
laws of God and man. That meant following fishery, boat, and pol-
lution regulations, and supporting one another in times of doubt and
duress. Many of the island's women signed a stewardship commitment of their own. Emmerich also fostered dialogue among islanders, regulators, and the CBF that made plain to all that they'd been
talking past one another.

The fifty-eight who signed the covenant constituted just over a
third of the island's licensed watermen, and they found themselves
harassed and ostracized by the majority. Emmerich herself encountered fiercer resistance. Behind her back, Tangiermen threatened to
kill her and to run her off the island. They were only slightly less brazen to her face: At a New Testament gathering she later compared to
a witch trial, she was castigated as an "Earth-worshiper," a "distorter
of scripture," and even as a "beguiler."

Nevertheless, a March 1998 conference Emmerich organized
on Tangier for watermen, state and federal officials, scientists, and
environmental activists was a success. It gave Tangiermen a venue
for sharing their "vision for environmental, economic, and cultural
stewardship—based on their faith, an integral part of who they are
as people, and how they relate to the natural world," she wrote in
her dissertation. As for environmentalists and officials, "hearing the
Tangiermen's testimonies of transformation enabled them to recognize, for the first time, the centrality of the Tangiermen's faith to
their view of their environment and of the world."

If one were to pinpoint when relations between the island and

CBF began swinging positive, it would be that conference. Same goes for the island's appearance. Though it's still pretty disheveled in places, Tangier is far cleaner and tidier today than when I first visited. Its people turn out in numbers for organized cleanups of the guts and marshes, and watermen are likely as not to bag their trash and bring it ashore. Small gains, perhaps, but they started with the work and the courage of Susan Emmerich.

Five years after Emmerich's departure, the Kayes arrived.

"Yeah, the Kayes," Ooker says when I bring them up, and although I expect him to say more, he merely sighs. Doctors Neil and Susan Kaye are a complicated subject on Tangier—even for Ooker, who was close to them, and whose four daughters were closer still; Devi even lived with the Kayes on the mainland for a year, and still refers to them as "a second set of parents."

Mention the Kayes to most islanders, and the response is almost always something like: They were good people who loved the island and who did a lot of good while they were there. They were generous with their money, their time, and their sweat. They midwifed positive change. "I'll tell you what," former mayor and town manager Danny McCready told me, "if Dr. Neil were still here, we'd have that jetty by now."

Likewise, ask islanders why the Kayes left Tangier, and their answers tend to follow a predictable course. "You can't come into a community and try to change it," said one of the Kayes' close neighbors, Hanson Thomas. "That weren't going to work with the island."

"You've got to adapt to our ways," Eugenia Pruitt said. "I think that's where the Kayes went wrong. When they first come, they were a lot like Mr. Jander—they wanted to help the island. But then they tried to change it."

"We loved the Kayes, then didn't," said Lisa Crockett, my own next-door neighbor. "That's how we'll do. If you do something to cross us, you won't stay here."

Cross Tangier is exactly what the Kayes managed to do, even

after they spent years making themselves all but indispensable to the place. They seemed unlikely émigrés, these New York natives. He Jewish, she Catholic, they'd spent most of their married life in Wilmington, Delaware, and first touched down on the island in September 2002. Neil, a psychiatrist, had just earned his private pilot's license in helicopters and was at the controls of their first of many choppers. Susan, a pathologist, was along for the ride.

"They were extremely friendly," Neil said of the islanders they met. "They were welcoming. The physical beauty of the place is unbelievable. You have a mile and a half of beach to yourself, and it's pristine. The town is cute. There are birds everywhere. The wildlife of that island is just outstanding."

"We were both taken by it," Susan recalled. "When you first go there you don't see the bad, underbelly stuff."

The Kayes had been in the market for a weekend home and found one on the Eastern Shore a few miles south of Onancock. After closing the deal, they flew to Tangier for celebratory crab cakes, and on a post-dinner stroll they passed a brick rambler for sale on Hog Ridge, just a few doors from the beach. The mainland place suddenly seemed a mistake; they changed their minds on that house and closed on the island property in August 2003. "Tangier was a terrible investment," Neil told me thirteen years later. "It was already sinking and disappearing. But we thought it would be an easy place to sneak off to on weekends. We had no intention of getting involved in the community at all."

That changed when Hurricane Isabel raked the Chesapeake a few days after they moved in. "We had a chain saw and lent the chain saw to people," Susan said. "We took the mayor up [in the chopper] so he could take pictures for FEMA." Neil flew to the Eastern Shore for more chain saws and handed them out to their new neighbors. Though grateful, the islanders were mystified by the pair. That didn't change.

That fall, the Kayes stepped up their involvement. "They have proposed an idea of doing a genealogy of the entire island," the min-utes of the town council's November 2003 meeting reported. "The end result would be to place the Tangier Island Family Tree on the side of the Rec Center as a conversation piece." The report included a line that would characterize much of the couple's island activity: "They have indicated that they would fund the effort."

Sure enough, the Kayes assembled an immense computer print-out of the family tree with the help of schoolteacher Donna Crock-ett, herself a come-here. The years-long effort culminated when they installed the printout around the inside of the rec center in 2010. Islanders still talk about it.

In the meantime, Susan led a drive to create a library. The couple bought a sixteen-by-twenty-foot shed, installed it on Hog Ridge, and filled it with books. It opened in December 2005. At about the same time, they launched and paid for an artist-in-residence program that drew painter Ken Castelli to Tangier. The Kayes also sought to remedy a problem they had noticed soon after their arrival: The is-land lacked public restrooms. When they learned the town had not followed through on a state grant to build a visitors' center, they won permission to commandeer the project. Early in 2007, they applied for money for a museum of island history and culture, along with a system of walking and paddling trails. With Castelli's help, they built out a closed King Street gift shop to house the new attraction. "We'd go down every weekend and work," Neil said.

"Work," Susan emphasized.

The museum opened in June 2008 to justified praise: It tells the story of Tangier's past and cultural quirks, but even more it effectively captures the island's ongoing struggle with the bay. Among the high-lights are Castelli's wonderfully detailed map of Tangier, occupying an entire wall, and a backlit exhibit of the island's shrinking footprint that is rendered in colored film on multiple layers of Plexiglas. "It be-gan with the quest for the bathrooms," Neil said, "and a museum just happened to go along with the bathrooms."

The Kayes designed and ordered trash cans shaped like the old Tangier Lighthouse and placed them around town. They created a walking history trail spotlighting the island's older homes. And their good works took more personal forms. Neil, who was licensed to practice medicine in Virginia, offered free psychiatric care to a few islanders, whose kin cite his intervention as lifesaving. He flew the sick to mainland hospitals at no charge, ferried equipment during emergencies, and offered helicopter joyrides during the annual homecoming celebration. When Ooker's son Joseph asked him to photograph his wedding, Neil said yes—and served as photographer at Tangier weddings for years after. The photo in Swain Memorial of the rainbow descending on the church is his, too.

By any measure, the Kayes were terrific neighbors and a boon to the community. Which makes what happened later all the more frustrating.

IN APRIL 2009, the couple found themselves the subject of, as they put it, a "classic Tangier rumor." That spring they were immersed in a new project: trying to establish a government-subsidized, year-round ferry link with Onancock. This made sense, they figured, because the town's reliance on Crisfield denied it ready access to the Virginia services it supported with its tax dollars, from substance abuse treatment to driving tests.

But the effort stirred a deep and long-standing fear among Tangiermen about their school and its drain on rural, cash-strapped Accomack County. In 2009 the school's thirteen teachers taught eighty students, giving Tangier Combined one of the lowest pupil-teacher ratios in the state, at 6.2—lower by half than any other school in the district.

Islanders had long assumed that the only thing keeping the county from transferring their kids to mainland schools was the expensive and difficult commute across Tangier and Pocomoke sounds. But if the state were already running a ferry and paying for it—what then?

"If the mailboat went to Onancock," Nina Pruitt told me, "maybe our schoolchildren would be on it."

So it was no small matter when word circulated not that the Kayes were trying to start a ferry service, but that they were trying to close the school. As is typical with island rumors, it spread like lightning and was treated as fact before the couple learned of it. "A lot of people got pissed off," Neil said. He and Susan responded with an open letter to the town, which they posted in the grocery store, the post office, and on telephone poles around the island. "This is not true," they wrote. "We are 100% in favor of children being in school on Tangier from K–12, of the school being fully funded, of grades being kept separate, and of Tangier students getting all of the same benefits and opportunities that all children in the state receive."

As the Kayes saw it, kids wouldn't be schooled on the mainland, with or without a ferry. No matter how big the boat, it would have to contend with the Chesapeake's darker moods, and the district couldn't well have its students getting seasick on their way to class or racking up absences due to rough weather and ice. So the couple continued with the ferry project for another three months, until the town council voted unanimously that it wanted no part of the idea.

Folks got exercised, too, when the Kayes "floated the idea that Tangier could go to the government and get bought out," Neil told me. From their helicopter he and Susan could see the effects of climate change unfolding week by week, and the island's eventual fate seemed plain. They suggested that the town make itself "a model for how you relocate a community. We might be able to move Tangier as a block." This idea upset many.

So islanders already had a few qualms about the Kayes to go along with their admiration and gratitude when, one Saturday afternoon in August 2010, Susan opened Facebook while at home in Wilmington and saw a photo of the newly repainted Tangier water tower. As with the old paint scheme, Tangier's name faced north and south. New to the tank's east side was a big orange crab. Facing west was a startling addition: an enormous budded cross, its four tips ending in trefoils. She showed it to Neil.

"I was pretty horrified," he said. It seemed clear to both that the cross violated the separation of church and state: It had been painted at the town's behest on a tower that had been built with government money and was maintained at public expense. And besides being unlawful, it was bound to make many tourists uncomfortable, as it did them. "When I was in first grade, I was the only Jewish kid in my class," Neil said. "One of my friends had a birthday party. It was at a country club. They had no Jewish members. So I was the only one of my friends who wasn't invited to the party. I am passionate in my opposition to this sort of thing—to religious bigotry and exclusion."

Neil banged out an email to the town council:

> I just saw the photo of the water tower painted with a cross. I am asking you to repaint this. As a Jew, it is offensive. Further, it may be illegal. As the Constitution requires a separation of Church and State, if any public funds were used to paint this, it would be a breach of that sacred separation and will bring a lawsuit against the Town as soon as others who are also upset take note.
>
> While I have deep respect for all religions and everyone's right to practice his/her own brand of religion, Tangier should not be taking a stance on religion in this manner. . . .
>
> Citizens are free to erect crosses in their yards, to build temples, churches, mosques, sweat lodges, perform rituals, sacrifices, or whatever else they want. But, Town sponsored public displays endorsing a particular religion are a different situation.

In hindsight, his use of the word "offensive" in the opening paragraph was probably a mistake, because when news of the dispute raced through town, it was distilled to *The Kayes called the cross offensive*, which was all that many Tangiermen needed to hear. Within hours, the island's greatest benefactors were personae non gratae. "It was pretty clear that they saw it as attacking their religion," Neil said. He and Susan tried to clarify their objections to the paint scheme and offered to pay to have the cross modified into an anchor. The islanders weren't interested. At a town council meeting a few days later, Ooker announced that "the cross would not be taken off unless ordered to by a judge," according to the minutes. The vote backing him up was unanimous.

"Verbally, we were told by at least one member of the town council that if we insisted that the cross come down, we were 'done.' We took that threat seriously," Neil said. "It also probably caused us to reflect a little more—and we reflected on the whole crazy experience of Tangier. We realized that going to Tangier was some sort of calling and that probably the cross thing was the signal that our calling was over."

A week after their Facebook discovery and seven years after setting up house on the island, the Kayes again emailed the town council. "Our time on Tangier has come to an end," they wrote. "We realize that we cannot live under the shadow of the cross, and we know we also cannot live under the shadow of the hard feelings if the cross were removed. We have chosen instead to move off and sell our home." The next day, they were packed up and gone.

The conventional wisdom on Tangier holds that a host of other, more complicated issues, unconnected to island government, provoked the Kayes to leave and that the cross was merely a handy excuse. As Anna Pruitt-Parks observed, the tower's new paint scheme could hardly have been much of a surprise, as "Tangier was notoriously famous in our Christian faith."

But not only do the Kayes say the cross, and only the cross, drove them away, their emails and journal entries from the period offer compelling testimony that the conventional wisdom has it wrong. Five days before the couple's Facebook discovery, Neil, a faithful diarist, wrote that they'd bought lamps for their Tangier house. Just hours before finding the cross online, he wrote that they'd bought a painting for the place. Neither seems the action of people already mulling a move. After that, Neil's diary abruptly shifts in tone. His entry for the day after the town council's vote reads: "Susan and I are struggling to comprehend the enormity of leaving Tangier." The next day's reads: "Devastated."

When I asked Ooker about the controversy, he told me that the town council never considered taking the cross down. "We heard from some who didn't like it, who said they thought it looked like a Catholic cross. But that one's staying there," he said. "Me and some other citizens, we decided that if they took us to court, they'd have to paint over it. And if they paint over it, we'd take turns climbing up there and painting a new one." Ooker also told me that the idea for the cross came from Dr. David Nichols, the town's beloved "Doctor Copter": "He said, 'I think you should have a crab and a cross. Those are the two things that make you who you are.'"

The Kayes, who counted Nichols as a friend, heard the same thing, and brought it up in a September 2010 email to the doctor. "We later learned that you had suggested the designs to represent the pillars of the community, faith and crabs," they wrote. "We in no way hold you responsible for any of what transpired. But, it freaked us out."

Nichols replied a few days later. "I did mention to one of the Tangier clinic staff, when asked, that the Crab and Cross would be representative of the community," he wrote. "I must admit the notion of separation of church and state never came to my mind, but of course you are correct in this fortunate constitutional fact. I also was remiss in not thinking how this would impact all the islanders including yourselves. I understand your concerns and for this I sincerely apologize."

Eight years later, the cross remains.

So does the museum. So do the library, the trash cans, the historical markers.

The Kayes have been back just once, for Dr. Nichols's memorial service.

A homemade sign on the Main Ridge pushes the island's favored ticket, November 2016. (Earl Swift)

TWENTY

——

S THE SUMMER ENTERS ITS FINAL WEEKS, THE UPCOMING presidential election looms ever larger on Tangier. Tourists travel a harbor channel lined with TRUMP FOR PRESIDENT signs, then step off the boats into a gauntlet of them. Golf carts display Trump bumper stickers at bow and stern. Four Brothers is so adorned with MAKE AMERICA GREAT AGAIN placards and Trump flags that it looks more like a campaign headquarters than a cart-rental business and café.

The election even snakes its way into church. "Now, I'm going to get political on you for just a minute," John Flood announces in the midst of his sermon one Sunday. "How you vote is between you and the Lord. When you close the curtain, that's up to you."

Soon enough, he gets to a *but*. "There is one party that believes that there should be same-sex marriage," he says. "How can a Christian vote for that? How can a Christian vote for *anything goes*?" Many heads nod. "Who would have ever thought that there would be a party pushing the point that you could go in any bathroom that you want to?"

The larger message, the pastor says, is that "The world is still rejecting Jesus Christ. The world is not mourning the pierced savior, or the pierced prince. Right now, the world thinks that it's doing great. The world thinks that it's in control. But one day, and I'm afraid that it's going to be one day soon, they're going to mourn, because they're going to see that they've made an eternal mistake." He wraps up with a tacit endorsement of Dr. Ben Carson, whose candidacy ended months ago: "The man, I believe, knows the pierced savior."

In the Situation Room, the election moves near the top of the agenda, right behind crabs, wind, and waves. One mid-September afternoon we arrive to find the water's been shut off over much of the island, leaving the old health center without an essential ingredient for coffee. With a muttered "Lord sakes," Leon drives over to Daley & Son for a gallon of water, and we await his return before taking up anything substantial. He's back and making a pot when Lonnie gets down to business.

"I'm at the point that I'm thinking about voting for Gary Johnson," he says, referring to the Libertarian candidate. "That tells you where I'm at, if I'm thinking about somebody like Gary Johnson."

Jerry Frank points at him. "You're throwing your vote away, voting for Gary Johnson."

"Well, you're throwing *your* vote away, voting for Donald Trump," Lonnie says. "He can't win, Jerry. It's impossible."

"I'm smart enough to not get into this today," Jerry Frank says.

"Completely impossible," Lonnie says. "There's no path for him to win."

"I'll take him before I'll take her," Jerry Frank replies.

"Well, you'll have to take her," Lonnie says. "She's going to be president."

"I hope you're wrong."

"I'm not wrong," Lonnie says. "And really, voting for him, it's the same as voting for her. They're the same. If either of them wins, nothing changes."

"I don't believe that," Jerry Frank says. He looks over at me. "He has a talent for pulling you into arguments. He knows the magic word to get you going. He knows just how to strike that chord."

I look at Lonnie, who's grinning. "I'll tell you what, Jerry," he says. "I feel better about voting for Gary Johnson than I do for either of them. And he has as much chance as Trump does."

Ooker enters the room as Jerry Frank sighs and says, "Well, that remains to be seen," then adds, cryptically: "Like Captain Otis Evans says, the longest pole knocks down the highest persimmon." A little research reveals that Jerry Frank has uttered a rural nineteenth-century aphorism meaning that the stronger party will carry the day.

Switching gears, Lonnie announces that Cook Cannon, who's been undergoing physical rehabilitation in Crisfield, is coming home tomorrow. Also, he says, he's learned that Sam the Greek, a renowned Eastern Shore businessman, has died.

"Alcohol?" Leon asks.

"That's what I heard," Lonnie replies.

"Whiskey?" Leon asks.

Richard Pruitt, nodding: "Whiskey."

Allen Ray: "He was as honest as the day is long."

"That he was," Jerry Frank says, then turns to me. "He had a lot of business and lost it all to alcohol, but he promised he'd pay back everybody he owed. He owed a lot of people."

Lonnie explains further: "He was Greek, and come down to Crisfield, and built up a right big business. But drugs and alcohol ate into it."

"And women," Allen Ray adds.

"And women." Lonnie nods. "Drugs, alcohol, and women. I don't know what order it'd be in."

"Well," Ooker says, sitting down with a coffee. "I know I'd rather die of women than die of drugs or alcohol."

The mayor's in a mood for more politics. He says he was talking to Lisa Crockett, my next-door neighbor, and she told him that the island ought to vote for Hillary Clinton, because if the Democrat wins "the Lord will come back sooner." This earns a laugh that retreats with several members of the group into the hallway and out the door. It's four o'clock. Suppertime.

IN EARLY OCTOBER, Tangier begins to batten down for the winter. Mark Crockett's boat to Onancock suspends operations until next May. Fisherman's Corner closes its doors, along with all the other restaurants but Lorraine's. The museum and gift shops close—on weekdays at first, but soon enough altogether. Hilda Crockett's Chesapeake House stops taking guests, leaving the island with a single inn.

For the coming seven months, only a trickle of hardy outsiders will visit: adventurous weekenders, the odd journalist or wildlife photographer, and goose and duck hunters, whom a handful of islanders guide in the marsh islands above Uppards. The winds clock around to the northwest, and stiff breezes carry a chill that foretells the icy blasts to come.

Leon, Short Ed, and the island's other scrapers lay up their barcats. Most potters will keep working through the month's end, bundled against the deepening cold under their oilskins. Some trim the number of pots they deploy, and the harbor is crisscrossed by deadrises stacked high with traps being retired until the spring.

One Saturday I sit with Beth Thomas in the mailboat office, watching as watermen straggle in to pick up their checks from the Crisfield crab buyers. It's the first stop in a weekly ritual: With checks in hand, the peeler crabbers will pay Beth what they owe in freight charges for shipping boxes of soft crabs on the mailboat, then head over to the oil dock to pay for the diesel or gas they've pumped into their workboats over the week, then make their way to Daley & Son to settle their accounts there.

This particular Saturday, they're also saying good-bye. Leon enters the room and shuffles over. "I'm done," he announces. "I'm finished."

Another old-timer follows him in. "Thank you," he says as Beth hands him a check. "That's it. See you next season. See you next spring."

"Yeah," Beth replies. "See you."

Out at Ooker's shanty, the birds who've summered with him have taken off for warmer climes. The mayor builds a nest warmed by a spotlight for his four cats. But the approaching end of crabbing promises him no rest. On the water, he sets out eel traps in addition to his peeler pots. He soon has several tanks filled with the writhing brown creatures, which he ships live to New York. Fried, pickled, or served in tomato sauce, they're a Christmas Eve staple in Italian and Italian American households. And right behind the eels and the last of the peelers comes the wintertime harvest.

In the nineteenth century, that meant oysters, and the bivalves continued to dominate the island's cold-weather work on the water even as harvests declined precipitously in the first half of the twentieth. But by the 1930s, the rocks on which Tangiermen had most relied were scraped barren, and the state's harvests were averaging only 10 percent of the peaks of fifty years before.

Then, in 1949, the already beleaguered oyster was set upon by a new and terrible scourge: a single-celled parasite, *Perkinsus marinus*, or "Dermo." Initially misidentified as a fungus, it remains a mystery even today. But this much was quickly apparent: It proliferated in hot weather and salty water, killing mature oysters by the millions in the Chesapeake and its tributaries.

Ten years later, another microscopic invader, *Haplosporidium nelsoni*, or "MSX," was accidentally introduced into American waters with the importation of Asian oysters. Within a few years, MSX was slaughtering native oysters alongside Dermo, and the decimation of the species was at hand. A stretch of hot, dry weather in the mid-1980s nearly finished the job: The pathogens survived the warm winters and exploded in number over the summers, ravaging the surviving oysters. In 1985, Virginia had little choice but to shut down oystering in Tangier and Pocomoke sounds.

THE OYSTER'S DECLINE did not idle Tangier's watermen. Many switched their energies to winter crab dredging, a fishery that a relatively small number of bay crabbers had pursued each December through March for generations. Dredgers pulled a bigger, heavier, toothed variation on a scrape, which bit into the bottom and uprooted crabs burrowed there. This was not work casually undertaken, for winter dredging was centered near the bay's mouth, and required island men to work out of Hampton Roads or Cape Charles, down at the south end of the Eastern Shore, sometimes for weeks at a stretch. Still, it provided an income for crabbers who would otherwise have gone without.

When summer came, the bay's hard-potters also redoubled their efforts to catch blue crabs, and before long that fishery was stressed. Scientists debated "as to the exact status of the blue crab stock," a Virginia state task force reported in 2000. Its members could say with certainty, however, that even though more crabbers were putting more pots in the water, they weren't catching any more crabs. Throughout the late 1990s, Virginia crabbers pulled a relatively constant number of crabs from the bay; at the same time, the number of hard pot licenses rose by 14 percent and the number of licensed peeler pots leaped by 82 percent.

For the catch to flatline while watermen's efforts were at an all-time high strongly suggested that the overall crab population was slipping. The task force worried that the crabs were shrinking in size, too. A crab processor in Hampton told me that a pile of crabs that once yielded twelve or thirteen pounds of meat was, by 2000, producing just eight pounds.

So in the 1990s the Virginia Marine Resources Commission required crabbers to cut cull rings in the sides of their hard pots, enabling undersized crabs to escape. The agency trimmed the number of pots a crabber could put in the water. It declared a chunk of the lower bay a sanctuary to protect spawning sooks and their hatchlings and, later, placed a huge swath of the middle bay off-limits. Virginia began requiring watermen to buy a separate license for each kind of gear they used. Finally came the most drastic steps yet: a lockdown on new crabbing licenses and a ban on most license transfers, the measures many Tangier old-timers blame today for the departure of the island's young men.

All of these fixes had their effects, but none adequately addressed one continuing issue, and it was a big one. The number of sooks was dropping fast, by roughly 70 percent in just six years beginning in 1994. And without those females, the crab population was doomed.

Of all the bay's crabbers, winter dredgers had the most to fear from this development. Though accounting for only 13.5 percent of the total Virginia catch, dredgers aimed almost exclusively for adult females—96 percent of the crabs they plowed from the bay's bottom were sooks. All of these females were waiting out the winter so they could fertilize their eggs and produce sponges. By targeting them, the dredgers were crippling the blue crab's ability to reproduce—and cutting deep into their own prospects for success as crab potters in the summer.

So even with the new regulations of the 1990s, the species continued to struggle. Scientists reckoned that harvesting should not claim more than 46 percent of the bay's crabs in any given year—an "exploitation fraction" that would enable crabs to produce sufficient young to replenish the loss by the following season. But the harvest regularly exceeded that threshold. Over the ten years beginning in 1998, crabbers caught an average of 62 percent of the bay's total number of blue crabs, and some years a far greater share—71 percent in 2001, according to the Chesapeake Bay Foundation, and 70 percent in 2004.

By 2007, the fishery was poised at the brink of collapse. The total population of blue crabs in the bay had plunged from 791 million in 1990 to 260 million. Of that 260 million, just 120 million were adult crabs capable of producing a new generation, well below the 200 million adults that experts deemed the rock bottom necessary to sustain the crab as a viable part of the Chesapeake ecosystem. That year, 2007, was the worst for crabbers since bay-wide record keeping began in 1945.

Virginia officials recognized that only a profound reduction of the adult crab harvest would reverse the animal's slide to oblivion, and that they had to act quickly. And so, in April 2008, they closed the wintertime dredge fishery. It has not reopened.

This did not have the drastic effect on Tangier that it would have just a few years before, because dredgers, like crabbers in general, had fallen in number with the island's population decline, and some older watermen had retired from the harsh wintertime work even as they continued to set pots the rest of the year. That was true all around the bay: In 1994–95, the Chesapeake had been worked by 346 dredge boats, each manned by a captain and one or two mates. Just four years later, the number dropped to 192. The state task force of 2000 had recognized the trend and surmised that if the crabs could hold their own for a few more years, they'd outlast their pursuers. And indeed, by 2007 only fifty-three boats were still at it, just a dozen from Tangier. For those few crews, the fishery's closure was a hardship, just the same, especially because it coincided with a national economic crisis. The island weathered some lean years.

But this is an anointed place, as Tangiermen will tell you, and they cite more than merciful weather as evidence. Even as dredging ended, the last survivors of the Chesapeake's once-boundless oyster population continued their struggle against microscopic predators and human abuse—and against all odds, the creatures began a tentative recovery.

So, after they've caught the last of 2016's crabs, and after Ooker has trapped his eels, the mayor and at least fifty other Tangiermen are readying to go to work as oystermen.

THE ATLANTIC HURRICANE SEASON runs from June 1 to November 30 and thus overlaps much of each year's crab-potting season. Tangier's closest brushes with cyclones have come in the middle of that six-month stretch, in August and September; among those were the Great September Gust of 1821, the August storm of 1933, and 2003's Hurricane Isabel, the three most destructive blows that wind and water have landed on the island.

But the fall's cooling temperatures don't signal an end to the danger, as I'm reminded when Tracy Moore, Lonnie's brother, opens the September 25 evening service at New Testament by taking prayer requests, then asks Carlene Shores, Ooker's sister-in-law, to say a prayer to lift them up. "I thank you, Lord God, for the seawall you're going to be providing and the sand you'll provide," she responds. "This is hurricane season, Lord God. Protect us. Put your protection around us, as you have so often in the past, Lord God."

As she speaks, a storm born as a knot of thunderstorms off the West African coast speeds across the Atlantic. Within three days it has tightened into the telltale bull's-eye of a tropical storm. Then, as it cruises west along the coasts of Venezuela and Colombia, it swells into a ferocious category 5 hurricane. No sooner has Tangier shut down for the winter than Hurricane Matthew turns its way.

It smashes into southwest Haiti with winds of 150 miles per hour, and twelve hours later it scours eastern Cuba. It weakens as it spins north, then muscles back up as it rakes the Bahamas. At that point, the forecast tracks for the storm include arcs over the Chesapeake and smack into Tangier. With Matthew still more than a thousand miles to the south, the island's tides go haywire, and water jumps the guts to flood the roads.

But by the time Swain Memorial's faithful convene for the evening service on Wednesday, October 5, most forecasts call for the hurricane to make landfall in Florida or South Carolina, then turn into the open ocean. "Did anyone have to tread water to get here this evening?" Pastor Flood says by way of a greeting. "I've been told that there will be a time of prayer at the school tomorrow at eight o'clock, around the flagpole, about the storm. I think the Lord has already answered prayers. From what I understand, we're going to get a little bit of rain out of it."

His sermon centers on a passage from the first book of Peter: "But the end of all things is at hand; therefore, be serious and watchful in your prayers." It strikes me as an ominous choice.

"How many's been praying for the storm to turn?" he asks. He gets a near-unanimous show of hands. "How many have been praying for Haiti, for eastern Cuba, for the people of Florida—all of those in harm's way?" Most of the congregation again responds. The pastor nods. "Prayer is a serious thing. We should be diligent in how we pray for others, as if we were praying for ourselves."

Matthew crawls up Florida's Atlantic coast, never veering closer than twenty miles to land, but drenching the state with rain and storm-borne tides. When it makes landfall near Myrtle Beach, South Carolina, on Saturday morning, it has weakened to a category 1 hurricane—dangerous, to be sure, but no longer packing cataclysmic winds. It bends back offshore, bounces up North Carolina's barrier islands, and, in the predawn of Sunday morning, makes its predicted seaward turn at Cape Hatteras.

Still, this is a big cyclone, and Tangier is only about 170 miles from the cape. Matthew's outer bands slug the island with fifty-mile-an-hour winds. Windows rattle, and wooden walls shudder and sigh. A disabled airplane out beside the runway, parked there for more than two months, flips onto its back. When I step onto the deck shortly after daybreak, waves pounding the seawall are throwing great blooms of water skyward, and the bay beyond is white with foam.

I pull on my rain jacket and set off on my bike for the ten o'clock service at New Testament. West Ridge Road is under several inches of water, and the north wind fires a stinging rain into my eyes. When I make the turn onto Wallace Road to cross the marsh, I can only guess where the asphalt is—the path is completely overwashed—and I have to lean ten degrees to my left to counter the wind. It catches at my jacket's hood, twisting it sideways over my face, and its gusts throw me so off-balance that I'm forced to dismount halfway over and splash the rest of the way to the Main Ridge.

The New Testament parking lot is empty of golf carts. The door is locked. The service has been canceled. Rather than attempt to retrace my journey, I pedal against the wind to Swain Memorial to wait for its eleven o'clock service. I find the church unlocked, but the congregation doesn't show up. The island's landline telephone network no doubt did its usual efficient work in getting the word out. Unfortunately, I have no landline.

Soaked through, I decide I might as well have a longer look

around. I head north through Meat Soup, encountering no one. The
storm has caused no visible mayhem to the business district: Tangier
had plenty of time to prepare, and I see nothing out of place. I cross
over the Long Bridge and, with a gale-force tail wind, travel the
length of the West Ridge without turning a pedal.

Down on the spit, the beach is invisible under incoming breakers,
and the dunes have been carved into high vertical bluffs. The air is
filled with spray and blowing froth, and the cymbals and timpani of
bay hammering shore. Foam leaps from the surf to cartwheel into the
saw grass at my feet, and a small twister forms for a moment over the
sand, spinning furiously before blowing itself apart.

But except for damage to the dunes, Matthew leaves Tangier
unscathed—and with further evidence that it is protected by the Almighty, that the faith and prayer of its people have anointed the place.
That evening at New Testament, Tracy Moore asks for prayer requests
to open the service. Carlene Shores utters the obvious: "Thank the
Lord for sparing us again."

THE FIRST SUNDAY of November: Pastor Flood advises his Methodists that sitting out the election is not a good idea. "It's up to the
Christians," he says. "It is our duty." He asks Chuck Parks, a Tangier-
born minister visiting from the Eastern Shore, to offer a prayer.

"We are at an anxious time in our country politically," the Reverend Parks says. "Lord, we pray for the upcoming election, and we
pray that you put in the candidate who would best put in what you
have in mind for this nation." Judgment, he says, "is so close."

Two mornings later, I bike to the school, where the voting machines wait. On the way I encounter Ooker parking his beach cruiser
outside his house. "I voted a couple of times this morning," he tells
me. "I'll wait awhile, then go over and vote again." He glances my
way and, detecting skepticism, adds, "Actually, I forgot my wallet.
They asked to see a photo ID, so I had to come back here to get it."

I pedal onward, taking a meandering route via the Main Ridge.
It bristles with Trump signs. Citizens have supplemented the usual

"Make America Great Again" fare with homemade tributes, scrawled in permanent marker on painted wood. I see no signs for Hillary Clinton. At the school, I watch a steady procession of voters arrive by golf cart. By the end of the day, 221 Tangiermen have cast their ballots for Trump—more than 87 percent of the island's 253 votes.

That Friday, three days after the election, I take the mailboat to Crisfield. I share the wheelhouse with Brett Thomas and Cook Cannon, who's headed ashore for a doctor's visit and is sprawled out on a low bench behind the helm. We're approaching the red bell buoy that marks the halfway point when Cook speak-shouts how happy he is, after these past eight years, that America is again going to have a Christian president.

I have borne mute witness to political debate on Tangier for six months, and have heard a lot of such talk—that Barack Obama and Democrats in general are enemies of Christian values and the American way. That Donald Trump, a man who's boasted about the impunity with which he sexually abuses women, whose campaign has been energized by racial and cultural division, and who's no stranger to deception, character assassination, adultery, and divorce, is nonetheless more godly than the comparatively straitlaced Hillary Clinton. Ooker's reaction to the infamous *Access Hollywood* tape, which captured Trump making decidedly unchristian comments about women, was illustrative: "The Left, they're acting all shocked at the language on that video," he announced in the Situation Room a few days after the 2005 tape surfaced. "Well, *they're* the ones that stand for a godless society."

To which Jerry Frank Pruitt replied: "So many people vote who don't know what they're voting for."

I've been mystified by it. Today, for once, I decide to correct the record. "President Obama's a Christian and Hillary Clinton's a *Methodist*," I tell Cook. "And I was under the impression that Trump's pretty much an atheist."

"He ain't no atheist," Cook replies.

"That's another media lie," Brett says.

"He's a Christian," Cook says. "He's been saved." He refers to reports that Paula White, a thrice-married millionaire televangelist who preaches the prosperity gospel, led the president-elect to Jesus.

I turn to face Cook. "Have you ever seen a picture of Trump in church?"

"That don't mean nothin'," Cook replies. "You don't have to go to church to be a Christian. And just because you go to church doesn't make you a Christian."

Not the kind of talk I've often heard from Tangiermen.

But true enough.

PART FIVE

THE SEA
IS COME UP

Matt Wheatley deploys a smoke while Lonnie Moore and Steve Watson keep their eyes peeled for other oystering boats on the Rappahannock River, November 2016. (EARL SWIFT)

TWENTY-ONE

——

DUANE CROCKETT HAS SPENT THE BETTER PART OF A YEAR preaching on First Corinthians, guiding the New Testament flock through the book verse by verse. One Sunday morning he homes in on the gift of prophecy. "A story I always share around this time of the year," he says, "is the story of Mr. Close and the oysters."

George "Buzzy" Close was Swain Memorial's pastor from 1987 to 1992. "In the late eighties, anyone who was alive at that time will remember, nearly every oyster in the bay was dead or diseased," Duane says. "Nearly every one of them was. It looked as if there would never be another oyster caught again, or that no one would ever be able to make their living off of the oyster industry, catching oysters, again. I remember that well, as you do.

"And I remember Mr. Close, who was a man of great faith and believing in prayer. He requested that the men take him out all over the sound to pray over the oyster rocks." Duane peers out over the congregation until he locks eyes with a waterman. "You were one of them, weren't you, Larry?" The man nods. "Yeah," Duane says. "And he prayed over every one of them. And he said that the Lord gave him a vision that one day the oyster business was going to come back full force, and people were going to be able to make a living off of the oyster business again.

"There's always those nags that want to put a stop to the works of God," Duane says. "There were people who told him, 'The Lord didn't tell you that. He did not give you that vision. You come up on that on your own.' They said it so much I believe Mr. Close even began to wonder if what he had been told was of the Lord . . .'"

"Was Mr. Close's vision from the Lord?" Duane wonders. "Now, let's ask some questions. Was he God's child? Yes. Did what he said come to pass? *Yes.*

"Who would have thought that the watermen of the sound would be able to make a living off of two oyster rocks? Who thought it?" he says. He looks around, letting the words sink in, before pressing on. "Was this prophecy glorifying to God? Ooker will say often—he said last night in his prayer—'I thank God for the oysters.' He says that a-many a time. So it was glorifying to God.

"Was it edifying to the church? Well, I ask this: Have the needs of our people been met through the oysters? Have the missionaries been supported through oyster money? Have the needs of the church been met through the oyster money? Yes.

"So," Duane concludes, "through that, I have to say, it was a prophecy of the Lord that he gave Mr. Close almost thirty years ago."

THIS MUCH IS SURE: George Close delivered his prophecy at a point when the Chesapeake's oysters were on the brink of extinction, and the island's oystermen had little reason to hope they'd recover. Over-fishing had decimated the bay's stocks, and Dermo and MSX had all but finished them off. From 1980 to 1984, Virginia's already-gutted oyster harvest fell by half. By 1990, even the catch on privately owned beds had collapsed. The state all but gave up on the oyster at that point. It slashed its spending for replenishing the bivalve and shied from investing serious money in the effort for twenty years.

Scientists from the Virginia Institute of Marine Science (VIMS), a division of the College of William and Mary, continued to monitor Dermo and MSX, however. The salt-loving, heat-reliant pathogens dwindled in years with heavy spring rains and stormed back when the weather turned hot and dry. In 1998, VIMS found that Dermo infected at least three in four oysters on the rocks nearest Tangier. In 2001, the infection rate was 96 percent or more. The surviving oysters were barely hanging on. As recently as 2006, the harvest in Virginia produced fewer than ten thousand bushels, a fragment of 1 percent of the nineteenth-century peak.

In the face of this bleak reality, the federal government stepped in. It began devoting money and energy to replenishing the public rocks with fresh shells, which larval oysters require to settle and grow, and seeding the rocks with healthy wild and hatchery oysters. And by 2009, VIMS was noting that while "it has not been widely appreciated that resistance to disease is developing among oysters in Chesapeake Bay, or that such development is even possible," long-term data suggested that just that was happening. The same rocks that had been ravaged by MSX now boasted hardy survivors that produced what appeared to be hardier offspring.

Virginia officials gave the oysters a boost by establishing a rotational system on the state's public rocks, mandating that each get a rest of one or two years between harvests. Although the bay's oysters have neither licked the pathogens nor rebounded to anything approaching their past populations, their numbers have stabilized for the first time in decades. Since 2011, Tangier oystermen have enjoyed increasingly fruitful harvests.

Which is what prompts me one Thursday in mid-November to drive three hours through the dark to Windmill Point, a marina at the tip of Virginia's Northern Neck. There I find Lonnie Moore's *Alona Rahab* tied up alongside a few other Tangier boats and twenty-odd deadrises from around the lower Chesapeake. All are getting a jump on the 2016 oyster season, which doesn't open until December 1 in Tangier Sound but started weeks ago in the broad, tidal Rappahannock River. A big, healthy oyster rock waits in twenty-eight feet of water just inside the river's mouth.

Most Tangiermen don't stray so far from home. Lonnie, however, brings to oystering the same far-ranging spirit he demonstrated all summer with crab-potting. He doesn't mind camping four nights a week on the boat through Thanksgiving. His crew today consists of two licensed oystermen: the experienced Matt Wheatley of Tangier and Steve Watson, a newcomer who commutes two hours every day from Norfolk. All are eager to get started when I climb aboard, stepping with care past the dredging rig that occupies much of the *Alona Rahab*'s open deck.

The VMRC carefully regulates the oyster harvest, and one of its hard-and-fast rules is that dredging starts at sunrise—which today comes at 6:48—and not a second sooner. Furthermore, the boats can't leave the dock until exactly thirty minutes before then. At about 6:10, engines start up and down the line, and the air fills with shouts and diesel smoke. At 6:15, crews throw off their lines. And at 6:18, two dozen boats pull from their slips and turn for the marina's opening into the river.

We strike south in a crowd five boats abreast and all racing flat out over foot-high swells, the boats up front creating heavy chop in their wakes. To run in such a flotilla in the predawn gloom is electrifying—there's a swashbuckling Battle of Britain feel to it, as if we're flying in tight formation. As we near the rock, three miles out, the squadron splits up, boats peeling away for favored spots. Reaching his own, Lonnie cuts the engine. "Okay, fellas," he says. "Let's gear up."

They step into their oilskins, tugging them over jeans and hooded sweatshirts, as Lonnie eyes the time on a GPS monitor bolted to the ceiling. A chill is brisking over the river from the west, and even bundled the men wait at their stations with clenched teeth—Lonnie at the aft steering console, Matt and Steve on opposite sides of a stainless steel culling board. The dredge hangs at eye level at the end of a rusty chain. It's twenty-two inches wide, with a toothed mouth and a trailing bag of knotted rope, and is rusted well past flaking. It looks a hundred years old.

The clock hits 6:48. The dredge goes over the side, its chain spooling from a hydraulic winch with a puff of atomized rust. It rattles and jounces as it plays out, tightens to quiet creaks when it hits bottom. Lonnie steers hard to starboard, into a curving lick, as the boats around us carve their own tight circles: The *Mariah Taylor* out of Tangier is a few feet to our right; Lonnie cuts across its stern. A twenty-foot Carolina Skiff off our port stern does the same to Lonnie. Other boats in wider orbit around us jockey for position, drawing closer as we clock in an ever-tightening circle. It feels like a dogfight.

After ninety seconds the winch kicks on, the chain winds back in, and the dredge breaks the surface and rises, dripping, alongside the boat—empty. After another ninety seconds on the bottom, it contains only bright orange sponges. A third time, more sponges. On the fourth lick, however, the bag comes up drooping with oysters. Matt and Steve grab it, pull it aboard, and dump the load onto the tray, then shove the dredge back over the side. The chain unwinds with a deafening clatter.

They cull the lick, shoving half shells aside, giving closed shells a rap to determine whether they're oysters or empty "boxes," using hammers fashioned from rebar to break away barnacles and bits of shell, or "wings," cemented to the creatures. The choreography of boats around us grows crazier. Lonnie holds the *Alona Rahab* in a sweeping right turn as *One More Set*, out of Saxis, cuts left to right across our stern, then is itself forced off course by another boat encroaching into its path. They miss each other by five feet. A little skiff that looks overwhelmed by its dredge rig plants itself just off our starboard beam. "He's gonna be in our way all day," Steve hollers. Lonnie solves the problem by yanking hard to starboard, gunning the diesel, and crossing four feet behind the interloper. The dredge rises again, heavy with shells slimed in mud and garlanded with sea grapes.

The low sun spins from port side to stern to starboard to bow as the *Alona Rahab* screws itself into the river. I glance into the cabin, to the GPS monitor recording our track, and see that it consists of loops piled atop one another in a thick black tangle. We gain fast on a twenty-foot open boat steered by a white-bearded Eastern Shoreman, who waves at Lonnie to get off his tail. Lonnie signals him to move aside. "He thinks I'm in his way. I think he's in my way," Lonnie yells over the chain's rattle. "Everybody's in everybody's way."

Again the chain winds in, spraying water that catches the early saffron light, and another load of shell hits the tray. The first bushel fills.

I look at my watch. Seven o'clock. We've been at it for twelve minutes.

THE CONTAINERS into which the *Alona Rahab*'s crew dumps its catch are not the wooden bushel baskets used for crabbing. Virginia tubs, as they're called, are built of molded orange plastic and closely resemble something you'd use to haul laundry. They're also slightly larger than the crab basket. Across the state line, oystermen use a Maryland tub, which is smaller than a Virginia tub but, again, larger than a crab bushel. Tangier watermen seem mystified but not particularly concerned by this variance in what should be a standard measure. "A bushel basket's a bushel," Leon told me. "A Maryland tub's a bushel. And a Virginia tub's a bushel. They're all different sizes. Now, *something* ain't a bushel. They're calling them bushels, but there ain't that many bushels."

Virginia law permits a licensed oysterman to take eight Virginia tubs of oysters per day. The boats on the Rappahannock all have a captain and one or two crew aboard, entitling some to a total of sixteen bushels and others, like the *Alona Rahab*, to twenty-four. In theory, the day should be half again as long for Lonnie and company than for some of the smaller boats circling near us.

Except that Matt and Steve are speedy cullers. They attack each load with sharp eyes and efficient movement to get oysters into the tubs and sweep the empty shells and detritus, or cinder, back overboard. On top of that, Lonnie's a tough boss: He takes no breaks and gives none. So by eight o'clock, seventy-two minutes into the day, we have six bushels filled—a rate of five tubs an hour. Matt never has time to single-task a smoke, so he keeps a Marlboro clenched between his lips as he works.

The westerly breezes up. The river grows spiky. In one lick, an old shell comes up with ten or more tiny oysters attached in a bunch. "All them struck on one shell," Matt says, showing off the cluster. "In a couple years, they'll be a right good size." The next lick brings up an old beer bottle, barnacles crowded tight on its brown glass—but only six oysters.

Matt is dumping oysters into our ninth bushel when the big, high-sided *Julie Ann* swings past our port side, so close that Steve pauses to glare at its skipper. The *Mariah Taylor* slides near us to starboard, and Lonnie heels the boat over hard to stay out of its way. We pass a mainland boat with a two-person crew, and I'm surprised to see that the mate is a woman. "There are at least five boats out here with women on 'em," Lonnie tells me. He need not add that none is from Tangier.

The *Mama Say* passes close by. Its stern advertises its home port as FALLIN AND CRAWLIN, VIRGINIA, but the boat itself is a Tangier product, built by Jerry Frank Pruitt. "Heeeeey," Lonnie yells to its captain.

"Heeeeey," comes the reply.

We near an old round-stern wooden boat, *Grace*. "Woman on this boat, too," Lonnie says, pointing to its mate, busily culling, her sweatshirt's hood cinched tight around her face. The boats around us draw close, *Mama Say* just a few feet behind, *Grace* alongside. Lonnie waits for his moment, then turns hard right, and we pass just aft of the *Grace*'s dredge line and across the stern of the *Julie Ann*.

Steve is assessing the last cull, which produced just two oysters. "Ain't much here, Lonnie," he yells over his shoulder.

"No," Lonnie agrees. "We'll have to find another spot."

We break out of the circle and shoot one hundred yards upriver, where the next lick brings up a mess of oysters. Before the cull's finished, the *Julie Ann* slides up on our port side. "They're in the way," Matt growls.

"More in the way than anybody else out here," Lonnie fumes. "He don't know what he's doing." Inside of two minutes, we're again in a crowd—*Mama Say*, *Mariah Taylor*, and the Eastern Shore skiff have joined us. The *Julie Ann* is glued to our stern. "I can't shake him," Lonnie complains. "Everywhere I go, he comes."

Ten o'clock. Sixteen bushels on deck. The *Julie Ann* hangs close, with its dredge line stretched thirty feet off its side—far too much line, and a barrier to other boats trying to pass on its right side. I watch its mates cull. It takes them four times as long as it does Matt and Steve. "We'll be gone before that boy gets ten bushels," Steve says.

Lonnie glares at the other boat. "He keeps cutting in on us."

"You want to pull alongside him?" Steve asks. "I know you don't like cussing, but I don't have a problem with it."

Lonnie breaks away from our pursuer and bolts upriver. The next lick brings up a pile of shell and fourteen oysters slimed in gray mud. "There we are!" Steve bellows.

At eleven the *Mariah Taylor* heads for shore. Its skipper has just one mate aboard, and they've reached their sixteen-bushel limit. Just a few minutes later we have twenty-three tubs stacked at the stern and the last fills quickly: twenty-seven oysters in one lick, twenty-three in the next. "I think we've got it!" Steve hollers, and as the mates secure the dredge, Lonnie opens the throttle and we bounce back to Windmill Point.

A few days later, back at Tangier, I find Lonnie on the *Alona Rahab*, tied up at his dock. He's shucking a small pile of oysters that his crew caught just hours before. I climb aboard as he forces one open, then passes it to me. I slurp the oyster from its shell. It's fat, firm, salty. I have never had better, and I've certainly not tasted fresher.

"Good?" he asks.

I nod, my mouth full.

He shucks another and hands it over.

BY THE TIME the season opens in Tangier Sound, the deepening cold has transformed oystering into a potentially deadly enterprise. Mid-December, daytime temperatures hang in the low forties on the island and dip considerably lower in the northerlies that breeze over the open water. A white-capped chop is normal. Heavier seas build quickly. Spray soaks through the cotton that watermen invariably wear under their oilskins and turns to ice on a boat's deck and dredge rig. A pursuit with little tolerance for mistakes thus becomes even more unforgiving: To go overboard is to confront hopelessness. In water so cold, a waterman has only fifteen to thirty minutes before he loses more heat than he can keep, and hypothermia sets in. His energy drains away. His limbs quit working. And if he isn't pulled from the water, dried off, and warmed, he dies.

One dark, frigid morning, I make the questionable decision to join Ooker and Allen Ray Crockett aboard Allen Ray's forty-two-foot *Claudine Sue*. We motor out to a rock five miles northeast of Tangier, where we find forty-one other boats waiting for the dawn. The crazy choreography of too many dredges working too small a piece of water doesn't differ much from oystering on the Rappahannock. We're often within five or six feet of Leon, who's brought Short Ed Parks aboard the *Betty Jane* as his mate, and at times sidle even closer to other deadrises competing for space.

What sets the day apart is the wind. It builds out of the northwest as soon as the sun has cleared the horizon. By eight o'clock, the seas have split into deep troughs and three-foot ridges, and waves smacking the hull send shivers through the entire boat. By nine o'clock, the wind is up to twenty-five miles an hour, and the seas run a hissing, frothy four feet. The deck rolls and lurches. Spray leaps over the gunwales. An occasional wave comes over the bow, vaults the cabin, and lands with a clap on and around us.

I brace against the cabin doorjamb, struggling to stay on my feet. From there I can see a framed picture mounted to the cabin's plywood interior. It's a print of a 1950 painting by Warner Sallman, a Chicago artist who specialized in religious images. This one's called *Christ Our Pilot* and depicts a young mariner clad in a muscle shirt and gripping the wheel of a wooden ship snared in a titanic storm at sea. Behind him stands a ghostly and outsized Jesus, left hand resting on the mariner's shoulder, right hand pointing the way to safety. The image is standard equipment on Tangier workboats.

Ooker and Allen Ray are too busy to pay the cold and wet much mind. No way are we going in before they have their sixteen bushels. "We've been out here when, really, we shouldn't have been out here—waves coming in the boat, shells blowing at you," Ooker says, offering what he apparently mistakes for reassurance. He nods toward Allen Ray. "He's right salty."

I watch the *Betty Jane*, twenty feet away, as it yaws, pitches, and rolls all at once. This is not the little barcat, the *Betty Jane II*, in which Leon scrapes for peelers, but a bigger, older, and notoriously leaky deadrise. As its dredge comes up, Leon and Short Ed climb onto its culling board to guide the device into position, release its load, then clamber back to the deck. Leon is closing in on his eighty-sixth birthday. Short Ed is eighty-one. Over and over, they do it.

"Kyowkin'!" Ooker yells over the motor, the clattering chain, the wind. "Breezy, breezy, lemon squeezy."

"Rough," Allen Ray says, as he guides the boat into a lick.

"Roughest day of the year," Ooker says. He looks over his shoulder at me. "We'd think twice before we bring *you* again!"

Tomorrow, Ooker says, the weather is supposed to be far nastier. Temperatures tonight will fall below twenty degrees, and it speaks of even more wind. "We ain't coming out," he says. "But the headhunters will, the crazy people. Like Lonnie, he'll probably be out here. They've got to do it. It's psychological."

I think about that the next morning, when I leave the island for Crisfield. I step out of the house with my luggage to find the bay black and churning and a cold northwest wind blowing steady at thirty miles per hour. At the mailboat dock, Beth Thomas tells me that Brett has opted not to make the morning's run. Going over wouldn't be bad, but the weather's expected to get meaner, and he's worried about the return trip. It's cold enough for spray to turn to ice. There's passenger safety to consider, and as high as the steel-hulled *Courtney Thomas* sits in the water, he's wary of the extra weight affecting the vessel's manners.

In the mailboat's place, Mark Haynie is running the *Sharon Kay III*, so I lug my bags down the dock road to Main Street and hike up Meat Soup to the county dock near its north end. The *Sharon Kay III* has yet to arrive. I huddle with a knot of other passengers out at the dock's tip, hunched against the cold and gusting wind. One is a solid fellow approaching middle age whom I don't recognize, but whose ravaged ball cap and chin stubble mark him as either a waterman or a tugboater. I ask his name. "Jason Charnock," he says.

"Who are your folks?" I ask—my standard question for establishing a new acquaintance's place in the island's family tree.

"What, like my parents? Ed Charnock is my dad," he replies.

"Ed Charnock? Eddie Jacks?"

He chuckles. "Yeah, Eddie Jacks."

I introduce myself, tell him I'm a fan of his dad, that I've found that the few words he has to say are usually worth hearing. He nods, smiles, thanks me. I realize that at this first meeting, I already know quite a bit about Jason. I know he's married to Carol and Lonnie Moore's daughter, Loni Renee. I've met his kids, whom Carol often babysits; Lonnie's boat is named for two of them. And I know that Ed's mate on the *Henrietta C.* is his son—they've worked together for more than twenty years—and while Annette has two sons on the mainland, I'm pretty sure Ed has only one boy of his own. So here he is: Jason Charnock, a waterman like his father—whose own forebears worked the water like their fathers, who were the sons of other captains still. Sorting through Tangier's web of family links is something like playing three-dimensional chess.

Over Jason's shoulder I can see the curving dock where Lonnie ties up the *Alona Rahab*. As Ooker predicted, it's not there. Lonnie's out on the rock. Mark Haynie pulls up in his boat, and his chilled passengers climb aboard. The *Sharon Kay III* is big as deadrises go—forty-six feet long, fourteen wide, with a spacious cabin and the forward half of its open deck shrouded in plastic. In all, there's protected seating for twenty or so and space for cargo and smokers aft of the shroud. Still, it's nearly twenty feet shorter than the mailboat, and tons lighter. The bundled passengers around me chat lazily as we pull away from the dock, but a nervous energy charges the air. We're in for an exciting ride.

It doesn't take long to start. Even before we've rounded the southeast corner of Uppards, waves are crashing heavily into the boat's nose. Halfway across the open water between Uppards and Smith Island, the seas top four feet, and the *Sharon Kay III* rolls into troughs, slides sideways, and draws circles with its corkscrewing bow. Gusts batter and crackle the tenting around us and drench the portside windows in spray.

We hug the Smith Island shore, but the low marsh offers little protection, and as the wind builds the deck underfoot tilts every which way without warning. I glance into the cabin. The captain is sitting alert but relaxed behind the wheel, which he guides with one hand. He looks utterly unworried. Then I turn my gaze to the boat's stern. The sound has become a maelstrom of heaving water, pushed one way by the tide, slammed the other by the wind—big, muscular waves colliding with great eruptions of spray, and parting to expose deep chasms that seem sure to swallow us up. The scene is all the more fearsome for its absence of color: All is black and dark gray, save for the foam jetting off the peaks. I struggle to look bored.

Ah, but then we make the turn east to cross the sound. The wind howls, and I recall the waves of a few minutes ago with nostalgia. The *Sharon Kay III* becomes a bad carnival ride, and unwelcome thoughts intrude: The boat's forward movement is its only defense in this craziness. If the engine quits now, we'll take a wave over the stern and we'll go down, and it'll happen *fast*. How long can one survive in forty-degree water?

I dare another glance aft and resolve not to repeat the mistake. Instead, I study the faces of my fellow passengers. Several appear to be dozing. A couple meet my gaze and smile. A few chat as if they're sitting at a sidewalk café. I'm the only one of us who seems the least bit unnerved.

I peer southward, down the sound. Somewhere out that way, invisible behind the spray, Lonnie Moore is oystering.

AND SO WINTER closes over Tangier Island. Nor'westers sweep in, cold and damp. The annual Christmas cantata brings a crowd to Swain Memorial, and Christmas Eve services fill both churches. New Year's passes with a scattering of fireworks and without public toasts to 2017. One Friday in early January, snow starts to fall. Swirled by high winds, it collects a foot deep on the roads and marshes and drifts waist-high against houses and fences. It is slow to melt and immobilizes golf-cart traffic for days.

Still, the watermen of Tangier take to their boats. "Thank the Lord for the good oystering season our men are having," Carlene says a week after the blizzard, during prayer requests at New Testament. "The work's getting harder, but they've been able to catch their limits."

Oystering closes at February's end, and the cycle of life on the water, a rhythm grown familiar over generations, starts anew. The island's watermen dismantle their dredging rigs, returning their boats to the simpler setups they'll use for crabbing. A good many pull their boats out of the water to scrape down and repaint their hulls.

The hard-potters are back on the water by mid-March, cotton layered thick under their oilskins—Lonnie and Woodpecker and their mates, Ed Charnock and Jason, plus dozens of others. Irene Eskridge and Stuart Parks prepare Fisherman's Corner for the coming hordes. Principal Denny Crockett and his wife, Glenna, do the same at Hilda Crockett's Chesapeake House. Orders are placed for souvenir Tangier ball caps and T-shirts to stock the gift shops. Tour buggy drivers and museum volunteers enjoy the final weeks of quiet.

At his crab shanty, Ooker builds eighty new crab pots, preparing for the peeler season, and gives interviews to a rising tide of visiting journalists fascinated by this disappearing island that voted for a president who's called climate change a hoax.

Jason and Ed Charnock aboard the *Henrietta C.*, May 2016. (ELI CHRISTMAN)

TWENTY-TWO

MANY AN ISLANDER WILL SAY IT WAS RIGHT KYOWKING THAT afternoon, that the water was as rough as they'd seen it. But Monday, April 24, 2017, dawned far less daunting, with winds out of the east-northeast at twenty to twenty-five—conditions that might draw a complaint from Leon in the Situation Room, but nothing the crabbers of Tangier hadn't experienced on too many mornings to count. All knew it was expected to breeze up, that the afternoon spoke of stronger wind. Most planned to pull their last pot well before the seas got high.

So it was that Ed Charnock backed the *Henrietta C.* out of its slip at the north end of Meat Soup at about five that morning, while his son, Jason, readied the bait and bushel baskets out on deck. They motored from the boat channel's wave-scoured mouth and turned southwest, toward hundreds of pots they'd set in six long rows, the nearest about seven miles away.

It was a daunting place to work, surrounded by so much water, such a long way from anywhere. But along with Lonnie Moore and Ooker's son Woodpecker, the Charnocks ranged far from home as a matter of habit: They'd never been much for following the pack, and went where they thought the crabs were. Besides, father and son knew their business. Eddie Jacks had worked the water for fifty-four years. At seventy, he was a captain's captain, an expert boat handler and among the island's most ambitious and successful crabbers. Jason, a fit and strong forty, had been his full-time mate for twenty-one years. And their boat was among the prettiest and most proven around: Among Tangier's deadrises, graceful all, the sublime arc from its forepeak to stern set the *Henrietta C.* apart. It looked longer, lower, and more delicate than it was. Fast and stable, built twenty-eight years before of fir and yellow pine, it might have been Jerry Frank Pruitt's masterpiece.

Mindful that the wind would pick up, the Charnocks planned to pull just three hundred pots—more if they had time, less if the weather turned early. At about six they reached their nearest pots, two rows running north–south near the submerged wreck of a navy target ship—the old battleship *Texas*, renamed the *San Marcos* and pounded into a rusty tangle southwest of the island. The thirty pots they pulled there were disappointing, so they abandoned those rows for the four they'd set another five miles out, at the edge of the channel used by big ships bound for Baltimore. On the way they passed Paul Wheatley in the *Elizabeth Kelly*. He was Ed's son-in-law, married to his oldest daughter, Kelly. Ed's grandson Jonathon was his mate. Everyone waved.

Better luck waited out by the channel, where their pots came up filled with crabs. But as they worked up one row and down the next, the wind notched upward. By midmorning it was blowing twenty-five to thirty, and the seas sprang to four feet. The Charnocks kept pulling pots. A little wind didn't bother them, especially that morning. All the previous week they'd been mired in the tedious business of moving their pots, fine-tuning their placement, and they'd looked forward to this Monday as the payoff.

And so it was: By late morning, they'd fished up only two-thirds of their pots, but the deck was crowded with thirty-six bushels of crabs. Now, with the weather souring by the minute, they decided to head in. They set off to the northeast, directly into the wind and five-foot seas. Ed pushed the boat hard, but it was slow going, especially with something like 1,400 pounds of crab aboard. They were halfway home, just past the *San Marcos*, when at about 12:45 P.M. they noticed the boat felt soft. It wasn't quick to answer the throttle or turns of the wheel, seemed lazy.

They knew what it meant. The *Henrietta C.* was taking on water.

That, by itself, was no cause for alarm. A wooden boat is semi-permeable, and water finds its way in as a matter of course. But the Charnocks knew they'd have to address it. In these heavy seas, with the boat already riding low under the weight of their catch, they could little afford the extra burden of water in the bilge. It settled the boat lower in the chop, forced the diesel to work harder, left them suscep-tible to high waves.

And the waves were high, indeed, and coming at them from all directions. Tangiermen would talk about that afternoon for a long time after, and on one point they'd all agree: As the day progressed, the bay became anarchic, the waves building to six feet high and crashing in from all sides. They'd say they were strange, corduroy seas, with high crests close aside deep troughs—seas that don't rock a boat so much as bludgeon it.

LEON MCMANN HAS TOLD ME that the key to safety at sea is pretty simple: "You have to be sure you're getting rid of more water than you're taking on," he said. "You get into trouble if you start taking on more water than you're getting off." The *Henrietta C.* was fitted with two pieces of equipment to achieve this. The first was a bilge pump, which acts just like a basement sump pump—it's tripped on by the presence of water and runs until the water's gone. Ed Charnock's boat had two of them. One had been busted for a while, but the other was running fine.

The second was an automatic bailer, a short brass pipe, 1.5 inches in diameter, that passes through a boat's bottom. The boat's forward motion creates a vacuum at the pipe's lower end, which sucks water out of the bilge and overboard. The pipe is capped when not in use and is usually reached through a hatch in the deck just aft of the engine box. Jason took the helm while Eddie Jacks opened that hatch and, down on his knees, reached almost shoulder deep into the bilge to uncap the bailer. Most of his arm was in water.

In the cabin, Jason listened to the marine radio, which crackled with watermen complaining about how rough it was. Most were in Tangier or Pocomoke Sound, far less exposed than the *Henrietta C.* to the effects of an east–northeast wind; one was Lonnie Moore, Jason's father-in-law, headed for home with his limit of crabs. Rocking and rolling out here, the chatter went. These are five-footers, for sure. She's a-blowing.

If you think it's rough there, Jason radioed, you ought to see how nasty it is over this way.

Out on deck, Ed had the bailer open, but it wasn't pulling any water. Slow as they were moving into the wind, no vacuum had formed beneath the pipe. The remedy: They'd turn away from the wind, get up some speed, get the bailer working. Once they had the water out, they'd turn back for home.

Rather than surrender all the progress they'd made so far, Jason elected not to run with the wind but to turn solid to it—to head northwest, with the wind hitting them abeam. It seemed to work. They got up some speed. The bailer pulled water. Lonnie, reaching the safety of the channel into Tangier, radioed: Are you okay?

Yes, everything's fine, Jason replied. We got some water in, but no concern. We're going to run solid to it until we got it out.

Except that it didn't happen. The bailer sucked water at a furious rate, and the sole bilge pump drew hard, but the bilge remained flooded—in fact, more water seemed to pile into the space with each minute. And now Jason noticed something he'd never seen before. Water was collecting on the cabin floor. He checked the windows, found no leak, and realized that it was coming up from below.

With that, the bilge pump failed. The bailer was now the only route for the flooding to leave the boat, and the little pipe was clearly no match for the load. Jason responded with an emergency maneuver. He put the boat's tail to the wind and opened the throttle wide. The

Henrietta C. surged westward, running with the storm. So it ran, right up to the moment the engine quit.

In the weird quiet that followed, Ed shouted from the deck: Jason, you better holler for somebody. And get your oilskins off. Get them off right now.

JASON SHOUTED FOR HELP over the VHF channel monitored by Tangiermen. It was near two o'clock, and few other crabbers were still on the water. It took a couple of tries before he raised Billy Brown, an islander who was bucking the storm near Crisfield, sixteen or seventeen miles away. We're in trouble, he told Billy. We're taking on water, a lot of it. We need help. Better get ahold of somebody.

Billy: Who should I call?

Jason: Lonnie. Call Lonnie.

He had no time to say more. The water leaped to his waist. His father, out on deck: Get out the cabin, Jason!

He scanned the cabin for the two life vests they kept on hand, spotted one in a storage locker a few feet away, waded in that direction. But as he reached for it the water jumped again, to chest-high, and he lost his grip on the vest, had to feel under the water to find it. His father's voice again: Get out the cabin! Jason turned for the door. Before he reached it, water filled the cabin to the overhead.

He had to swim for the surface. His father was already there, treading water, thin hair plastered to his balding head. So was the *Henrietta C.* The boat stood straight up on its tail, about four feet of its bow rising from the water, suspended there by air trapped in its forepeak. Jason was dumbfounded by the sight, numbed by the cold, unable to move. His father snapped him out of his stupor. Come on, Ed yelled, and he led the way, swimming to the boat. They crawled onto the cowl beneath the front windows, searching for handholds. Ed latched on to the bowline. Jason wrapped an arm around the stem.

The boat's topside faced east, flat to the wind. Both men wore the standard cool weather attire for Chesapeake watermen: cotton blue jeans, cotton socks, cotton underwear, cotton shirts. Ed wore an additional layer, a cotton sweatshirt. The air temperature hung in the midfifties, the water about sixty. They were soaked, and there was no way to gain cover from gusts that tore at them and doused them with spray, and great wind-driven waves that pounded them against the wooden decking. Hold on, Jason, Ed shouted over the roar. He put an arm around his son, pulled him close.

Lord, it was cold. Jason had never felt so cold. His wet hair, trimmed short, felt like ice against his scalp. His neck and arms and hands ached. His shirt clung sopping to his back, and with each gust he was nearly robbed of breath. His father was feeling it, too. I ain't going to be able to take this, Ed told him. He repeated it several times.

The *Henrietta C.* bobbed with the waves, and as it dropped into troughs a heavy, shuddering thud would run through the boat. It was hitting bottom, Ed said. They were in forty feet of water. The stern was bouncing off the mud and sand on the bay's floor.

Every time it hit, a little air would burp from the boat's nose, and it would settle a few inches lower. The bay covered its front windows now and, with each thump, crept higher up their legs.

BILLY BROWN'S CALL came into the oil dock, where watermen gather after work to chew over crabs, boats, the weather, and wrongheaded government regulators. By the time the phone rang on this afternoon, they'd been and gone—only one crabber was on hand when Sandra Parks, working the counter, received Billy's report that the *Henrietta C.* was in trouble.

That one crabber was Andy Parks, Stuart Parks's husband and one of Ed's closest friends. When Sandra passed on what Billy had told her, Andy was alarmed. When an experienced captain like Ed Charnock calls for help, you know he's already done all he can do for himself. And as good a captain as Eddie Jacks was, the situation had to be very bad indeed.

Andy grabbed the phone and called Ed's daughter Kelly. Listen, your dad's boat is taking on water, he told her. Is Paul home? Yes, she replied. Her husband, Paul Wheatley, the crabber who'd exchanged waves with Ed and Jason earlier in the day, had just come in from his boat.

Tell Paul he has to go back out, Andy said. He's got to go help your dad. Andy's voice was cracking. Kelly had never heard him sound so agitated. She thought he might be crying.

Down in the boat stalls lining Tangier's harbor, Freddie Wheatley, first cousin to Paul, was about to step off his boat, the *Cynthia Lou*, when Jason's call to Billy Brown came over the radio. He heard Jason's plea for help, and could hear Ed's voice in the background. He was shouting, "It'll soon be too late!"

Freddie yelled over to Dean Dise, who was tidying up his own boat in the next slip. Eddie Jacks is going down, he said. Both men started their engines and threw off their lines. Minutes later, Paul Wheatley and Ed's grandson Jonathon arrived at the *Elizabeth Kelly*. They fired it up and struck west through the boat channel. Word reached Tangiermen aboard other boats around the harbor, and they pulled out, too. As the little flotilla left the island's protection, news of the emergency was spreading by landline among the island's homes.

It missed Lonnie Moore. Exhausted after eight hours crabbing in Pocomoke Sound, he'd changed into pajamas and fallen fast asleep on his sofa, the ringer on the house phone turned off. But it reached his daughter, Loni Renee, who was teaching special ed at the school. When Nina Pruitt, Ed's first cousin, pulled her out of class, Loni Renee couldn't imagine the *Henrietta C.* was in any real trouble. The boat took on water all the time, and her husband and father-in-law knew how to deal with it. They'd seen plenty of rough days.

Then she heard that no one had been able to raise the boat by radio, and VHF chatter made clear that the bay was in chaos and the wind still rising, and her calm turned to worry, then to dread, and finally to panic. Unable to reach her father by phone, she sprinted the 250 yards to her parents' house. By the time she burst into the room where Lonnie lay sleeping, she was winded and crying. Dad, she told him, Jason's boat is sinking. You have to go out and look for him.

Lonnie was up off the sofa and throwing on his clothes, firing questions. Where were they? How much water was aboard? His daughter knew nothing, but Lonnie, like Andy Parks, had to assume the worst. He was out the door before Loni Renee caught her breath, running for his dock. Bring Jason home to me, she called after him. Please.

Lonnie had just reached the *Alona Rahab* when Michael Parks, a stout tugboater and the island's volunteer fire chief, yelled across the water from the county dock: You want me to get a pump from the firehouse?

No time for that, Lonnie hollered back.

Do you need help?

Yes, Lonnie said. Come on.

Michael leaped aboard as Lonnie started the boat's diesel, threw off the lines. The *Alona Rahab* tore down the channel, nose high, and ran full throttle past the island's western edge and into the storm.

IT'S AT SUCH MOMENTS that the unique place Tangier Island occupies on the American landscape comes into focus. A treacherous sea had already claimed a storied boat, skippered by one of the most respected watermen on the bay: Eddie Jacks, the last man anyone could imagine in this situation. Yet his neighbors, in their squadron of too-small craft, didn't hesitate to thrust themselves into harm's way. About twenty boats left Tangier that day to find the *Henrietta C.*, most of them with two or three islanders aboard. Fifty or more Tangiermen. A significant share of the island's adult male population.

The Coast Guard dispatched two cutters, a smaller boat, and a rescue helicopter. The navy scrambled a chopper out of its Patuxent River air station. All set out with only guesses as to where the Charnocks might be. Eddie Jacks, like most Tangier crabbers, had not fitted his boat with an emergency position-indicating radio beacon, or EPIRB, a device that automatically sends a distress signal from a sinking boat and pinpoints its location for rescuers. They cost a few hundred dollars. Islanders aren't inclined to spend the money.

Freddie Wheatley and Dean Dise, knowing that Ed had been

fishing pots far to the southwest—and figuring that he was inbound from there when he ran into trouble—made for "the spar," a buoy out toward the *San Marcos* wreck. Several other boats headed that way, too. Other boats ran blindly for points north of the *San Marcos*, informed by little more than instinct.

Lonnie Moore, meanwhile, was replaying his last radio conversation with Jason in his head. He knew the *Henrietta C.* was homebound when it started taking on water. He knew that Jason turned the boat solid to the wind. With the gale coming from the east–northeast, that meant he had tracked northwest. The time between that conversation and Jason's call to Billy Brown had been—what, an hour? Eddie Jacks and Jason could have covered miles in that time.

The *Henrietta C.* was a good five miles to the northwest of the *San Marcos* wreck, Lonnie decided. It was likely up near a red nun buoy about six miles west–southwest of the island. The "corn buoy," watermen call it, because it occupies a spot close to an old marker, long since removed, that warned mariners away from the wreck of a barge that sank with a load of corn aboard.

Lonnie steered the *Alona Rahab* for the corn buoy, and urged the armada of workboats around him to head that way, too.

THE *HENRIETTA C.*'S BOW continued to slip into the water. Ed and Jason rode it down. Look for a helicopter, Ed told his son. That's our only hope. Keep looking for a helicopter.

They saw no helicopter. They saw no boats. They could see little at all, except seas that seemed to defy every norm. The tide was outbound, running with the wind, but waves seemed to be raging without direction, without pattern. They collided in fizzing geysers, parted to open deep canyons, and ambushed the men from all sides. With every passing minute, their soaked clothes leached more of their heat and strength.

You better get right with the Lord, Ed told Jason. Call unto the Lord, because he's the only one who can get us out of this situation.

A while later, as Ed looked to the east, he said: I don't guess nobody's a-coming.

A little after that he said: I didn't think I would go like this.

Then: I'm scared.

Jason realized his father was crying. He was too stupefied to do so himself. A blank. None of this seemed real. And the cold was so intense that it erased any but the simplest, lizard-brain thought. The bow was almost submerged, the water up to their chests, and Jason released his hold on the stem long enough to slip the life vest off his arm. He handed it to Ed. A wave ripped it from his father's grasp and carried it off.

Forty-five minutes after the men took hold of the bow, the last few inches of the *Henrietta C.* slipped under. In the water now, untethered in the chop, they were almost immediately torn in different directions. Eighty feet of water opened between them. Across the divide, Jason could see Ed staring at him.

An hour after Jason's call to Billy Brown, the *Alona Rahab* and several other Tangier boats converged on the corn buoy. Pressing farther to the northwest, they came upon a debris field floating on the water: the big plywood lid from a workboat's engine box; empty bushel baskets, others still full of crabs; and a tote labeled with Ed Charnock's name.

So they weren't searching for a boat, Lonnie realized, but for two men in the water. As the searchers split up, he began to appreciate just how high the seas were running. Five to six feet now, and there was no sense to these waves. They manhandled the boat.

Lonnie saw that wind and tide were carrying the debris to the southwest, and fast. He pressed the "Man Overboard" button on his GPS console. It would enable him to cover the surrounding water in loops, always returning to this spot to start again.

He knew, from his own close call in 1991, that a man treading water doesn't drift as fast as flotsam. That meant that Ed and Jason were likely upwind of the debris. He turned to the northeast. Out on deck, Michael Parks climbed onto the engine box. The *Alona Rahab* was tossing like a cork, but the big firefighter somehow kept his feet while he scanned the surrounding water.

Lonnie ran up the debris field until it ended, then circled back to the point he'd marked on his GPS. Another Tangier waterman was talking by radio to the crew of the Coast Guard chopper. Get him to find the debris field and fly against its flow, Lonnie told the waterman. Ed and Jason, they're going to be lagging behind the debris.

Lonnie made another looping venture to the northeast. Light and lively as his boat was, the deck swerved, dipped, and tilted under his feet. The storm was strengthening. Tangier boats had fanned out to the north and south of him, and he could see they were getting tossed. Lonnie reckoned it was blowing thirty-five, gusting to forty.

The Coast Guard chopper roared low overhead, circling. Flew over again. Now Carol's voice came over the radio. "Lonnie, do you see anything?" Like many Tangier families, the Moores kept a marine VHF in their house.

"Nothing yet, Carol," he replied.

"Do you have hope?"

He did not answer. As the chopper made another pass, a terrible thought began to grow in his mind. He might have to tell his daughter that he couldn't find her husband, that his own grandchildren had lost their father. He did not share this worry with the other boats. To the contrary, as he made another circle in the *Alona Rahab*, and another, always returning to that spot he'd marked on his GPS, Lonnie tried to encourage their crews. I know they can still be alive, he said into his radio. I know because I've been there.

Still, an hour and a half had now passed since Jason's call to Billy. Time was getting short. Carol called again. "Anything yet, Lonnie?"

"Nothing yet, Carol."

He made another circle, and suddenly Michael shouted, "I see something!" From his unsteady perch he was peering over the cabin to the water ahead of the boat. "It's right smack over your bow!" Lonnie turned the wheel and slid open his window as Michael yelled, "I think it's Jason!"

Dead ahead, the waves had parted to reveal a man neck-deep in the water. He'd stripped off a pale red T-shirt and was waving it over his head. Michael heaved a life ring to him and pulled him to the *Alona Rahab*. He came up ghostly pale, almost blue.

"Which way is your dad?" Lonnie asked him, looking around.

Jason was trembling, unable to walk, barely coherent. "I don't think Dad made it," he said.

FROM JASON CHARNOCK'S STATEMENT to the U.S. Coast Guard:

> Dad kept on floating away staring at me. I was looking for a helicopter to come. I kept looking, and then I looked back to see where my dad was, he wasn't there and must have went under.
>
> After that I felt like I was there for about an hour. I had to gain my strength. Eventually, I had to relax to conserve my energy. I kept looking east to the Coast Guard. Eventually a helicopter flew right over top of me. There was a boat coming from the east to the west, but he passed right by me. Once it passed me, I thought they would mark off this location [as already searched]. I kept fighting.
>
> I saw another boat, and remember it started to rain on me again. I saw another boat from the west, who was Lonnie. I took off my shirt and waved it.

Two nights later, when John Flood opens Swain Memorial's Wednesday evening service, Ed is still lost at sea. "Dear Heavenly Father," the pastor prays, "as we gather together as your people this evening, we are hurting. And, Father, we're trying to comprehend everything that has happened this week."

Nancy Creedle is off the island, so the congregation sings "Amazing Grace," all five verses of it, a cappella, lending a humble, beseeching air to the proceedings. On hearing of the tragedy, Pastor Flood tells the assembled, his first reaction was to give thanks that Ed had been saved just the previous month. "It was March the first," he says. "He hadn't been saved that long, but he was saved, just as if he had been saved when he was twenty years old."

He peers at his flock. "You have to wonder, where is God in all this? What *happened* here?" he asks. "I found out that there were two blessings right immediately. That Ed—in that crisis time, in that stormy sea, hanging on—he told his son: 'This don't look good. You have to get right with the Lord. You need to accept him as your savior right now.' And then Loni Renee told me that she made a deal. She said, 'God, if you will only bring Jason home safe, I'll serve you the rest of my life.' And you know, I had to think how that would have put a smile on Ed's face. You know that had to be his heart's desire, to see his son and his daughter-in-law saved.

"And I believe that out there, as rough as those seas were Monday . . . I still believe that in all of that, that God was there."

With that, the pastor has outlined how Tangier will process the tragedy. Ed Charnock had long been a churchgoer—he accompanied Annette to Swain every Sunday, wearing his best. Islanders regarded him as honest and upright, and kindly and good-humored. But that he was *saved* mitigates the pain of his loss. And that the events surrounding his death have led two other Tangiermen to salvation is a cause for rejoicing. "God is here with us," Pastor Flood tells his flock. "And I believe that God's glory is already being shown in Jason and Loni Renee."

Members of the congregation share their memories of Ed. Hoot Pruitt says Ed was "always a joy to be around." Marlene McCready calls him "a good man." Jack Thorne says Ed was honest, always honest, and that he didn't die in vain. "I thought, 'Oh Lord, how come Ed, he's drownded?'" Jack says, then answers his own question: "If he hadn't got drownded, maybe Jason and Loni would have got saved. But *next day*?"

Pastor Flood nods. "There are a lot of opportunities to see God's glory in this," he says. "Satan didn't take Ed. Jesus came and got him."

The entire congregation stands and, holding hands, forms a circle. It encloses most of the sanctuary. Marlene leads a prayer. "Please, Lord, watch over our watermen, our men who make their living on this great big bay," she says, "and protect them.

"And, Lord, if it is your will, please bring Ed's body back to his family."

THE NEXT DAY I STOP IN at the Charnock home to visit with Annette. I'm one of a procession of well-wishers who crowd into her living room, joining her sons Brock and Travis, who've hurried to the island from their homes in western Virginia and the North Carolina coast. Annette greets me with a pillowy hug. She's distracted with shock and exhausted after three sleepless nights, but somehow she maintains her usual chatty warmth. The house is filled with food, she tells me. Eat something.

Jason is present as well, standing near the door and embracing his neighbors as they enter. "I ain't a huggy person," he says, "but I've hugged a lot of people, and I ain't minded it."

Stuart Parks steps in and wraps her arms around him. "Hey, Jace," she whispers.

"Yeah," he murmurs. "Got my second chance."

Fred Pruitt, Eugenia's husband, comes through the door. He clutches Jason for a long moment, patting his back, and tells him that both churches devoted their services to his father last night. "Ed was the only subject," another islander agrees, adding that Carlene led the New Testament service and "did a wonderful job."

Jason is as laconic as his father, and the exchange visibly taxes him. "She was always my favorite teacher," he manages. Fred asks how he's holding up. "I've been spending as much time as I can away from home," he replies, retreating to a sofa. "The morning is the hardest, when I first wake up and I'll start thinking."

The conversation turns to his radio call for help. What are the

odds, someone wonders, that a VHF call could cover so much distance, in conditions so foul? Jason says he didn't realize where Billy
Brown was when he made the call. "I hollered to him one time and
didn't get anything back," he tells us. "I had no time to call for anybody else. If he hadn't answered, then I'd have been drownded. Nobody would have known where we were, and we were always one of
the last ones to come in anyway. They wouldn't know we were missing, even."

Annette is in constant motion, checking with her guests to ensure
that their needs are met, soothing their grief as much as they're attending to hers. I'm talking with her sons when, about an hour after
my arrival, Ooker appears at the door with a friend, a private pilot
from Suffolk, Virginia, named Kenny Carpenter. The mayor tells Annette they're going up in Kenny's Cessna 172 to look for Ed. There's
room for a second spotter on the plane, and I volunteer to join them.
Thirty minutes later, after unloading a pile of stale hamburger buns
that Kenny's brought Ooker to use as bird food, we take off and circle
to gain altitude.

A thousand feet up in Kenny's airplane, the view is revelatory.
Tangier appears so tiny, so fragile, so water-rasped and ponded, so
utterly insubstantial, that even after a year of studying the island's
vulnerabilities I'm surprised by how conspicuously it is losing its fight
with the sea. The ridges look even more slender and defenseless than
they do on the ground. The south end of Tangier proper, where the
Big Gut empties, is dissolving. Uppards is a loose macramé of marshy
strands enclosing great tidal pools, and the breach most emphatically
splits the island in two.

Minutes later we pass over a yellow can buoy, and Ooker announces we're getting close. Kenny extends the flaps, and slowing,
we drop to four hundred feet. The water is furrowed by a light breeze
and dotted with crab pot markers. "Can see a lot of buoys out here,"
Kenny says. "If you can see buoys, we should be able to see a body."

"Oh, yeah," Ooker replies, and as he says it a red nun buoy
appears—the corn buoy. "Okay, that's it," he says. "Jason said that

after the boat went down they drifted to the southwest, so over here."
He points off to our left, and as we bank that way, Ooker and I com-
mence our duty as lookouts, he peering off to starboard, I to port. It's
a lovely, sunny spring day, and the bay glitters benignly. "On a day
like today," Ooker says, "it's hard to imagine a boat going down."

In a widening circle we fly, over dark green water that seems in-
calculably vast. Perspective is elusive: I'm not sure how large a body
would be from this height or whether I'd be able to make out the
colors of Ed's clothes, so I look for anything: any anomaly in the
surface, any odd breaker, any discordant patch of bright or dark. We
search, with few words exchanged, for forty minutes, seeing nothing,
until Kenny suggests that we fly to the mainland to get fuel. Ooker
notes that one piece of wreckage—the lid to the *Henrietta C.*'s engine
box—washed up yesterday at Milford Haven, on the western shore,
which is more or less on the way to the airfield, so we head that way,
still flying low and slow, still scanning the water below.

Sandbars lurk beneath the surface, coloring the bay a bright, almost Caribbean teal. Buoys of red and blaze orange and electric blue
dot the shallows, and I can see gulls riding the waves. If we fly over
Ed, it seems we'll be able to see him. But our route past Milford
Haven yields nothing, and we press southward to the Rappahannock
River and fly into its mouth, and over the oyster rock Lonnie dredged
five months ago, to land.

THE FOLLOWING SUNDAY MORNING, twenty-six islanders gather in
Swain Memorial's South Room for the weekly class meeting. "They're
supposed to go out again today to look for the body," Hoot Pruitt announces from the lectern, "so let's pray."

"We know, oh Lord, that his soul's already in Heaven with you,"
Marlene says, "and with his mom and his family and everyone he
knew when he was a little boy. And we're thankful for that. We ask,
Lord, that all the Christians on the island pray hard that the body
be found."

"Loni Renee came to me and said, 'Regardless of what happens,

I'm going to trust in Jesus for the rest of my life,'" says Carol Moore's mother, Grace. She wipes her eyes. "I prayed for my grandchildren for so long of a night. I'd think, 'Neither of them are saved.' And now Loni Renee is."

"That body is going to be found," Marlene says. "Grandma Sadie lost a little boy in the ditch, out in the marsh there. It was in the cold wintertime, and his body drifted out. But she prayed, and that body came right back into that ditch."

An hour later I'm aboard Mark Haynie's *Sharon Kay III* with Richard Pruitt and crabber Allen Parks, headed into the bay to join a flotilla of boats dragging for Ed's body. It is another lovely, near-cloudless day, with a light and steady breeze out of the south. "Very different from when we were all out here Monday, looking," Allen tells me. "Now, *that* felt like a big body of water."

The *Henrietta C.* lies on the bottom thirty-eight feet down, so Allen and Richard fashion two especially long lines for their drags. Allen's line is tied to a steel bar from which dangle seven large and evil-looking treble hooks. Richard's is studded with giant bent nails. Allen tosses his drag over the starboard stern, and Richard, his over the port. They play out the lines, which are visible fifteen feet down through the crystalline water, and we begin a three-knot search of the bottom west of the sunken *Henrietta C.*

Thirty minutes later they pull the drags back up. Richard's has snagged nothing. Allen's treble hooks have collected bundles of algae and clusters of sea grapes, which he rips away before carefully throwing the device back into the water. A Coast Guard boat arrives, its deck crowded with crew in blaze-orange life jackets. Its skipper radios that we're to stay two hundred yards clear, as he plans to anchor to the wreck and send down a dive team.

An hour passes. The drags come back aboard. Again, nothing but eelgrass and red moss and sea grapes. We eat lunches we've packed and sweat under the hot afternoon sun. Allen, one of Tangier's younger watermen, talks about Ed. "Pot for pot, he was the best fisherman on Tangier," he says. "When Ed shined the best in his waterman's career was when crabs and oysters were scarcest. That's when he caught a boatload. He could *think* like a crab or oyster."

TEN DAYS AFTER THE SINKING, a Smith Island waterman comes upon Ed's body floating in the bay not far from the wrecked *Henrietta C.* The body passes through the hands of the medical examiner, then is prepared for burial by a mainland funeral home. He finally comes home on the mailboat. I watch a crew from the volunteer fire department load his casket into Tangier's ambulance and carry it over to Swain Memorial.

Later that afternoon, a few hours before the visitation, the regulars convene at the Situation Room. We sip our coffee through a lengthy silence. Leon breaks it. "Ol' Ed's gonna be missed."

"Yes, he will," Jerry Frank murmurs.

"If there was a group of watermen and Ed was in the group, he'd be telling stories," Ooker says.

Jerry Frank: "I'll tell you something else about him. There weren't no more honest a man in all the world."

"That he was," Richard Pruitt agrees, nodding.

We sit in silence for fifteen seconds. Again, Leon kick-starts the conversation: "Seems like he was all fixed up to die."

"Yeah, it seems like everything was working so that Ed wouldn't come back," Ooker says. "Couldn't get to the life vests. Ran away from the island to get water out of the boat, which made him harder to find."

"Well," Leon says, "I mean being saved and all."

Everyone nods. Talk turns to the *Henrietta C.* A reporter from Norfolk, here to write a story about Ed, asks Jerry Frank whether there was anything unusual about the boat. Jerry Frank, who has spent seventy years living "within three or four hundred feet" of Ed, shakes his head. "When I first heard about it I went over in my mind, wondering if I did anything different on that boat," he says. "But no, I did it exactly the same as I did the others." Then, without waiting for a follow-up question, he offers up a piece of backstory that's been the subject of rumor for three weeks, but on which he is one of the few islanders who can speak with authority.

"About four year ago," Jerry Frank says, "he had the boat to the boatyard and up out of the water, and they found a place in the bilge, up under the cabin, where the wood was eaten with worms—not the kind you get overboard, but the kind that get in trees, that kill trees."

"A dry wood worm," Jerry Frank would call it later. Park your boat or a piece of onboard gear under the trees, and the worms would drop in. They ate a boat from the inside out. He suspected they came from trees over Reedville way, on the bay's western shore, where Tangiermen had been known to store equipment.

"I've taken them out of quite a few boats around here," he said. "They're hard to see. That hole's so small you'd better look very close. The worm gets bigger as he eats. When he first goes in, he's a little thing." The boatyard might not have noticed the infestation, except that the affected boards wouldn't hold paint, he said. So it scraped out the damage it found, filled and painted it, and told Ed he'd need to swap out that wood.

Point being, Ed had known for years that those boards were bad.

He'd loved that boat. He'd been a stickler for maintenance. He'd replaced several boards since, at the starboard stern where he fished up his pots, a high-wear region of the hull. The previous November he'd had the *Henrietta C.* in a Smith Island boatyard, getting the hull sanded and filled and painted, zinc applied to its rudder and prop shaft, a new seat installed in the cabin.

But he had not gotten around to replacing that weakened section of the bottom. The work would have kept the boat out of the water, would have kept them from working. And that, he knew, would be a hardship on Jason—who, as Ed had done, was raising four children.

That Monday, they'd likely cracked one of those worm-tunneled boards early in the run home. And after one went, so did another, and another, like a zipper pulled, until the *Henrietta C.* was no more.

The *Joyce Marie II* leaves Tangier harbor. (EARL SWIFT)

TWENTY-THREE

————

MAY 18, 2017. A YEAR TO THE DAY AFTER MY ARRIVAL ON Tangier, I stand on the mailboat dock in the predawn dark, looking out over the harbor and waiting for Ooker. A light breeze out of the south riffles the water as black sky brightens to deep purple, then—in the course of ten minutes—blooms red, orange, and gold, and the sun, not yet risen, filigrees a scattering of dark clouds off to the east. It's against this riot of color that I make out the *Sreedevi* backing into the channel from the mayor's crab shanty at 5:30. It putters slowly across the dark water and halts alongside the dock. "Hey, Ooker," I say.

"Hey, Oral," he replies.

I jump aboard. We slide three boxes of soft crabs onto the mailboat's aft cargo deck. I ask whether they're bound for New York. Ooker grunts and shakes his head as we pull away from the bigger vessel. "They're all going to Crisfield now," he says. "The price ain't good anywhere, and it isn't worth shipping them." Jumbos are fetching just fourteen dollars a dozen—a third of what he's seen them reach in the past. Mediums are a paltry four bucks a dozen. There have been times when a single whale brought that much.

The market's flooded, he explains. A big shedding house reportedly spent $500,000 buying up doublers a week or two back, so now restaurants and stores all along the Eastern Seaboard have all the soft crabs they need. It'll likely be June before prices rebound. We cruise at six knots through the harbor's shanties. Workboats are tied up alongside a good many of them. "A lot of 'em are laying in, because of the funeral," Ooker says. "I won't stay out too long myself. Check a hundred pots or so."

The tall pines on the P'int loom black against the sky. We cruise past the CBF's darkened campus, 150 yards off to starboard, and round the island's northeast tip. The scent of honeysuckle wafts strong and sweet over the water.

The first pot holds nine doublers, the second, two. The third brings up just a single lemon. In the fourth: ten doublers. "There's no doubler run down here anymore to speak of, but what few you do see are right here around Port Isobel," Ooker says. "Used to be you could put pots here, when the doubler run was happening, all up thisaway, and you'd catch six or seven bushels." Those days appear to be fading deeper into the past, and this year, lemons seem to have dropped off, too. "I don't know what changed or why they changed their habits," he says. "People always talk about how high the tides have gotten. There are abnormally low tides, too. Things are messed up—crab habits have changed and the tides are all out of whack."

The sun pulls free of the horizon. From the water off to starboard juts a thin metal pole, to which is attached a rumpled flag, its orange fabric faded to dull pink. It marks a hazard to navigation: Years ago, Randy Klinefelter built a bulkhead there, hard against the shoreline, in an attempt to halt Port Isobel's shrinkage. It didn't work. The bulkhead survives, but it's underwater and well out to sea.

We finish the pots off the P'int and motor south to four long rows Ooker has set off the east side of the spit. Watts Island, four miles to our east, is silhouetted against the low sun, and as often happens when I have a good view of it, I find myself struggling to comprehend its appearance. On my 1994 kayak trip around the bay, I paddled out of Onancock Creek one morning and sighted Watts miles away across Pocomoke Sound, and on a whim decided to visit. It was a dumb move. A west wind rose when I was halfway across, and I had to push hard for three hours before I was able to land the kayak on a sandy beach lining the island's east side.

Watts was about a mile long, a slender crescent of sand and upland forest. The woods rose in two dense thickets tangled with poison ivy, so I kept to the perimeter. I recall feeling both exhilarated and marooned as I walked the beach. I could find no sign that anyone had stepped on this shore in years, stumbled on no relics of its long human occupation. And I did not see, hidden in the trees, a big electrical junction box: In 1977, Tangier had retired the diesel generators installed by Henry Jander in favor of an underwater cable from the mainland, and here the lines from island and Eastern Shore met. My only company was birds—hundreds of them, high in the trees, shrieking as I passed.

Tangier looked close enough to touch. I gazed on Swain's steeple and the water tower during my hour's walk and watched a scattering of deadrises plying the same water Ooker and I are working today. I was sorely tempted to keep paddling to reach the town. But the breeze was still coming strong and steady. I thought it better to turn back.

Now, from the *Sreedevi*, I can see that Watts has shrunk to a third of its size of twenty-three years ago. One of its forests is gone without a trace, along with the ground it stood on, and the survivor is falling, several trees a month, into Tangier and Pocomoke sounds. Nothing remains of the beach I walked. And the junction box hasn't moved, though the ground beneath it has; once well inland, it now rests on a platform out in deepening water off the island's south end.

Up one row of thirty pots, down a second. A few big jimmies, a great many lemons, a handful of doublers. Ooker baits each pot with a number two, first breaking its claws with a sideways tug. "It seems to attract more females," he explains, "and this way the jimmies can't hurt any soft crabs in the pot."

An osprey flies out to the boat and hovers, flapping loudly, just a few feet overhead. Ooker plucks a menhaden from his bait box and offers it with an outstretched arm. The osprey eases closer, until it's just three feet from the meal, then two—then spooks and wheels away. Ooker heaves the fish into the air after him. "I ain't ever had an osprey take one out of my hand like a gull, but he's come close," he says. "I'm going to try to get him to do it this summer."

It's a pleasant beginning to the day, which is otherwise blue. A few hours later, I'm at the packed funeral at Swain Memorial for Ed Charnock. I sit in a rear pew, listening to a eulogy from Twin Denny Crockett, who lost his grandfather Donnie to the island's last accident at sea, and who is married to Ed's daughter Danielle. Hoot Pruitt sings a solo. Duane Crockett tells us that God was with the *Henrietta C.* when it went down: "He was out there to save Jason's life, and he was there to save Jason's soul, and he was there to take Ed home." John Flood wraps up the service by observing that "Ed has received a good and full measure."

WHEN I RETURN to Tangier in early June, the island is enjoying a burst of press attention, thanks to a most unusual sook that Ooker's son Woodpecker found in his pot: a crab with two oysters that had attached themselves to its face in almost perfect symmetry. A photograph of Ooker holding the creature has just appeared in the *Washington Post* and has been picked up by newspapers around the world.

But aside from that little thrill, the island's crabbers have little to cheer about. Peelers have all but vanished in the shallows around Tangier. The underwater grasses that typically grow so thick in the island's lee have been slow to show up, and crabs about to molt have sought safe territory elsewhere. Some of the potters and all the scrapers have little to show for their efforts of the past three weeks—Leon complains of catching no more than a dozen peelers total over three days on the water and has even considered parking his boat. "I know Dad's a scraper," his daughter Carlene says during prayer requests at New Testament, "but there are a lot of other scrapers, and there ain't no crabs, and they're hurting.

"We will praise him," she says. "The Lord knows we have need of crabs, and he'll provide."

Leon's woes are compounded the next day, when the *Betty Jane II*'s hydraulic steering fails while he's out on the water, leaving him adrift until another boat tows him in. "I went out," he tells me later in the Situation Room, "but I had a lot of trouble."

"Like crapping your pants," says Ooker, pouring himself a coffee. "Ain't that right, Leon?"

Leon nods, unfazed. "That's one thing you can't do when you get some years on you. You can't say, 'It's okay—I'll wait awhile,'" he tells Jerry Frank, Cook, Bruce, and me. "Everything down there, you don't have the strength you once had."

"The check valves don't work as good," Bruce suggests.

Leon nods. "That's right."

We hear the outside door open, movement in the hall, and into the Situation Room step four people from CNN: two videographers laden with enormous cameras and tripods, a producer, and one of the network's on-air meteorologists, Jennifer Gray. Ooker has been talking to the network for weeks about this visit and invited the crew to join us here. What's about to happen will propel Tangier into the news to such a degree that everyone will soon forget about the weird crab.

Jennifer Gray, several months pregnant, surveys the room. "It's cool," she says, taking in the cracked tile, the tacked-up pictures, the fruit flies. "I like it."

"So do we," Cook tells her.

"The more I see of the outside world," Jerry Frank says, "the more I like it here."

"I know, right?" Gray says. "Your own little place." She looks to Ooker. "So, is this like the city council?"

"This probably has more authority than the city council," the mayor replies.

The producer has everyone switch seats, so that Bruce, Leon, Jerry Frank, and Ooker are lined up beneath the big publicity still of the *Harry S. Truman*. Cook, who says he doesn't want to participate, takes a seat next to the coffee maker and behind one of the cameras. I sit off-camera about two feet from Ooker.

The cameramen fit their subjects with lavalier mikes—first Ooker and Bruce, who wear T-shirts and need only drop the cords inside them, then Jerry Frank, who wears a button-up shirt tucked into his pants. A cameraman unfastens the buttons one by one, threading the mike cord under his shirt. Jerry Frank is visibly uncomfortable. Leon, watching from the next chair, looks horrified. Finished with Jerry Frank, the cameraman approaches Leon, who shakes his head. "I don't need that," he says. "I'll yell into his." He leans toward Bruce to demonstrate.

Gray takes a seat before the four and, with the cameras rolling, asks whether they've noticed the island shrinking. "Yes, I have, and they're letting us go," Jerry Frank says. "It could have been prevented years ago." He pauses before adding, "It's never too late."

Gray asks whether they've noticed signs of the bay rising. Jerry Frank again speaks for the group. "I've been living here for seventy-two years, and it hasn't changed much at all," he says. "This here argument about sea level rising, I don't go along with it at all, because your sea level depends on the way the wind is blowing." He launches into an explanation of how the wind affects tides and water levels.

Ooker jumps in. "What we focus on is erosion," he says. "Sea-level rise is at such a slow pace that erosion will get us long before sea-level rise does. . . ."

"The island's problem, the erosion problem, it's on your mind when you wake up in the morning," he continues, "and it's on your mind when you go to bed at night. It's something we think about all the time. It's always in the back of your mind."

"Always," Jerry Frank says.

"It causes people to hold back from investing here," Ooker says. "You find yourself thinking, 'Should I invest here, when it might not even be here?'"

Gray: What makes the place special?

Jerry Frank: "The history alone is worth saving."

Bruce: "That it is."

Ooker: "Lots of history here. Lot of military service. And now we need help from Washington."

Gray shifts the focus to Donald Trump. "If you could say anything to him or his administration today, what would you say?" Gray asks.

"I would say . . ." Ooker begins. He pauses for an instant, and Leon, who has been silent until now, blurts out: "Build us a wall!" No microphone necessary.

"Yeah, build us a wall," Ooker says, as Cook chuckles off-camera. "They talk about a wall—*we'll* take a wall. We'd like to have a wall all the way around Tangier."

The subject turns to the island's affinity for the president. Ooker says he likes the man's willingness to slice through red tape: "He's gonna cut down on the time it takes to study something. We've been studied to death. We just need something done."

He does not stop there. "I love Trump," Ooker declares without the hint of a smile, "as much as any family member I got."

Even after fourteen months of exposure to Ooker's strident political views, I'm goggle-eyed. Two thoughts immediately cross my mind. First: There's no way CNN isn't using *that*. And second: Trump is bound to take notice.

Gray wraps up the interview by asking again what her panel would like to say to the president. "Donald Trump, if you see this, I mean, anything you can do—we'd welcome any help you can give us," Ooker says. "If Donald Trump would come out here, I'll take him out crabbing and take him out for a good crab cake dinner."

THE SEGMENT AIRS four days later, on Friday, June 9. Gray's report juxtaposes the island's steady inundation, its support for Donald Trump, and Trump's characterization of climate change as a hoax—and includes a brief interview with me toward the end. In the first minutes after the story runs, several Tangiermen post Facebook comments lauding the piece and congratulating Ooker for his comments.

But almost immediately CNN's Twitter account blows up with comments from viewers astounded that the island voted overwhelmingly for a man who derides the science behind sea-level rise. "I have NO sympathy for the people of Tangier Island," one writes. "If they voted 87% for that IDIOT, they are getting what they ASKED FOR! GOOD LUCK."

"Dear Tangier Island, Va: Be swallowed by the sea," another reads. "You're all #Trump supporters and deserve what Nature gives you: submersion."

So they go, screen after screen: "What do you call a sinking island of Trump supporters? A good start." And: "Hard to be empathetic for residents, who are objectively stupid and proud of it." And: "Hope they know how to swim." And this: "Tangier Island in MD?? Seriously. It's a test tube for inbreeding."

The islanders I speak with, and those whose Facebook accounts I follow, are dumbstruck by the ferocity of this invective from strangers, and confused—heartsick—disgusted—alarmed, even—that their support for Donald Trump has moved fellow Americans to wish them dead. Barb Baechtel, who moved to Tangier with her husband, Rob, in 2013, was furious when I spoke with her in the midst of this storm. "They have been asking for twenty years or more for help with a natural disaster," she said of her neighbors. "It's a slow-moving natural disaster, but it's a natural disaster. If it was a fast-moving natural disaster, like a wildfire, everyone in the country would be saying, 'Oh, those poor people! They need our help!'

"We've actually got people sitting around debating whether *these* people are worth saving. How is that okay?" she said. "I don't care if you want to call it erosion or sea-level rise or Aunt Sadie's butt boil. It doesn't matter what's causing it. The point is that this disaster is happening, and these people need help."

WE SOON LEARN that fallout from the CNN report is just getting started. Three days after the story airs, Ooker is out potting when he sees Woodpecker racing across the water toward him. His son pulls his boat alongside the *Sreedevi* and says, Dad, you have to come in. The president's trying to call you.

The president of what? Ooker asks.

It takes his son a minute to convince him that the White House has been calling the oil dock. Ooker motors into port, waits for a call, decides to buster up his peelers—his crabs won't wait—and when he again comes ashore, the phone rings. A woman tells him the president wants to know whether he can speak with the mayor. Yes, Ooker replies. He sure can.

Donald Trump comes on the line, introduces himself, and, as Ooker later recounts it, tells him: You've got one heck of an island there.

This is Trump Island, Ooker replies. We really love you down here. He tells Trump that he believes the president stands for the workingman, and that he wants to put people back to work. That he's for the military, and Israel, and protecting religious liberties. That he's the right man at the right time for the country.

Trump tells Ooker he appreciates that. He comments that Tangier looks like a beautiful place, and that if Ooker's ever in Washington, he should stop in for a visit. He calls Ooker his kind of guy. He says that he and his family love the citizens of Tangier.

And he provides fodder for headlines around the world. As Ooker describes it to a reporter, "He said not to worry about sea-level rise. He said, 'Your island has been there for hundreds of years, and I believe your island will be there for hundreds more.'"

Many islanders are reassured by this.

The world at large finds it preposterous.

Foreign reporters descend on Tangier—I encounter Norwegians, Russians, French, and Brits on the roads. Stephen Colbert makes the island the centerpiece of his *Late Show* monologue. "Now in the unlikely event that Donald Trump's words didn't calm residents of the soon-to-be-lost city of Tangier," Colbert says, "their mayor believes there is a solution to coastal erosion. They need a jetty, or perhaps even a seawall around the entire island, and . . . Trump will cut through red tape and get them that wall.

"Yes!" he says over laughs and jeers. "Trump is going to get them that wall and then make the ocean pay for it!"

This does not go down easy on the island. But it's forgotten soon enough, because Tangier is again propelled into the news: The mayor is invited to appear at a televised "town hall" moderated by CNN anchorman Anderson Cooper and featuring former vice president Al Gore, whose efforts to sound the alarm on global warming earned him a Nobel Peace Prize in 2007.

Cooper introduces Ooker about a third of the way into the hour-long program. The camera closes in. The mayor is wearing his Sunday clothes: a fresh-looking Tangier souvenir ball cap and a button-down plaid shirt. "Vice President Gore, Mr. Cooper, I'm a commercial crabber, and I've been working the Chesapeake Bay for fifty-plus years," he says into a handheld microphone. "And I have a crab house business out on the water, and the water level is the same as it was when the place was built in 1970.

"I'm not a scientist, but I'm a keen observer, and if sea-level rise is occurring, why am I not seeing signs of it?" Those familiar with Ooker's usual easygoing manner can detect that he's nervous. But he has great presence. He comes across as unpretentious and intelligent, and the slight tremor in his voice lends him an endearing humility. "Our island is disappearing, but it's because of erosion and not sea-level rise," he says. "Unless we get a seawall, we will lose our island. But back to the question: Why am I not seeing signs of the sea-level rise?"

Sounding supremely confident, the former vice president asks, "What do you think the erosion is due to, Mayor?"

"Wave action," Ooker replies. "Storms."

Gore: "Has that increased any?"

Ooker: "Um, not really—"

Gore: "So you're losing the island even though the waves haven't increased."

"Yes," Ooker tells him. "This erosion's been going on since Captain John Smith discovered the island and named it. It's got to our doorstep now, and we focus on it more."

Gore nods. "Won't necessarily do you any good for me to tell you that the scientists do say that the sea level *is* rising in the Chesapeake Bay. . . . And that the forecast for the future is another two feet of— what, if there was another two feet of sea-level rise, what would that mean for Tangier Island?"

"Tangier Island, our elevation is only about four foot above sea level," Ooker says. Which is true of only the high spots.

"Yeah," Gore says.

"And if I see sea-level rise occurring, I'll shout it from the house-top."

Gore, nodding: "Okay."

"I mean, we don't have, you know, the land to give up," Ooker concludes. "But I'm just not seeing it."

"Yeah. Okay," Gore says. He holds a finger to his chin, seemingly deep in thought. "Well, one of the challenges of this issue is taking what the scientists say and translating it into terms that are believable to people, where they can see the consequences in their own lives. And I get that, and I try every day to figure out ways to do that."

At this point, Gore might have explained that day-to-day eye-balling, even over decades, provides an unreliable perspective on in-cremental change. That by contrast, hard scientific evidence leaves no doubt that the water is climbing. That just because Ooker has not noticed the water coming up doesn't mean it isn't happening.

Instead, he takes a tack that shows he understands something about Tangier culture but addresses Ooker's question only obliquely. "Reminds me a little bit of a story from Tennessee," he says, turning to Cooper, "about a guy who was trapped in a flood. He was sitting on the front porch and they came by in an SUV to rescue him, and he said, 'Nope. The Lord will provide.' And the water kept on rising. He went up to the second floor, and they came by the window in a boat, and they said, 'Come on, we're here to rescue you.' And he said, 'Nope. The Lord will provide.' Then he went on up to the rooftop as the water kept rising, and they came over in a helicopter and dropped a rope ladder. He said, 'Nope. The Lord will provide.'

"Well, he died in the water," Gore says, "and went to Heaven and said, 'God, I thought you were going to provide.' And he said, 'What do you mean? I sent you an SUV, a boat, and a helicopter.'"

So ends the exchange. The consensus on Tangier is that Ooker has "won" the face-off, and I, watching at home on the mainland, draw the same conclusion. The world's preeminent spokesman for the perils of climate change has failed to answer a question compelling in its simplicity, asked by a plainspoken crabber with a high school education. In so doing, he has enabled anecdotal folk wisdom to persevere as a competitor to bona fide science. Mighty unfortunate, this.

Islanders are also angered by Gore's joke, which they believe mocked their religious faith. I don't read it that way—he was making a point, even if it was lost on his intended audience. The Lord has provided the islanders with minds for recognizing the danger that faces them. That might be the sum of what the Lord plans to provide them with, this time around. Denying that the danger exists—or expecting a miracle to chase it away—might not be what the Lord has in mind.

FOR THE REST OF THE SUMMER and into the fall, the global media maintains its fascination with Tangier. A British network sends a team ashore, and Lonnie raises some eyebrows by telling its reporter, "I don't care if ISIS supports us and puts a seawall around here. Put any name you want on it, just so we get the seawall." In the same report, Ooker questions the infallibility of science by noting: "Scientists also say we evolved from apes, too." Over those same months the island is alert to the appearance of helicopters out on the runway. Rumors swirl that choppers from Andrews Air Force Base are casing the place for a presidential visit. It doesn't come, which does nothing to temper the general enthusiasm for Donald Trump.

George "Hambone" Thomas dies. The population drops by one. Both Ricky and Nick Laird, father and son watermen, move to Crisfield. Another two down. Caleb Cooper, a young waterman, dies unexpectedly, and Gail Smith, a volunteer at the museum, passes. The head count continues to slip.

My landlady, Cindy Parks, marries a mainlander and moves off the island. Another loss. Cindy sells her house to her son, Jared. She's fortunate: Some Tangier houses linger on the market for years. Fourteen months after moving in, I give up my quarters on the West Ridge.

In time, life settles back into a routine dictated by the seasons and the blue crab. Jason Charnock goes to work as Lonnie's mate on the *Alona Rahab*. The new school year starts with sixty students at Tangier Combined. Leon quits scraping for the winter. Ooker catches a mess of eels. The jetty project grinds its slow bureaucratic course through the Corps of Engineers, as it has for more than twenty years.

As for the big, expensive concept for saving the island as a wildlife sanctuary, it's early yet. One might assume that Donald Trump's reported feelings for the place can't hurt its chances, but these days, assuming anything about Washington seems rash. And the idea would face rigorous testing in Congress—and, ultimately, before the American people.

I'M SOMETIMES ASKED by mainland friends whether, after more than a year among the islanders, I have an opinion on the prospect of taxpayers spending millions of dollars to save Tangier. My standard reply is that I hope the island is saved, that I *want* it saved. Tangier has worked itself into my pores. I have come to love the breezy openness of the marsh, its whispers and sways, the subtle play of sunlight and shadow on its grasses. I've treasured solitary hours progging on the spit and contemplating my own impermanence at Uppards. I've reveled in life without need of a car, even when it's meant pedaling hard to stay ahead of the flies. I have learned to savor the smell of creosote, diesel smoke, and old crab on the docks and shanties. I'm excited every time I board a boat.

All those delights are nothing, I probably need not add, next to my feelings about the island's people. Many are confounding in their beliefs and hidebound in their ways. They can be harshly judgmental, contemptuous of authority, dismissive of book learning. God knows, they are gossipy. But they're also remarkably resilient, hardworking, and courageous. They are willing to die for one another. Their faith is unshakable, and their optimism, wondrous.

They are bound by blood and history to a degree most modern Americans cannot fathom. They are creatures of place, so rooted to their tiny lump of mud in the Chesapeake that they seem products of another age. And they are warm, loving, generous. Once they figured out I wasn't a day-tripper, they treated me as one of their own.

But saying I hope the place is saved is not answering the question. Whether it *should* be saved is a far more difficult matter, because it requires some national consensus on how we address the changes wrought by sea-level rise in the coming decades. Those changes are already happening—the good people of Tangier will see that, in time—and as they accelerate they will affect thousands of American communities.

We will not have the money, the physical means, or the time to save them all. So we as a people will have to develop a rubric for deciding which towns and properties we save and which we surrender to the sea. Should we restrict our efforts to the places with the greatest populations? Some of the country's biggest cities face inundation, and government spending to protect them would certainly be put to its most efficient use. If that's our yardstick, however, we'll have to agree on a baseline population for intervention. A million people? Half a million?

Whatever the number, that criterion would doom Tangier and thousands of localities important to our history and culture. Assuming we're unwilling to sacrifice smaller places we hold dear, we'll have to come up with other measures of value. We must devise means to quantify the intangibles that make a place special.

In the vast sea of styles and experience that constitute American culture, Tangier is an island both literal and metaphorical. We must decide whether such cultural outliers are as worthy of salvation as the mainstream, whether a circle's circumference is as essential as its middle. Or, to couch it in everyday mainland terms, whether the mom-and-pop restaurant that the locals cherish should be preserved along with the much busier chain eatery out by the interstate.

Tangier forces us to confront some of these questions now. How we choose to respond to its plight will speak to what we hold important and how we tackle the more complex rescues and retreats to come. In one respect, it's a perfect candidate for so key a role, in that it is a place of insignificant population and is unquestionably unique—a word that's tossed around a lot but here really means something. In another respect, it's a rotten candidate, as its demographic issues raise the possibility that, ultimately, it cannot be saved, even by heroic intervention.

We don't get to pick our first test, however. Nature has done that for us. So in answer to whether we should save Tangier, the responsible citizen in me says we have to decide how to decide and proceed accordingly.

But as one who's been allowed, for however brief a time, to pretend he's a Tangierman, it would pain me deeply to see the place disappear. I'd hate to see its people forced to the mainland. I'd mourn losing this direct connection to the past. The world would be a little less interesting without it.

So again, I hope we intervene.

That I do.

LATE AFTERNOON, EARLY AUTUMN. I'm in Carol Moore's skiff, idling beside the dock it shares with Lonnie's *Alona Rahab*. Carol is at the stern and consulting her husband, who stands above us on the decking.

"Lonnie, what do you think?" she asks. "East side or west side?"

Lonnie cocks his head as if listening hard and holds the pose for a couple of seconds. "Way the wind's blowing, I'd go east."

"Then we'll go east," Carol says, and with a twist of her hand, the outboard's putter turns to a loud buzz, and we jet away into the harbor. Out past the docks and parked boats we race, into the channel, among the shanties. The town recedes behind us—the steeple and water tower, the clustered houses, all balanced on a wafer of green so slender that they appear already part of the bay.

We sweep wide into Tangier Sound, bucking on its riffled surface. At Canaan, Carol noses the boat against the peaty shore, and we step onto ground turned gelatinous by the recent high tide. Every footstep leaves a puddle. Carol carries the skiff's anchor ten feet from the water and drives its blade into the boggy turf. It gulps the steel down.

Uppards is silent under a sky blazoned with bowl-shaped clouds, the breeze too mellow to tousle the marsh or to waken surf. The shore here has changed profoundly since my first visit seventeen months ago. The land that held the anchor that day has been carried away by the Chesapeake. The beach on which the headstones lay has retreated fifteen feet, and in places, twice as far. A dense tangle of bleached tree limbs, which in the spring of 2016 blocked our eastward progress along the water's edge, has been overtaken by the bay. New beach has taken shape behind it.

For the first time in our many visits to Uppards, Carol lets me join her as she progs. We tread east from the boat along the tide line, alert for arrowheads and bottle glass. "There's got to be something good in here," she murmurs, scanning the ground. "Got to be."

The sand is soon replaced by ancient oyster shells, and we crunch over monsters that dwarf those I've seen pulled from the bottom. In time the shells give way to the pulpy roots and stems of drowned water bushes, then to a thick mat of rotting marine grasses. "My nana used to always tell me, 'Don't walk in the sea oats, because you don't ever know what's under them,'" Carol says. "Boards with nails in them, that sort of thing." She veers clear of the stranded plants and splashes ankle-deep in the shallows, pausing to scoop a half-inch-wide jimmy out of the water. She opens her hand to show it off.

"A baby," I observe. "Hey, little fella."

"He's dead," she says. She slings him into the drink.

We press on. The shoreline transitions to soupy mud littered with water-smoothed brick and old timber veneered in slime and tunneled by worms. The mud is loose and gluey underfoot and emits a flatulent stink. Unbothered, Carol slops through it to a knee-high escarpment where beach meets marsh. "Look there," she says, pointing.

Beyond a narrow strip of spartina and sand is a huge pond in Uppards's interior. It is separated from the sound, here and at numerous other points, by slivers of ground that barely clear high tide. "That's just happened recently," Carol says. A year ago, the pond was smaller, shallower, farther inland. Another breach, this one on Uppards's east side, seems imminent.

We turn around, having found nothing in our beachcombing but the broken necks of a few old bottles. As Uppards retreats, the wreckage of Canaan fades farther offshore, as do its relics. "Not finding anything up here anymore," Carol says, and sighs. "I thought we would, with all the wind we've had. But no.

"One day last winter I came up, and I think I found seventeen bottles. The water was clear and the tide low. That it was." She shakes her head. "It makes me sad, because that part of history is gone. It's hard to tell somebody about Uppards if you can't bring them up here and find something. But it's just *gone*."

Maybe progging here runs in cycles, I suggest. Maybe she'll return in a month and find the beach littered again with remnants of her forebears.

"Maybe," she says. "Maybe. But I've been coming up here religiously for twenty years."

"And you've never seen it like this?"

"No. Never," she says. "I mean, the land loss is just unreal."

We've backtracked to the oyster shells. I spot a dark sphere among the piled white and come up with a fat nugget of coal, from a cookstove in an Uppards home, perhaps. Carol gazes over the water as we walk. "You've lost the will to hunt," I tell her. "You're not even looking."

We complete our crunching traverse of the shells and return to the quiet of the sand. "When you come up here," I say, "do you think about people coming to Tangier in the future, and progging the beach for signs of you?"

"Well, if we get a seawall, they'll be able to look," she says. "If we don't, there won't be an island left to prog on." She sighs again. "I don't know why I come up here. I mean, it's peaceful. And it's pretty. Just look."

We stop. The breeze has stilled. Tangier Sound is slick calm and deep blue. The sinking sun tinges the shore a warm amber, burnishes the water's placid surface, reflects off the condos at Crisfield, twelve miles away. The Chesapeake seems so deceptively benign on this tranquil evening, so lovely.

"It don't get much better than this," Carol says.

I look down at my feet. We stand on a shallow crescent of beach. Where it meets the water, the sand is discolored by deposits of granulated wood, seagrass, seeds, and fine black soil, all left by the retreating tide and forming a mash the color and consistency of coffee grounds.

Pieces of Tangier Island, minced by the bay to near nothing.

ACKNOWLEDGMENTS

This book is the product of the many months I spent in one of the most isolated, tightly bound communities in the Lower 48, and as such owes much to the people of that place: Had they decided they wanted no part of me or this project, I'd have little to show for my time there. As it happened, all of Tangier made me feel welcome, and more than sixty islanders agreed to sit for interviews with me. I thank all for their hospitality and patience and for trusting me with their stories.

A few played key roles. James "Ooker" Eskridge put up with my company on close to a daily basis—on his boat, at his crab shanty, in the Situation Room, at Fisherman's Corner, and in his living room. He must have grown sick to death of me over fourteen months, but he always treated me well. His help was vital.

Likewise, Carol and Lonnie Moore were essential to my research. Carol let me tag along with her to Uppards on numerous occasions and impressed me with her thoughtfulness and intelligence. Lonnie took me crabbing and oystering and answered my wearying questions about both, proving himself a smart, funny straight shooter in the process.

Cindy Parks Walter gave up half of her West Ridge house to me, shuttled me and my luggage back and forth to the docks, and let me tap her filtered water supply. Logistics are a real challenge to any reporter on Tangier, and without Cindy's help I'd have had a much harder time of it. I'm mighty grateful to her.

I looked forward every day to the ninety minutes I spent in the Situation Room. I thank the regulars for letting me play fly on the wall during their sessions: Leon McMann, Jerry Frank Pruitt, Allen Ray Crockett, Ernest Ed Parks, Richard Pruitt, Cook Cannon, John Wesley Charnock, Hoot Pruitt, Danny McCready, and Bobby Crockett.

Bruce Gordy, another of the regulars, introduced me to the group. He and his wife, Peggy, also provided a refuge of calm and good conversation during my frequent visits to their home, and treated me to a couple of crab boils that were highlights of my stay.

I thank John Flood of Swain Memorial and Duane Crockett and the other elders of New Testament for making me welcome in their congregations. Captains Mark "Mooney" Crockett, Mark Haynie, and Brett Thomas delivered me safely to and from the island over the course of my stay. After my car was destroyed at the Onancock town wharf in July 2016, Devi Eskridge drove me 250 miles home, and from that day on, Jack and Carolyn Chandler let me park my replacement car in their yard, high on a bluff over the creek.

I owe as much or more to numerous people on the mainland. My agent, David Black, supported this book from its conception, helped shape it, advocated its merits with ferocity, and encouraged me through the reportage and writing that followed. He is the best in the business, and our every interaction reminds me just how fortunate I am to have him.

This is the second book on which I've worked with my editor at HarperCollins, Peter Hubbard, and on both we've enjoyed the kind of collaboration a writer can only hope to have with his publishing house. Peter brings a big brain and ravenous curiosity to his projects, and he has invested himself fully in this one. It's his baby as much as mine.

The Virginia Foundation for the Humanities at the University of Virginia was my professional home while I reported and wrote

Chesapeake Requiem, as it has been for five years. My fellowship there has surrounded me with a smart, nurturing community—and in the solitary business of writing, that's a gift of incalculable value. I thank everyone there, but especially the foundation's founding president, Rob Vaughan; its executive director, Matthew Gibson; and the assistant director of the Fellows Program, Jeanne Siler.

At the *Virginian-Pilot*, research librarians Maureen Watts and Jakon Hays opened the paper's files to me, made for good company as I pored through hundreds of old stories about Tangier, and suggested strategies for tracking down more elusive source material. At the Corps of Engineers, Mark W. Haviland let me loose in the Norfolk District's library, which contained a wealth of useful information. I'm grateful to all. I also owe a huge debt to the corps' Dave Schulte.

I owe thanks to several writers who read behind me. John Pruitt, a Tangier native and old colleague of mine at the *Virginian-Pilot*, read and corrected an early draft, and his insights were invaluable. He also brokered my housing arrangements and had me over to his family's homeplace for food and conversation whenever he was on the island.

Kyle Langston and Cindy and David Fuller read pieces of the story and offered helpful comments.

And without five others, this book would not have developed as it did, and might not have happened at all. Maria Carrillo read and critiqued the manuscript, as she has every one I've written. This time she did it while changing jobs, moving halfway across the continent, and wrestling with hurricanes at both ends of the trip. I can't thank her enough.

Laura LaFay has been my dear friend for more than thirty years, over which I've benefited from her keen intellect, boundless heart, and fierce loyalty. I've long admired her prowess as a storyteller. Over the past year, I've come to recognize that she's an equally skilled editor. This is a very different and far better book for her contributions to it.

I met Mark Mobley, to whom this book is dedicated, at about the time I did Laura, and he's been a wonderful friend to me since. His roles in this project go far beyond his careful hand as an editor, though that was considerable: He's been my cheerleader, strategist, and confidant throughout the adventure.

My daughter, Saylor, managed our mainland home, wrangled our pets, and generally kept everything running during my long stints on the island. She also kept my spirits buoyed wherever I was. I could not have attempted, let alone finished, this project without her.

Finally, my fiancée, Amy Walton, recognized this as a story worth another, deeper look, fifteen years after I'd last written about Tangier. She encouraged me to propose it, was my unflagging supporter throughout my months of reporting, and several times brightened my stay by venturing across the bay to join me on the island. She has believed in this book, and in me, from the start.

As they say on Tangier, I ain't lucky none.

NOTES

The bulk of this story is based on my firsthand reporting on Tangier Island over a period of nearly two years, beginning with reconnaissance visits on December 24–26, 2015, and February 10–12, 2016. On May 18, 2016, I took up full-time residence on the island, which continued into November; during this period I made only occasional, brief trips home to see family, cut the grass, and catch up on mainland business. From December 2016 through March 2017, I spent about a week per month on the island, and from April through June 2017, two weeks per month. After that, I made another two reporting forays to the place, each of several days, in September and October 2017.

The result is a roughly chronological account of life on Tangier during the 2016 crab and oyster seasons, and of events that drew international attention in the first half of 2017.

I've relied most heavily on scenes that I witnessed. In those, I recorded what was said electronically or in handwritten notes and typed them into my computer very soon after. Those scenes that I did not see and hear for myself are drawn from multiple sources.

My long stay on the island put me in several conversations on the same topic with some of the story's main players. I've tried to distinguish those comments made "live"—that is, within a particular scene—by writing them in present tense. Those comments that inform the subject, but were spoken at other times, are in past tense.

Where I have placed quote marks around a comment, I'm confident that I have captured verbatim what was said, either because I heard and recorded it, or I reconstructed it with the help of multiple participants who agree on the particulars. If quoted speech is included in a larger quote—if I quote a character recalling a conversation, for instance—I've punctuated the interior quote, regardless of my confidence in its veracity, simply for clarity's sake.

In spots I've recounted exchanges without the use of quote marks, meaning that I'm reflecting the gist of what was said, rather than the exact phrasing. In some cases I've chosen this device to distill a long back-and-forth into its essence; in others, I've done so because the participants did not remember their exact wording but agreed on the content and character of what was said; and in a few places I've done so because corroboration was impossible, and I've thus relied on a single source.

All the characters had at least one birthday during the course of my reporting, and many have had another between my last trip over and the publication of this book. Whenever possible, I've thus used years of birth in place of ages. When that approach hasn't fit comfortably into the narrative, I've used a character's age at the time of the scene in which it's mentioned. So it is that Leon McMann is eighty-five early in the book and eighty-six later in the story.

INTRODUCTION

1 A day after the storm: My description of Carol Pruitt Moore's October 31, 2012, visit to Canaan is based on interviews I conducted with her on February 11 and October 30, 2016.

4 Skip ahead to a clear: This scene unfolded during a visit to Uppards on September 27, 2016, on which I was accompanied by Carol Moore, James "Ooker" Eskridge, and a team of researchers from the Virginia Foundation for the Humanities.

CHAPTER ONE

9 Then, starting about 1900: My account of the abandonment of James and Sharps islands was informed by William B. Cronin, *The Disappearing Islands of the Chesapeake* (Baltimore, Md.: Johns Hopkins University Press, 2005), and Michael S. Kearney and J. Court Stevenson, "Island Land Loss and Marsh Vertical Accretion Rate Evidence for Historical Sea-Level Changes in Chesapeake Bay," *Journal of Coastal Research* 7, no. 2 (Spring 1991): 403–15.

9 They abandoned Holland Island: I here relied on Cronin's *The Disappearing Islands of the Chesapeake*; Sheila J. Arenstam Gibbons and Robert J. Nicholls, "Island Abandonment and Sea-Level Rise: An Historical Analog from the Chesapeake Bay, USA," *Global Environmental Change* 16, no. 1 (February 2006): 40–7; and Irving M. Parks Sr., "Vanishing Island: A True Story of Hollands Island," an unpublished pamphlet written by a former islander in 1972, which I found in typescript form at https://sites.google.com/site/taleof2shores

/enjoys-seeing-hollands-island-through-the-memories-of-irving-m
-parks-sr/vanishing-island-a-true-story-of-hollands-island-by-irving
-m-parks-sr (retrieved November 7, 2017). Population figures are from
the U.S. Census.

10 Worrisome ghosts, those places: I was present on Ooker's boat on
May 24, 2016.

12 To look at it: My physical description of the bay is from numerous
published and online sources, as well as maps, but a nice summary
can be found on the Maryland Sea Grant's website at "Chesapeake
Bay Facts and Figures," http://www.mdsg.umd.edu/topics/ecosystems
-restoration/chesapeake-bay-facts-and-figures (retrieved November 7,
2017).

13 Even on calm days: Crisfield, Maryland, stakes its claim to the title
via a big sign on its city dock.

16 I left impressed: I've used the measurements supplied in David M.
Schulte, Karin M. Dridge, and Mark H. Hudgins, "Climate Change
and the Evolution and Fate of the Tangier Islands of Chesapeake Bay,
USA," *Scientific Reports* 5 (2015). The paper estimates the island's area
at 789 acres (vs. Central Park's 843) and the town's area, minus the
airstrip, at 83 acres (vs. the reservoir's 106).

16 That's what everyone called it: For an interesting discussion of
the term "global warming," see NASA, "What's in a Name? Global
Warming vs. Climate Change," https://pmm.nasa.gov/education
/articles/whats-name-global-warming-vs-climate-change (retrieved November 14, 2017).

18 I took a sea kayak: Carol Moore's uncle Kenny R. Pruitt Sr. operated his Pruitt's Paradise bird-hunting camp from Canaan for years. It
had a troubled relationship with federal game authorities.

18 Sixteen years later: Author interview with Carol Moore, February
11, 2016.

18 Ooker Eskridge thus: The passage on Holland Island was informed
by Cronin, *The Disappearing Islands of the Chesapeake*. The population
figures are from Gibbons and Nicholls, "Island Abandonment and
Sea-Level Rise."

19 Count the mayor: Ooker said this on his boat on May 24, 2016.

19 Carol Moore shares: Author interview with Moore, February 11,
2016.

19 Perhaps just as daunting: The 2000 population figure is from the
U.S. Census. The 2016 population figure is from an annual island demographic study compiled by islander Donna Crockett, who shared it with
me in July 2016. Tangier Combined School enrollment figures were
supplied to me by Principal Nina Pruitt during a July 13, 2016, interview.

20 In fact, a February 2016 report: The article I quote is Robert E. Kopp, Andrew C. Kemp, Klaus Bittermann, et al., "Temperature-Driven Global Sea-Level Variability in the Common Era," *PNAS* 113, no. 11 (March 15, 2016): 1434–41. The article was released in advance of publication on February 22, 2016.

20 Before I made that trip: Schulte, Dridge, and Hudgins, "Climate Change and the Evolution and Fate of the Tangier Islands."

CHAPTER TWO

25 Cars and trucks remain: This vehicle inventory reflects the situation in June 2017.

26 Witness Ooker Eskridge: This scene is built around my day with Ooker on the water on May 24, 2016.

27 A couple of lemons: The blue crab's life cycle is described in the all-time classic of Chesapeake Bay nonfiction works, William W. Warner's *Beautiful Swimmers: Watermen, Crabs, and the Chesapeake Bay* (Boston: Little, Brown, 1976). I also drew from Curtis L. Newcombe, *The Biology and Conservation of the Blue Crab, Callinectes sapidus Rathbun* (Richmond: Commonwealth of Virginia, 1945); Robert Aguilar, Eric G. Johnson, Anson H. Hines, et al., "Importance of Blue Crab Life History for Stock Enhancement and Spatial Management of the Fishery in Chesapeake Bay," *Reviews in Fisheries Science* 16, nos. 1–3 (2008): 117–24; and the website of the Virginia Institute of Marine Science, "Blue Crab Life Cycle," http://web.vims.edu/adv/ed/crab/cycle.html (retrieved November 14, 2017).

29 Ooker and his kin: My estimate of a crabber's maximum haul is based on 47 bushels (each bushel containing 72 crabs) multiplied by 196 days crabbing—which in 2016 covered the period from March 17 to November 1, minus 33 Sundays.

30 Ooker is keenly aware: The mayor's quotes are from a telephone interview on February 25, 2016.

31 And there's the matter: Ooker's quote is from our day on the water, May 24, 2016.

33 We chug to the next: My reference to Ooker's genealogy, and similar references throughout the book, was made possible by a Johns Hopkins University asthma study conducted on the island in the 1990s. Islander Donna Crockett allowed me access to her copy.

33 Two of his older brothers: "Tangier War Victim 1st Since World War II," *Virginian-Pilot*, February 5, 1969. Warren Eskridge died of gunshot wounds in Vietnam's Tây Ninh Province on January 28, 1969. He'd been in the country for about six weeks.

37 A typical session might begin: I was present to hear Leon's comments at the Situation Room session of May 21, 2016.

37 The group might mull: Ibid.

37 They might discuss: Situation Room session, June 27, 2016.

38 Talking "over the left": I heard the Mustang exchange as the mailboat prepared to leave Crisfield on December 24, 2015.

38 On some days, too: Situation Room session, July 14, 2016.

CHAPTER THREE

42 The unnatural schedule: Situation Room session, July 11, 2016.

45 Indeed, much of Tangier: The figures I mention are from the demographic study of the island that Donna Crockett shared with me in July 2016.

45 In sum, about half: My conversation with Carol Moore and the tour described in subsequent paragraphs took place on May 19, 2016.

45 Carol replies: Crockett, demographic study.

48 Down at the end: I found a draftsman's rendition of the earlier Heistin' Bridge in the library of the U.S. Army Corps of Engineers' Norfolk District headquarters. Titled "Sketch of bridge at Tangier, Va., complained of by boatmen of the locality," the December 1925 drawing depicts the bridge with a three-foot-wide deck, a fifteen-inch gap in its middle, and fifty-two inches of clearance at high water. USACE Baltimore District File No. H-50-20-09.

52 We negotiate: The statistics are from Crockett, demographic study.

53 The first Englishmen: Captain John Smith's journal of his 1608 exploration was published in 1612 as part of his *A Map of Virginia: With a Description of the Countery, the Commodities, People, Government and Religion.* Numerous editions of the book are available. I used Lyon Gardiner Tyler, ed., *Narratives of Early Virginia, 1606–1625* (New York: Charles Scribner's Sons, 1907), 142, which is available online at www.americanjourneys.org/aj-075/.

53 In 1666, it's said: The specious overcoat and first settler stories are described, and refuted, in Kirk Mariner, *God's Island: The History of Tangier* (New Church, Va.: Miona Publications, 1999). Mariner's well-researched book is the most complete history of the island, though it suffers a bit for its strictly chronological structure. Still, for readers seeking greater detail on Tangier history, I can recommend it.

54 The earliest published account: Thomas Crockett, *Facts and Fun: The Historical Outlines of Tangier Island* (Berkley [Norfolk], Va.: Berkley Daily News Print, 1891). Crockett's book is arranged in two parts, the first a questionable but entertaining history and the second a Tom Sawyer–esque memoir of his boyhood on the island. So scandalized were his

neighbors by the book's latter half that they destroyed every copy they could lay their hands on. I read it first in 2000, when a librarian on the Eastern Shore snuck me a photocopy, first making me promise not to reveal where I got it. Nowadays you can actually buy a typescript of the book at the island museum. It excises the original's use of the N-word.

54 But with all due respect: Mariner, *God's Island*, and David L. Shores, *Tangier Island: Place, People, and Talk* (Newark: University of Delaware Press, 2000). The latter was written by a Tangierman who went on to a career as a linguist and professor at Norfolk's Old Dominion University. That Joseph Crockett, not John, was the island's first settler is affirmed by the Johns Hopkins asthma study cited earlier, insofar as few native Tangiermen are not descended from Joseph; no John Crockett appears.

55 In fact, the island: The marooning, which occurred in August 1645, is briefly chronicled by Martha W. McCartney, "Narrative History," in *A Study of Virginia Indians and Jamestown: The First Century*, ed. Danielle Moretti-Langholtz (Williamsburg, Va.: Colonial National Historical Park, 2005). Sugar Tom's quote is from Crockett, *Facts and Fun*, 2.

55 Mean though island: Thomas's settlement on Tangier is described in the classic work on his life, Adam Wallace, *The Parson of the Islands: A Biography of the Rev. Joshua Thomas* (Philadelphia: Methodist Home Journal, 1872). I used the fourth edition, published eleven years after the original.

55 Thomas had been born: Ibid.

55 The young Thomas apprenticed: Ibid., 21.

56 One summer's day: Thomas's visit to the camp meeting is described in Wallace, *The Parson of the Islands*, and in Andrew Manship, *Thirteen Years' Experience in the Itinerancy* (1856; repr. Ann Arbor: University of Michigan Library, 2005). The quote I use, however, is from Robert W. Todd, *Methodism of the Peninsula: or, Sketches of Notable Characters and Events in the History of Methodism in the Maryland and Delaware Peninsula* (Philadelphia: Methodist Episcopal Book Rooms, 1886), 87.

56 John Crockett, Thomas said: The quote is from Wallace, *The Parson of the Islands*, 27. My characterization of early Methodism relies on Wallace; Mariner, *God's Island*; and S. Warren Hall III, *Tangier Island: Study of an Isolated Group* (Philadelphia: University of Pennsylvania Press, 1939).

56 Later in the summer: Wallace, *The Parson of the Islands*, 31. My description of the first island service was informed by Mariner, *God's Island*.

56 From then on: Ibid.

57 The Methodism here was: My description of Tangier Methodism's Victorian character was informed by my December 15, 2016, interview with the Reverend Alexander Joyner, the Eastern Shore district superintendent, UMC Virginia Conference. The faith healing I describe took place on May 29, 2016.

58 Likewise, it's not: Duane Crockett delivered the sermon on May 29, 2016.

CHAPTER FOUR

61 In June 1812: Passages about the British occupation draw on Mariner, *God's Island*; Manship, *Thirteen Years' Experience in the Itinerancy*; and Wallace, *The Parson of the Islands*.

62 And Joshua was surely rough: The quote about Joshua's reading comprehension is from the Reverend James A. Massey's introduction to Wallace, *The Parson of the Islands*, 7. The "wretched poverty" quote is from Bruce Gordy's "History on the Beach," a short manuscript he prepared for use by patrons of the island museum. My description of Thomas's gifts relies on Massey.

62 "I told them they had better": The passages about Thomas's famed sermon rely most on Wallace, *The Parson of the Islands*, 54–6, and Manship, *Thirteen Years' Experience in the Itinerancy*.

63 Alas, no plaque marks the place: The disappearance of the site by 1900 is chronicled in C. P. Swain, *A Brief History of Tangier Island* (1900; repr. Coram, NY.: Peter's Row, 1993). Sugar Tom's quote is from Crockett, *Facts and Fun*, 2.

63 Earth has seen: My passages on ice age effects on the bay rely on Benjamin D. DeJong, Paul R. Bierman, Wayne L. Newell, et al., "Pleistocene Relative Sea Levels in the Chesapeake Bay Region and Their Implications for the Next Century," *GSA Today* 25, no. 8 (August 2015): 4–10; and Simon E. Engelhart, Benjamin P. Horton, Bruce C. Douglas, et al., "Spatial Variability of Late Holocene and 20th Century Sea-Level Rise Along the Atlantic Coast of the United States," *Geology* 37, no. 12 (December 2009): 1115–8.

64 At the same time: A wealth of information is available on ice age sea levels. A particularly succinct summary can be found in a NASA briefing by Vivien Gornitz, "Sea Level Rise, After the Ice Melted and Today," https://www.giss.nasa.gov/research/briefs/gornitz_09/, January 2007 (retrieved November 15, 2017). See also the U.S. Geologic Survey's science summary prepared by Thomas M. Cronin, "Sea-Level Rise and Chesapeake Bay," https://chesapeake.usgs.gov/sciencesummary -sealevelrise.html, May 2013 (retrieved November 15, 2017).

64 So most of what we know: The ancient Chesapeake is described in Michael W. Fincham, "Channeling the Chesapeake: In Search of Ancient Estuaries," *Chesapeake Quarterly*, April 2011, http://www.chesapeakequarterly.net/v10n1/main/; and Cronin, "Sea-Level Rise and Chesapeake Bay."

65 The changes wrought: DeJong et al., "Pleistocene Relative Sea Levels in the Chesapeake Bay"; and Engelhart et al., "Spatial Variability of Late Holocene and 20th Century Sea-Level Rise."

65 Sugar Tom had never heard: Ibid.

66 By the time Sugar Tom: National Geographic Society, "Sea Rise and Storms on the Chesapeake Bay," at https://www.nationalgeographic.org/news/sea-rise-and-storms-chesapeake-bay/ (retrieved January 2, 2018).

67 The 1850 map: I was supplied an electronic scan of the Coast Survey map by Dave Schulte. Readers can peruse other nineteenth-century Tangier maps on NOAA's Coast Survey Historical Map and Charts Collection page at https://historicalcharts.noaa.gov/historicals/search.

68 The map testifies: My interview with Will Eskridge took place in March 2000 in Ooker's childhood home in King Street. Leon McMann and Jerry Frank Pruitt recalled Oyster Creek on March 14, 2017.

68 As word of Joshua: The 1820 reference derives from Charles B. Cross Jr., "Camp-Meeting Tradition Rooted in Tangier Sands," *Norfolk Ledger-Star*, January 26, 1976. Henry Wise's quote is from his *Seven Decades of the Union: The Humanities and Materialism* (Philadelphia: J. B. Lippincott, 1876), 95.

69 Renowned preachers: Miss Crippin's glowing complexion was described in "A singular display of the goodness and power of Almighty God, at a Camp Meeting held at Tangier Island, August 15th, 1824," originally published in the *Norfolk Beacon* of August 19, 1824. I found the story reprinted in the *Boston Herald*, September 1, 1824.

69 Better documented were spirits: Wise, *Seven Decades of the Union*, 95–6.

69 By the time the map: Swain, *A Brief History of Tangier Island*, 6-7.

70 In the late spring: The New Testament service and Wheatley funeral occurred on May 22, 2016.

73 A couple of weeks later: The graduation took place on June 9, 2016. Trenna Moore, Nina Pruitt, and several students supplied me with scripts of their speeches.

CHAPTER FIVE

79 I found him outside: My visit with Ed Charnock occurred in early March 2000.

80 I've since heard: A wealth of insight into crab claws can be found

in H. J. Pynn, "Chela Dimorphism and Handedness in the Shore Crab *Carcinus maenas*," *Field Studies* 9 (1998): 343–53; and David L. Smith, "Patterns of Limb Loss in the Blue Crab, *Callinectes sapidus Rathbun*, and the Effects of Autotomy on Growth," *Bulletin of Marine Science* 46, no. 1 (January 1990): 23–36.

80 A Tangier crabber: For an incredibly detailed description and photos of both sexes, I'd recommend Austin B. Williams, "The Swimming Crabs of the Genus *Callinectes* (Decapoda: Portunidae)," *Fishery Bulletin* 72, no. 3 (1974): 685–92.

80 As invertebrate romances go: My passage on crab mating relied on Newcombe, *The Biology and Conservation of the Blue Crab*; and "Blue Crab" on the website of the Chesapeake Bay Program, at https://www.chesapeakebay.net/S=0/fieldguide/critter/blue_crab (retrieved January 2, 2018).

81 Crabbers know: For more on this nasty little trick, read Peter Hess, "Crabbers Use Sex to Catch Naked Soft-Shell Crabs," *Washington Post*, June 1, 2016.

83 Ooker's own courtship: Ooker described his romance with Irene on June 13 and September 13, 2016. I found copies of *Harbor Light*, the school's now-discontinued yearbook, in the island museum. He made the comment about marijuana aboard his boat on July 30, 2016.

83 He was eighteen: The "light comes on" quote is from May 24, 2016. The "hard to get" quote is from September 13, 2016.

84 "It gets long, like brown hair": Ooker's quotes in this and subsequent paragraphs are from June 13, 2016.

85 "It seems to be a water quality issue": Author's telephone conversation with John Bull of October 2017.

87 Plus, there's the allure: Leon made his "right dangerous" comment in the Situation Room on March 15, 2017. He described pulling up the pole on July 25, 2016.

88 Comes a Thursday: The wedding took place on June 16, 2016.

90 "I thought I was destined": Author's interview of Annette Charnock, October 6, 2017.

91 Then, after thirty-six: Ibid.

91 Now I take a seat: Jerry Frank's quote is from the Situation Room session of November 10, 2016.

CHAPTER SIX

95 Dolphins are routine sights: This and the following paragraph were informed by Tim Prudente, "Researchers Finding More Dolphins Than Expected in Lower Chesapeake Bay," *Baltimore Sun*, June 11, 2017.

95 Whether this is a new: The bay's rising temperature is described by the Chesapeake Bay Program on its website at "Climate Change," https://www.chesapeakebay.net/issues/climate_change (retrieved November 10, 2017), and in Haiyong Ding and Andrew J. Elmore, "Spatio-Temporal Patterns in Water Surface Temperature from Landsat Time Series Data in the Chesapeake Bay, USA," *Remote Sensing of Environment* 168 (October 2015): 335–48.

96 Whatever the case: I conducted the dump inventory on July 23, 2016.

97 When I mention: My conversation with Ooker on his boat and at the shanty occurred on July 30, 2016.

98 Two cats prowl: Ooker introduced me to his cats on May 24, 2016.

100 He thinks for a moment: Ooker told me the "howard" story on July 30, 2016. I later consulted Allen Parks, who provided additional details.

102 Among its early: Canaan's settlement is described in Mariner, *God's Island*. The quote is from Crockett, *Facts and Fun*, 23-24.

102 Another family: The surname tally is from the demographic study.

102 But in 1820: The population figure is from Hall, *Tangier Island*. The hurricane quote is from Crockett, *Facts and Fun*, 20-21.

104 Then, around 1840: The Yankee invasion is described in Mariner, *God's Island*, and David M. Schulte, "History of the Virginia Oyster Fishery, Chesapeake Bay, USA," *Frontiers in Marine Science* 4 (May 9, 2017).

104 The Yankee oystermen: The quote is from Crockett, *Facts and Fun*, 28.

104 Indeed, it did: Mariner describes the island's alignment with the Union and Civil War years in the Chesapeake in *God's Island*.

105 Or at least: This scene in the Situation Room unfolded on May 27, 2016.

CHAPTER SEVEN

111 Fifty yards beyond Uppards: I rode with Cook on May 25, 2016.

112 Later, I'll examine: The map is available for inspection at https://historicalcharts.noaa.gov/.

112 Cook steers the boat: Ooker spoke of Hearn aboard Allen Ray Crockett's *Claudine Sue* on December 14, 2016.

114 We cross the state: The chimney's origins are described in Jason Rhodes, *Crisfield: The First Century* (Charleston, Sc.: Arcadia, 2006), and Don Beaulieu, "Plain Janes," *Washington Post*, June 12, 2002.

114 It's a reminder: Scott Dance, "At Blackwater Refuge, Rising Sea Levels Drown Habitat," *Baltimore Sun*, December 31, 2016; DeJong et al., "Pleistocene Relative Sea Levels in the Chesapeake Bay"; and Daniel Strain, "The Future of Maryland's Blackwater

Marsh," https://www.climate.gov/news-features/features/future
-marylands-blackwater-marsh, January 14, 2015 (retrieved November 15, 2017).

114 The northernmost: I visited Saxis during my 1994 kayak circumnavigation of the bay and again on September 28, 2016. I can recommend the coffee and company at Martha's Kitchen.

115 The region's relative: The report I mention is Samuel S. Belfield, *Sea Level Rise and Storm Surge Impacts to May 2016 Roadways in Hampton Roads*, Hampton Roads Transportation Planning Organization, May 2016. Available at http://www.hrtpo.org/uploads/docs/Sea%20 Level%20Rise-Storm%20Surge%20Impacts%20to%20Roadways%20 in%20HR%20Final%20Report.pdf (retrieved November 11, 2017).

115 Cook throttles back: Leon made his comment on May 26, 2017.

116 We tie the boat: CBF's history is described on the group's website at http://www.cbf.org/about-cbf/history/ (retrieved January 3, 2018).

117 At first inspection: Cook made his remark on September 27, 2016.

117 But consider: Ibid. His experience at the sewage treatment plant was chronicled in Cyril T. Zaneski, "Island Cleans Up Its Waste Problem," *Virginian-Pilot*, March 6, 1989.

121 Erosion devoured: Jerry Frank's quote is from the Situation Room session of March 14, 2017.

122 This would have been: Wallace, *The Parson of the Islands*, 13.

122 War's end only intensified: Mariner, *God's Island*; Hall, *Tangier Island*; and Rhodes, *Crisfield*. See also Maryland Historical Trust documents at https://mht.maryland.gov/secure/medusa/PDF/Somerset /S-517.pdf (retrieved November 11, 2017).

123 That would soon: The cholera epidemic is described in Crockett, *Facts and Fun*; Mariner, *God's Island*; and a brief item in the *New York Times*, October 27, 1866.

124 "The people began to die": Crockett, *Facts and Fun*, 35.

124 The wonder is: Peggy Gordy's remarks are from a September 2016 telephone conversation. See also Peggy Reynolds, "New Tide Washes Old Tangier Isle," *Washington Post and Times Herald*, April 28, 1957, which includes the line: "Only half a dozen homes have bathrooms."

124 The situation wasn't remedied: I witnessed the Situation Room exchange on March 15, 2017.

125 On an afternoon: I made this trip to Uppards on June 27, 2016.

CHAPTER EIGHT

129 Tangier found itself: The passage on steamboat service was informed by Mariner, *God's Island*; Hall, *Tangier Island*; Shores, *Tangier Island*; C. P. Swain, "Tangier Island: A Protest Against a Recent Let-

ter in Relation to It," *Richmond Dispatch*, July 30, 1899; and letter from the chief of engineers, *Transmitting Report of the Board of Engineers for Rivers and Harbors on Review of Reports Heretofore Submitted on Tangier Channel, Va., with Illustration* (H.R. Doc. No. 51, 72nd Congress, 2nd Session), which I found in the library of the corps' Norfolk District headquarters.

130 They had one now: The passages on the border dispute were informed by *Report and Accompanying Documents of the Virginia Commissioners Appointed to Ascertain the Boundary Line Between Maryland and Virginia* (Richmond, Va.: R. F. Walker, 1873), and *Final Report of the Virginia Commissioners of the Maryland and Virginia Boundary to the Governor of Virginia* (Richmond, Va.: R. F. Walker, 1874). The documents make fascinating reading for their description of bay erosion. Much discussion centered on the location of a point of land cited in seventeenth-century documents delineating the boundary— a place called Watkins Point—Maryland commissioners arguing that the point had moved in the two centuries since. "It is scarcely necessary to remark," they wrote, "after the many proofs we have seen and heard of the great changes in the '[P]oints,' shores and islands of the Chesapeake bay and its tributaries, which have taken place especially in that vicinity, and within the memory of living witnesses, and which are still going on, that the place called Watkins point, in the charter to Lord Baltimore, has long since been washed away, or does not now exist as it then was." It was "proved that the main land originally extended in one unbroken point to Watts island," they continued, so "as to confirm the implication that Watkins point was at the southern end of what has since become known as Watts island." They offered as evidence that "very many large stumps of trees are in the waters and marshes adjacent" to Watts, which surely suggested the existence of a "continuous neck of land as far south as that island." Needless to say, this argument did not prevail—Watts and Tangier islands would otherwise be part of Maryland. Still, it recognizes the effects of sea-level rise in the bay more than 140 years ago, before it had shifted to overdrive.

130 The arbitrators: "Appendix C: Opinion of Arbitrators—1877 Opinion Regarding Boundary Line Between Virginia and Maryland," http://www.virginiaplaces.org/pdf/mdvaappc.pdf (retrieved November 11, 2017).

130 The oyster industry: The figure is from Mariner, *God's Island*. It was confirmed by Dave Schulte, author of "History of the Virginia Oyster Fishery," in a June 16, 2017, email exchange with the author.

130 That worry, quickly realized: Schulte describes the decline of har-

vests in "History of the Virginia Oyster Fishery." The oyster wars are described in John R. Wennersten, *The Oyster Wars of Chesapeake Bay* (Centreville, Md.: Tidewater Publishers, 1981), and Mariner, *God's Island*.

131 Tangier had been transformed: The quote is from "Tangier Island Colony: Poverty and Crime Almost Unknown in the Community," a *New York Times* report published in the *Washington Post* of June 7, 1903.

131 An hour before one June: The scene in Ooker's shanty occurred on June 13, 2016.

132 This is a reality: Leon delivered this quote to me on October 30, 2016.

137 Some nicknames: The quotes are from the Situation Room session of March 15, 2017.

137 That was Half-Ass: Author interview with Kim Parks, November 7, 2016. The back-and-forth in the Situation Room is from the session of February 15, 2017.

138 During another Situation: This exchange occurred on March 14, 2017.

138 One evening at Swain: This scene unfolded on September 26, 2016.

139 We fish up Ooker's pots: The crabbing described in this section occurred on June 25, 2016.

141 "Bad," Leon declares: Situation Room session, June 27, 2016.

142 Aging is a topic: Situation Room session, February 15, 2017.

142 More often, he introduces: Situation Room session, March 12, 2017.

CHAPTER NINE

145 Mary Stuart Parks: I hung around the kitchen at Fisherman's Corner on September 2, 2017.

147 It's also a family business: Ooker told me of his contributions on August 20, 2016.

147 Stuart also scoops: My passage on the hepatopancreas was informed by Rob Kasper, "To Cut, or Not to Cut, the Mustard," *Baltimore Sun*, June 21, 1992.

148 Down the road: Denny Crockett explained the sourcing of the Chesapeake House's crabmeat in a September 2017 telephone conversation.

148 Off the island, however: For an interesting study of imported crabmeat, see Michael Paolisso, "Taste the Traditions: Crabs, Crab Cakes, and the Chesapeake Bay Blue Crab Fishery," *American Anthropologist* 109, no. 4 (December 2007): 654–65.

149 I'm not alone: Author's phone interview with Sydney Meers on November 10, 2017.

149 Such nuance is lost: Ed Charnock made the remark during an interview in March 2000.

151 Irene and Ooker: Leon's remark is from an October 30, 2016, conversation.

152 The orphanage also: Devi Eskridge's adoption, and the details of her life before and after, are from my interview with her in North Garden, Virginia, on August 23, 2017.

152 Irene homeschooled: Devi told me about the Post-its in our August 23, 2017, interview. Nina Pruitt made her comment to a group of visiting mainland principals on July 14, 2016.

152 Once she had tentative command: Both Devi and Ooker explained the sequence of events—Devi on August 23, 2017, and Ooker during a conversation at Lorraine's on October 7, 2017. The Alabama couple's experience is related at "An Adoption Gone Wrong," *Morning Edition*, NPR, July 24, 2007, https://www.npr.org/2007/07/24/12185524 /an-adoption-gone-wrong; and "Adoption Victim Meets Her Mother After 9 Years," ACT, December 25, 2005, http://www.agains tchildtrafficking.org/2005/12/6160/ (both retrieved November 14, 2017).

153 And so it was: Leon made this comment on October 30, 2016.

153 So, yes: Devi supplied me with the graduation dates via text message on September 18, 2017.

153 Nowadays, all the girls: Ooker described his daughters' lives and locations in a February 25, 2016, telephone conversation.

153 And Devi, who visits: Author interview with Devi Eskridge, August 23, 2017.

157 When I ran this: Ibid. Annette Charnock's quote is from our interview on October 6, 2017.

CHAPTER TEN

159 To a point: I approached Ooker after the church service of June 12, 2017.

160 No, it's the weather: Mark Crockett and I watched the distant storm on June 30, 2016.

161 Storms like that: The squall that hit the Eastern Shore occurred on July 1, 2016.

161 "A squall can blow": This exchange took place on July 25, 2016.

161 Some violent storms: Ooker was uncertain about the year, let alone the date, of his close encounter with the waterspout. He told me the same story during my 2000 stay, so it was evidently before then.

162 I can attest to the power: I camped on Honeymoon Island on the evening of July 3, 1994.

163 Heavy weather: My description of William Henry Harrison Crockett's death draws from Mariner, *God's Island*. See also "Four Men Drowned," *Peninsula Enterprise* (Accomac, Va.), February 29, 1896, preserved at http://chroniclingamerica.loc.gov/lccn/sn94060041/1896-02 -29/ed-1/seq-3/ (retrieved November 13, 2017).

164 "She sank at once": Ibid.

164 Tangier felt "its loss": "Resolutions of Respect," *Peninsula Enterprise* (Accomac, Va.), February 29, 1896.

164 Another hard loss: Mariner, *God's Island*, and "Dies After Rescue," *Richmond Times-Dispatch*, February 23, 1914, preserved at http://chroniclingamerica.loc.gov/lccn/sn85038615/1914-02-23/ed-1 /seq-4/#date1=1900&index=0&rows=20&words=Asbury+Crockett+T angier&searchType=basic&sequence=0&state=Virginia&date2=1920& proxtext=asbury+crockett+%2B+tangier&y=11&x=16&dateFilterType =yearRange&page=1 (retrieved November 13, 2017).

164 The years since: Several islanders told me of Puck Shores's death, among them his daughter-in-law, Carlene Shores; Leon McMann; and Jerry Frank Pruitt. Harry Smith Parks's disappearance was documented in "Officials Halt Boat Search," *Daily Press* (Newport News), April 7, 1989; and "Watermen's Bodies Found After Separate Accidents," *Daily Press* (Newport News), May 2, 1989.

165 And then there's: My re-creation of Donnie Crockett's sinking relies on Joshua Partlow, "Tangier Island Aches for Lost Waterman," *Washington Post*, March 14, 2005; and Joanne Kimberlin, "An Island Waits for Its Lost Soul," *Virginian-Pilot*, March 16, 2005.

165 By the time: Author interview with Leon McMann, October 30, 2016.

166 Indeed, the mailboat's: Beth Thomas, Rudy's wife, told me of the mailboat run and Rudy's comment to his passengers in an October 8, 2016, interview. I confirmed the mailboat's partial passenger list in a January 4, 2018, phone interview with Danielle Crockett.

166 Donnie, meanwhile: Leon offered his comment in the Situation Room on July 4, 2016.

166 He had company: Partlow, "Tangier Island Aches for Lost Waterman."

167 Dorsey Crockett, running: Kimberlin, "An Island Waits for Its Lost Soul." The "weren't no letup" quote is from Partlow, "Tangier Island Aches for Lost Waterman."

167 Word reached: Lonnie's quote is from an interview on April 29, 2017.

168 When William Henry Harrison Crockett drowned: This favorite island story is related in Mariner, *God's Island.*

168 When William Henry Harrison Crockett died: My passages on Swain and the church named for him were informed by John I. Pruitt, *Beacon of the Soul: A Centennial Remembrance* (Tangier, Va.: Centennial Committee of Swain Memorial United Methodist Church, 1997); Mariner, *God's Island*; and two unpublished typescripts—the undated "The History of Swain Memorial Methodist Church" by Hattie Thorne, Carol Moore's grandmother, and "This Is the Lord's Doing," a 1954 compilation of church facts by former pastor Oscar J. Rishel marking Methodism's 150th anniversary on Tangier. See also Swain, *A Brief History of Tangier Island.*

168 The year after: "Lead Simple Lives," *Washington Evening Star,* July 15, 1899, preserved at http://chroniclingamerica.loc.gov/lccn /sn83045462/1899-07-15/ed-1/seq-15/#date1=1898&sort=relevanc e&rows=20&words=ISLAND+Island+Islander+TANGIER+Tangi er&searchType=basic&sequence=0&index=4&state=&date2=190 0&proxtext=%22Tangier+Island%22&y=16&x=9&dateFilterType =yearRange&page=1 (retrieved November 13, 2017). It was picked up in the *Richmond Dispatch* of July 19, 1899. Swain's reply, dated July 26, appeared in the *Dispatch* of July 30, 1899, under the headline "Tangier Island: A Protest Against a Recent Letter in Relation to It."

169 The incident for which: Swain's death was detailed in Mariner, *God's Island*, and Pruitt, *Beacon of the Soul.*

169 Swain Memorial's bell: Population figures from 1880 and 1900 are from Hall, *Tangier Island.* The Heistin' Bridge is depicted in "Sketch of bridge at Tangier, Va., complained of by boatmen of the locality," USACE Baltimore District File No. H-50-20-09, December 1925. The footbridge to Uppards is remembered by islanders Jack Thorne and Ginny Marshall and depicted in another Corps of Engineers document I found in the Norfolk District's library: *Report of June 22, 1928, by Lt. Col. C. R. Pettis, USACE, on preliminary examination of Tangier Sound, Va., with a view to securing a channel to the foot of County Road on the south end of Tangier Island.*

170 "The road that they made": Author interview with Jack Thorne, June 26, 2016.

170 A few features of daily life: Mariner, *God's Island.*

170 The gasoline engine: My description of trotlining is from personal observation. The 1913 population figures are from a letter from the secretary of war, transmitting, "Preliminary Examination of Channel from Tangier Island, Va., to the Mainland," January 24, 1913,

included with a letter from the chief of engineers, *Reports on Preliminary Examinations and Survey of Channels to Tangier, Va.* (H.R. Doc. No. 107, 63rd Congress, 1st Session). It's interesting that early twentieth-century corps documents consistently spelled Canaan as Canane, which prompted me to ask older islanders whether the settlement was pronounced that way—that is, *kuh-NANE*, as West Virginians pronounce their Canaan Valley. None had heard it said that way.

171 The Army Corps: The reference to Tangier's fleet was contained in a letter from the secretary of war, transmitting, "Preliminary Examination of Channel from Tangier Island," with a letter from the chief of engineers, *Reports on Preliminary Examinations and Survey of Channels to Tangier.* The reference to the channel's depth is from *Report of the Chief of Engineers, U.S. Army, 1923* (Washington, D.C.: Government Printing Office, 1923).

171 During World War I: The description of the work and the quote are from *Report of the Chief of Engineers, U.S. Army, 1923.*

172 August 1879 brought: Author interview with Jack Thorne, June 26, 2016. Storm information in this and subsequent paragraphs was informed by *The Eastern Shore of Virginia Hazard Mitigation Plan* (Accomac, Va.: Accomack-Northampton Planning District Commission, 2005). The document is accessible at http://a-npdc.org/wordpress/HazardMitigationPlan.pdf (retrieved November 13, 2017).

173 In August 1955: I interviewed Duane Crockett at the firehouse on July 14, 2016.

173 "I think of this little island": I interviewed Dewey Crockett at the school in March 2000.

174 Iris Pruitt, at eighty-eight: Our interview took place on December 12, 2016.

174 Carol Moore's late father: Author interview with Carol Moore, October 30, 2016.

175 Damaging though Sandy was: The damage figures I list are from *Eastern Shore of Virginia Hazard Mitigation Plan.*

175 In all the fuss: Ibid.

175 Almost as soon: Ibid., and Mariner, *God's Island.*

176 The track put Tangier: Author interview with Jack Thorne, June 26, 2016.

176 Ginny Thorne Marshall: Author interviews with Ginny Marshall, November 10, 2016, and Jack Thorne, June 26, 2016.

176 Folks on the mainland: Author interview with Will Eskridge, March 2000.

177 Consider that in 1930: Population figures are from the U.S. Census.

CHAPTER ELEVEN

179 Hours before daybreak: I spent the described day aboard the *Alona Rahab* on June 28, 2016.

183 The economics of hard potting: Lonnie explained his expenses when we talked on the boat and via text message on April 22, 2017.

187 In the days before reliable: The effects of the 1893 freeze are captured in "Seven Dropped on the Ice," *Washington Post*, January 24, 1893; and "Frozen Oystermen," *Los Angeles Times*, January 25, 1893, from which I quoted. My passage on the 1936 freeze relied on "Island Group Claims Food Needed Badly," *Washington Post*, February 7, 1936; and "Life Lost in Mercy Dash: Aid Tragedy Draws Fire," *Los Angeles Times*, February 9, 1936.

187 Nowadays a frozen sound: "Our Heritage: Pungoteague River Lighthouse," U.S. Lighthouse Society, Chesapeake chapter, http://www.cheslights.org/heritage/pungoteague.htm (retrieved November 13, 2017).

187 A long roster: The lives and deaths of the Janes Island, Solomons Lump, Sharps Island, and Hooper Strait lighthouses are described by the Chesapeake chapter of the U.S. Lighthouse Society at https://cheslights.org/category/heritage-maryland/ (retrieved November 14, 2017).

188 There was plenty: The balky first electric plant is described in Mariner, *God's Island*, and Anne Hughes Jander, *Crab's Hole: A Family Story of Tangier Island* (Chestertown, Md.: Literary House Press, 1994).

188 That same year: My account of the crab pot's invention relies on James Wharton, "The Pot at the End of the Rainbow," *Baltimore Sun*, June 3, 1956; and James Wharton, "Of Time and the Dipnet," *Rappahannock Record* (Kilmarnock, Va.), August 11, 1983. Available at https://virginiachronicle.com/cgi-bin/virginia?a=d&d=RR19830811.1.3# (retrieved November 7, 2017).

189 Lewis received a patent: Larry S. Chowning, *Barcat Skipper: Tales of a Tangier Island Waterman* (Centreville, Md.: Tidewater Publishers, 1983), 140–42.

190 Another big change: "FCC Authorizes Radiotelephone to Bay Islands," *Washington Post*, October 16, 1940.

190 Over the next twenty-six years: The phone booths and their 1966 replacement with home phones are described in "Dial Cuts Distance of Island," *Virginian-Pilot*, October 13, 1966; "1st Phones for Homes on Tangier," *Virginian-Pilot*, October 17, 1966; "Telephones in Tangier," editorial, *Virginian-Pilot*, October 18, 1966; and "Changes Coming to Tangier I.," *Washington Post*, October 27, 1966.

190 But then, Tangiermen: Jander, *Crab's Hole*, and Mariner, *God's Island*.

192 The wind persists: The Situation Room scene unfolded on July 11, 2016.

192 So it seems: My car was totaled on July 28, 2016.

192 Between blows: Paulie McCready made his comment on July 24, 2016.

CHAPTER TWELVE

195 The first Sunday in July: The New Testament service took place on July 3, 2016.

197 The yoke of bondage: This and the following passage on observance of the Sabbath are ibid.

198 The island has never been: Author interview with Jack Thorne, June 26, 2016, and Iris Pruitt, December 12, 2016.

198 Such an exception: "Boy Who Defies 'Go to Church' Blue Law Shot," *Chicago Tribune*, April 19, 1920; Mariner, *God's Island*; and author interview with Annette Charnock, April 28, 2017.

198 The boy survived: Mariner, *God's Island*, and author interview with Annette Charnock, April 28, 2017.

199 A great many: Author interview with Annette Charnock, April 28, 2017.

199 Which brings us to a saga: My description of Richardson's past and early years on Tangier was informed by Mariner, *God's Island*; James C. Richardson, *7 Acres: The Story of the New Testament Church on Tangier Island* (Shippensburg, Pa.: Companion Press, 1997); and interviews with Grace Kimpel (October 31, 2016), Iris Pruitt (December 12, 2016), Ginny Marshall (November 10, 2016), and Jack Thorne (July 3, 2016).

200 So it came to pass: Author interview with Grace Kimpel, October 31, 2016.

200 Richardson didn't leave alone: "Trouble in Tangier," *Newsweek*, October 13, 1947.

200 Stella Thomas's sister: Ibid.

200 So the lawlessness began: The passage on the post-split mayhem on Tangier was informed by Richardson, *7 Acres*; Mariner, *God's Island*; Shores, *Tangier Island*; my interview with Grace Kimpel, October 31, 2016; and ibid.

201 Richardson met these trials: Richardson, *7 Acres*, 80.

201 Incredibly, the town: Ibid., 81.

202 Embarrassed by the story: While researching this section in June 2017, I was contacted by former islander Sarah Newton Palmer, whose father served as principal of Tangier's school during the troubles. Ms. Palmer told me via email that prior to the *Newsweek* story, her father

wrote the governor to warn that if left unaddressed by the state, the split would end in violence. I've been unable to verify any role he might have played in the state's intercession.

203 Piety and lawlessness: The shirt factory is described in James Marinus, "A Visit to Tangier Island," *Peninsula Enterprise* (Accomac, Va.), May 7, 1927. Its destruction was described to me in numerous interviews with islanders and referenced in Harrison Smith, "Tangier Island Is Sinking. Its Population Is Shrinking. And These Guys Want to Make It the Oyster Capital of the East Coast," *Washingtonian*, March 6, 2016.

204 The news reaches: Situation Room session, July 27, 2016.

204 A few hours later: I attended the Swain Memorial service of July 27, 2016.

CHAPTER THIRTEEN

209 Asbury Pruitt was that: My portrait of Asbury Pruitt was informed by interviews with Jack Chandler (July 17, 2016), John W. Charnock (January 12, 2017), Grace Kimpel (October 31, 2016), Danny McCready (October 10, 2016), Connie Parks (September 26, 2016), Kim "Socks" Parks (November 7, 2016), Inez Pruitt (September 24, 2016), and Jerry Frank Pruitt (June 27, 2016).

210 In 1958, Asbury had: The target ships and the spotting operation are described in "Bombs to Drop Off Tangier Isle," *Washington Post*, June 2, 1957; "6 Operate Navy Bomb Range," *Virginian-Pilot*, November 3, 1967; and John Stevenson, "Business Is Explosive at Navy's Island Range," *Virginian-Pilot*, April 6, 1970. I also relied on my interviews with Jack Chandler, July 17, 2016, and Ed "Short Ed" Parks, October 11, 2016.

210 At some early point: Asbury's quote is from Jack Dorsey, "Eastern Shore Islands Yield to Sea as Men Argue Peril," Our Vanishing Shoreline, *Norfolk Ledger-Star*, October 1979.

210 On January 8, 1964: Author interview with Jerry Frank Pruitt, June 27, 2016.

211 Asbury repeated: Asbury's methodology is described in Dorsey, "Eastern Shore Islands Yield to Sea," and Donald P. Baker, "Tangier Island: 17 Feet Lost to Bay in Last 3 Months," *Washington Post*, April 20, 1979. Jerry Frank's quote is from our June 27, 2016, interview.

211 Rightly so, because: Morris Rowe, "Erosion Threatens Tangier," *Virginian-Pilot*, February 15, 1973; and John Pruitt, "Tangier Islanders See Land Washing to Sea," *Virginian-Pilot*, January 14, 1974.

211 By January 1975: Asbury's measurements were reported in "Tiny Tangier Island Losing Erosion Fight," *Virginian-Pilot*, January 9, 1975.

Asbury's quote is from Pruitt, "Tangier Islanders See Land Washing to Sea."

211 Its most immediate: The channel's construction is described in *Report of the Chief of Engineers, U.S. Army, 1965* (Washington, D.C.: Government Printing Office, 1965); and *Report of the Chief of Engineers, U.S. Army, 1966* (Washington, D.C.: Government Printing Office, 1966).

212 The project was bundled: My passage on the airport's development was informed by "Tangier's Isolation Due to End Soon," *Norfolk Ledger-Star*, January 28, 1966; "Airport for Tangier: Accessibility to Grow," *Virginian-Pilot*, February 22, 1966; John Pruitt, "Tangier Airport Tentatively Granted $102,000," *Virginian-Pilot*, April 30, 1968; Don Hunt, "A Link to Tangier Takes Shape," *Virginian-Pilot*, February 24, 1969; "First Airfield Is Opened on Island off Virginia," *New York Times*, August 10, 1969; John C. Stevenson, "Control Tower Begun for Tangier Airport," *Virginian-Pilot*, March 17, 1970; and "Tangier Airport Upgrade Sought," *Virginian-Pilot*, August 3, 1973.

212 Five years later: "2nd Barge for Tangier Project," *Virginian-Pilot*, December 2, 1975; and Morris Rowe, "Island Gets 13,000 Tons of Riprap to Fight Erosion," *Virginian-Pilot*, February 22, 1976.

212 Meanwhile, the state: Robert J. Byrne, *Shore Erosion at Tangier Island*, Virginia Institute of Marine Science, College of William and Mary, 1976. See also Paul G. Edwards, "Bay Waters Rapidly Erode Tangier Island," *Washington Post*, July 4, 1976.

213 Whatever remedy: Byrne, *Shore Erosion at Tangier Island*, 23.

213 It so happened: The land loss statistics are from "Saving Tangier," editorial, *Virginian-Pilot*, April 12, 1979; and Baker, "Tangier Island: 17 Feet Lost."

213 Down below Hog Ridge: Author interview with Jack Chandler, July 17, 2016.

213 Little more than two years: Baker, "Tangier Island: 17 Feet Lost." I quoted the report in "The Tangierman's Lament," a two-part story resulting from my 2000 stay on the island and published in the *Virginian-Pilot* of June 11–12, 2000.

214 That grim forecast: Claudia Turner Bagwell, "Seawall for Tangier Draws Official Eyes," *Virginian-Pilot*, April 20, 1979. The telephone lines are mentioned in Baker, "Tangier Island: 17 Feet Lost."

214 His visit seemed: The agreement was described in Claudia Turner Bagwell, "Tangier Island Seawall Given Officials' Accord," *Virginian-Pilot*, July 13, 1979. The rescue's early fate in Congress was reported in "Panel to Discuss Tangier Seawall," *Virginian-Pilot*, July 24, 1979; Felicity Barringer, "$3.5 Million for Seawall at Tangier Backed on Hill,"

Washington Post, July 27, 1979; "House Panel Okays Tangier Seawall Bill," *Virginian-Pilot*, July 31, 1979; Claudia Turner Bagwell, "2 Attempts to Block Seawall Project Fail," *Virginian-Pilot*, January 29, 1980.

214 The House passed the bill: "House Approves Tangier Seawall," *Virginian-Pilot*, February 6, 1980. Warner's letter was included in Claudia Turner Bagwell, "Warner Seeks to Jar Loose Tangier Funds," *Virginian-Pilot*, June 3, 1980.

215 The bill failed: "Warner's Bill Seeks Seawall," *Virginian-Pilot*, February 8, 1983; Linda Cicoira, "Seawall Is Urged for Tangier Island," *Virginian-Pilot*, September 28, 1983; and Linda Cicoira, "Panel Recommends Erosion Bill," *Virginian-Pilot*, May 25, 1984.

215 In the early 1980s: Dewey Crockett made the comment in Linda Cicoira, "Residents of Tangier Push Road Development," *Norfolk Ledger-Star*, September 17, 1987.

215 The centerpiece: Linda Cicoira, "Erosion Worsens on Tangier Island," *Virginian-Pilot*, January 11, 1984. Asbury's quote is from Cicoira, "Panel Recommends Erosion Bill."

215 Congress wrote and passed: Linda Cicoira, "New Tangier Island Seawall Likely," *Virginian-Pilot*, June 28, 1984; "Funding OK'd for Seawall off Tangier Island," *Virginian-Pilot*, September 19, 1984; and Linda Cicoira, "Bay Nearing Landing Strip at Tangier," *Norfolk Ledger-Star*, January 9, 1985. "Save Tangier Island," editorial, *Virginian-Pilot*, October 26, 1985.

215 Tangier had been: Author interview with Duane Crockett, July 14, 2016.

216 The island had reached: Junior Moore was quoted in Jean McNair, "Tangier Fights to Keep Head Above Water," *Virginian-Pilot*, October 22, 1985. Asbury's measurements were included in Linda Cicoira, "Tangier Loses 42 feet to Erosion," *Virginian-Pilot*, January 14, 1986.

216 Whatever the case: Linda Cicoira, "Senate OKs $5.4 Million for Tangier Seawall," *Virginian-Pilot*, March 28, 1986; "House OKs Tangier Seawall," *Norfolk Ledger-Star*, June 25, 1987; Linda Cicoira, "Accomack Pledges Its Share of Tangier Seawall," *Virginian-Pilot*, April 22, 1988; and Patrick K. Lackey, "Keeping Bay at Bay: Tangier Building Wall to Save Island," *Virginian-Pilot*, June 27, 1988.

217 Just as vexing: Leon's "good harbor" comment came on May 26, 2016. His "crab floats" quote is from the Situation Room session of August 18, 2016.

217 The channel's unhappy: Author interview with Ed "Short Ed" Parks, October 11, 2016.

217 The Corps of Engineers: 153 Cong. Rec. 12,039 (2007). The jetty's

glacial progress is reflected in Tangier Town Council minutes of June 5, 2001; July 23, 2002; November 19, 2002; February 10, 2004; September 28, 2004; August 17, 2009; February 2, 2010; and July 23, 2012.

218 That might explain: Susan Svrluga, "Harboring Hope on Tangier Island," *Washington Post*, November 21, 2012.

219 Islanders were convinced: Author interview with James "Ooker" Eskridge, May 30, 2016, and Denny Crockett, February 11, 2016.

219 Those in attendance: Author interview with Denny Crockett, February 11, 2016.

219 The stormy meeting: Ooker made this comment during our crabbing foray of May 30, 2016.

219 "They do studies": Ooker's quote is from September 13, 2016.

219 One late afternoon: This venture to Uppards and Carol Moore's comments are from undated notes I recorded in summer 2016.

220 The gut's expansion: Gibbons and Nicholls, "Island Abandonment and Sea-Level Rise," 42.

220 But back to this breach: William B. Mills, Chih-Fang Chung, and Katherine Hancock, "Predictions of Relative Sea-Level Change and Shoreline Erosion over the 21ˢᵗ Century on Tangier Island, Virginia," *Journal of Coastal Research* 21, no. 2 (March 2005): 36–51.

221 The Tangier Town Council: Council minutes of September 12, 2011.

CHAPTER FOURTEEN

223 His skiff's outboard: My outing with Cameron Evans took place on October 8, 2016.

227 Though older Tangiermen: Cindy Parks's quote is taken from my undated interview notes from the winter of 2016–17.

227 Principal Nina Pruitt: Author interview with Nina Pruitt, July 13, 2016.

227 In the spring of 1914: J. W. Church, "Tangier Island," *Harper's Magazine*, May 1914.

228 But not because: Author interview with Jean Crockett, October 29, 2016.

228 The first such outfit: My passage on the grocery store is based on personal observation and interviews with JoAnne Daley, September 22, 2016, and Terry and Lance Daley, October 10, 2016.

230 Another must-have: My passage on the mailboat is based on my October 8, 2016, interview with Beth Thomas, and on Thomas Ferraro, "Neither rain, nor snow, nor broken rudder stops the mail," a UPI report carried in the Galesburg (Il.) *Register-Mail*, April 13, 1977.

231 And there are no painless: Rudy Thomas Jr.'s comments are from *Voices of the Chesapeake, 2010-2013*, a podcast of interviews conducted by Michael Buckley of WRNR-FM in Grasonville, Md., and available free on iTunes.

232 Then there's the big one: Author interview with Cindy Parks, winter of 2016–17. My subsequent passage on the health of Tangier Combined School is based on my interview with Nina Pruitt, July 13, 2016, and telephone interview with Rhonda Hall, September 2017.

232 Islanders have watched: Nina Pruitt supplied me with the projections during our July 13, 2016, interview.

233 Even if enrollment: Author interview with Denny Crockett, February 11, 2016.

233 But islanders do have: Nina made the "older than dirt" comment to a contingent of visiting mainland principals on July 14, 2016. Her comment about teacher replacement is from our July 13, 2016, interview.

234 On a Friday in mid-August: The session took place on August 19, 2016.

CHAPTER FIFTEEN

238 But nowhere: John Charnock's quotes are from the Situation Room session of September 12, 2016.

238 Patricia Stover was a native: The pastor's biography prior to her arrival on Tangier is from our telephone interview of May 3, 2017.

238 So Pastor Stover: The pastor's comments regarding her tenure on Tangier are from our telephone interview of May 4, 2017.

239 The parsonage: Pastor Stover's feeling that something wasn't right at Swain was backed up by the Reverend Robbie Parks, an island native who was serving the UMC Virginia Conference on the mainland. He told me in an August 29, 2016, interview that he detected unrest in the congregation years before Stover's arrival.

239 Some of the faithful: Author interview with Cook Cannon, September 27, 2016.

239 Duane Crockett, who: Author interview with Duane Crockett, January 14, 2017.

240 Pastor Stover agreed: Author interview with Nancy Creedle, January 12, 2017.

240 "When you're elected": Author interview with Eugenia Pruitt, October 11, 2016.

241 Titled "Effective and Constructive": The 2011 *Book of Reports* is available online at http://www.vaumc.org/ncfilerepository/AC2011/2011BOR.pdf (retrieved November 14, 2017). Resolution 13 appears on pages 59–60.

241 "Well, when I read": The votes on the resolutions are available at "Results of 2011 Annual Conference Resolutions," http://www.vaumc .org/ncfilerepository/ac2011/2011ResultsofResolutions.pdf (retrieved November 14, 2017).

241 "I don't think anybody": Author interview with Jean Crockett, October 29, 2016.

241 No matter that: Author interview with Duane Crockett, January 14, 2017.

242 One might argue: Author interview with Cook Cannon, September 27, 2016.

242 Duane, who knew: Author interview with Duane Crockett, January 14, 2017. At least one congregant, John I. Pruitt, declined to stand. I've spoken with others who said they were confused by Duane's entreaty but stood anyway.

242 The subject came up: This and subsequent quotes from Duane Crockett on the split are from our interview of January 14, 2017.

243 The letters appeared: The letter is preserved at http://www.progress -index.com/progress-index/news/1.1178887/archive (retrieved November 14, 2017).

244 Building or no: Author interview with Jean Crockett, October 29, 2016.

244 "I was hurt": Author interview with Eugenia Pruitt, October 11, 2016.

245 Like the earlier: Author interview with Nancy Creedle, January 12, 2017.

245 In 2013, the time: Author interview with Robbie Parks, August 29, 2016.

245 John Flood remembers: The pastor's comments are from undated notes of our interview in early June 2016.

246 One evening I sat: The session with Alex Joyner took place on September 26, 2017.

CHAPTER SIXTEEN

250 And then comes: Situation Room session, July 14, 2016.

250 Eleven days later: Situation Room session, July 25, 2016.

250 The carnage is still: I joined Ooker on his boat on July 30, 2016.

251 The barges are a sore spot: The barge project was described in Scott Harper, "Rigell: Use old barges to stem Tangier Island erosion," *Virginian-Pilot*, June 28, 2011. A video produced by Rigell's office captures a meeting between the congressman, islanders, and a salvage company official; it is available at https://www.youtube.com /watch?v=tqYR92YO0_Q.

252 At Tangier, the three men: Rigell's quote is from a press briefing released by his office, and available at https://votesmart.org/public-statement/620760/delmarvanowcom-rigell-proposes-sinking-barges-off-tangier#.Wk6OQEtG3BI. Ooker's made his comment on July 30, 2016.

253 One morning I drive: I visited Dave Schulte on June 22, 2016.

254 Schulte stayed alert: Kearney and Stevenson, "Island Land Loss and Marsh Vertical Accretion Rate Evidence."

254 In 1995, another paper: Rachel Donham Wrayf, Stephen P. Leatherman, and Robert J. Nicholls, "Historic and Future Land Loss for Upland and Marsh Islands in the Chesapeake Bay, Maryland, U.S.A.," *Journal of Coastal Research* 11, no. 4 (Autumn 1995): 1195–203.

254 The following year: Michael S. Kearney, "Sea-Level Change During the Last Thousand Years in Chesapeake Bay," *Journal of Coastal Research* 12, no. 4 (Autumn 1996): 977–83.

255 Clearly, the middle: Gibbons and Nicholls, "Island Abandonment and Sea-Level Rise."

255 And the literature: Raymond G. Najjar, Christopher R. Pyke, Mary Beth Adams, et al., "Potential Climate-Change Impacts on the Chesapeake Bay," *Estuarine, Coastal and Shelf Science* 86 (2010): 1–20.

255 What would that look: William B. Mills, et al., "Predictions of Relative Sea-Level Change and Shoreline Erosion over the 21st Century on Tangier Island, Virginia."

255 How Tangiermen: W. Neil Adger, Jon Barnett, Katrina Brown, et al., "Cultural Dimensions of Climate Change Impacts and Adaptation," *Nature Climate Change* 3 (2013): 112–7.

256 Schulte undertook his own: To refresh your memory, the paper is David M. Schulte, Karin M. Dridge, and Mark H. Hudgins, "Climate Change and the Evolution and Fate of the Tangier Islands of Chesapeake Bay, USA," *Scientific Reports* 5 (2015).

259 If there are two words: My passage on Poplar Island's past was informed by Stephen P. Leatherman, *Vanishing Lands: Sea Level, Society, and Chesapeake Bay* (Washington, D.C.: U.S. Department of the Interior, 1995); and Jon Gertner, "Should the United States Save Tangier Island from Oblivion?" *New York Times Magazine*, July 6, 2016.

259 But unlike so many: The Poplar Island project is described in Gertner, "Should the United States Save Tangier Island from Oblivion?"; in Nevin Martell, "In Chesapeake Bay, Poplar Island Is Man-Made Miracle," *Washington Post*, September 24, 2015; and on the website of the U.S. Army Corps of Engineers' Baltimore District, at http://www.nab.usace.army.mil/Missions/Environmental/Poplar-Island/ (retrieved November 14, 2017).

260 Here's the part: Jerry Frank's quote is from the Situation Room session of February 15, 2017.

CHAPTER SEVENTEEN

263 Late on the morning: I have reconstructed the firehouse meeting through interviews with Gregory Steele (September 15, 2016), John Bull (October 2017), Renee Tyler (August 18, 2016), and Anna Pruitt-Parks (October 31, 2016).

263 The business at hand: Both Steele and Bull described their lunch meeting to me—Steele in our interview of September 15, 2016, and Bull in an October 2017 telephone conversation.

264 Representing Tangier: Ooker related his feeling about email when we were aboard his boat on August 20, 2016.

264 "I was excited": Author interview with Anna Pruitt-Parks, October 31, 2016.

265 Whatever the case: I reconstructed the August 17, 2016, town meeting with the help of a digital recording made by islander Barb Baechtel.

268 The following day: This Situation Room session occurred on August 18, 2016.

270 A month after the meeting: My meeting with Gregory Steele and Susan Conner took place on September 15, 2016.

CHAPTER EIGHTEEN

278 Across the years: Church, "Tangier Island."

279 The studio called the film: My account of the *Message in a Bottle* saga relies on Marylou Tousignant, "Tangier Island Gives Film Script a Thumbs Down," *Washington Post*, March 12, 1998. I also relied on an October 31, 2016, interview with Anna Pruitt-Parks, and a September 2017 telephone conversation with Beth Thomas.

280 The vote attracted: The quote is from "Immorality vs. Immortality," *Baltimore Jewish Times*, March 20, 1998. See also Tousignant, "Tangier Island Gives Film Script a Thumbs Down."

280 A good many: Author interview with Anna Pruitt-Parks, October 31, 2016.

280 The following week: Beth Thomas telephone conversation, September 2017.

281 In their homes: Author interview with Jack Thorne, July 3, 2016.

281 "That's when people": Author interview with Iris Pruitt, December 12, 2016.

281 Dozens of people: Author interview with Jean Crockett, October 29, 2016.

282 In all, more than: Duane Crockett delivered this sermon on September 25, 2016.

282 Another clue: Ernest Ed Parks and I explored the Sunset Inn on or about August 30, 2016. My fiancée spotted the light on September 9 or 10, 2016.

284 "That was the big shocker": Author interview with John Flood, early June 2016.

284 But not all: The meeting took place after the Sunday evening service of September 11, 2016.

285 Twenty-four hours later: The Tangier Town Council meeting took place on September 12, 2016.

286 The meeting reveals a host: In our October 31, 2016, interview, Anna Pruitt-Parks told me that the sewage treatment plant "dollars the town to death." The failing wells and the poor quality of Tangier water were discussed at a Tangier Town Council "special water meeting" on February 1, 2005.

287 But no difficulty: The first quote is from the meeting cited in the text. The second is from the meeting of December 4, 2006.

287 When I raised the subject: Author interview with Inez Pruitt, September 24, 2016.

287 I also broached: Author interview with Nina Pruitt, July 13, 2016.

287 The man with the most: I rode patrol with John Wesley Charnock on January 12, 2017. The quotes in subsequent paragraphs, as well as details of his biography, are from our conversation in the car.

290 Wearied by the job's: Ooker's quote is from July 30, 2016. The cop's complaint to the town council took place at the meeting of April 4, 2006.

CHAPTER NINETEEN

293 In mid-September: This scene occurred on September 13, 2016.

294 Henry Jander was a: Anne Hughes Jander, *Crab's Hole: A Family Story of Tangier Island* (Chestertown, Md.: Literary House Press, 1994).

294 But Henry Jander turned: Ibid.; "Approval Seen for Islands' Power Plans," *Washington Post*, October 8, 1947; and "Virginia Isle Again Has Electric Light," *Washington Post*, December 25, 1947.

295 That it took: Author interview with Anna Pruitt-Parks, October 31, 2016.

295 I heard much the same: Author interview with Nina Pruitt, July 13, 2016.

296 Denny Crockett, the former: Author interview with Denny Crockett, July 15, 2016.

296 I saw it again: This scene unfolded on July 4, 2016.

297 One has to wonder: Oscar Rishel's contributions are described in Peggy Reynolds, "New Tide Washes Old Tangier Isle," *Washington Post*, April 28, 1957. Also see Mariner, *God's Island*.

297 Rishel also played: Ibid.; Peggy Reynolds, "Doctor's Welcome to Tangier Isle Marred by the Death of a Villager," *Washington Post*, April 17, 1957; "Dr. Kato's Island Domain," *Washington Post*, April 18, 1957; and Jeff O'Neill, "Trouble Again on Tangier: 'Copter Plucks Heart Attack Victim Off Island That Has No Doctor," *Washington Post*, March 30, 1959.

298 Jander and Rishel: My section on Dr. Nichols was informed by a September 24, 2016, interview with Inez Pruitt and by "Dr. David Nichols, Tangier Island's Angel, Dies," *Richmond Times-Dispatch*, December 30, 2010; Angela Blue, "Treating Tangier Island," *Coastal Virginia*, August–September 2015; and Bill Lohmann, "Five Years Later, Tangier Island Still Feels Presence of Doctor," *Richmond Times-Dispatch*, December 20, 2015.

299 Finally, in 1988: "Donor of Island Education Left a Legacy," Chesapeake Bay Foundation, January 23, 2008, http://cbf.typepad .com/chesapeake_bay_foundation/2008/01/donor-of-island.html (retrieved December 14, 2017).

299 Perhaps no come-here: Susan Emmerich, "Faith-Based Stewardship and Resolution of Environmental Conflict: An Ethnography of an Action Research Case of Tangier Island Watermen in the Chesapeake Bay" (Ph.D. diss., University of Wisconsin–Madison, 2003).

301 Mention the Kayes: Danny McCready's quote is from our conversation on Mark Crockett's *Joyce Marie II* on August 17, 2016.

301 Likewise, ask islanders: Hanson Thomas made this comment on October 10, 2016.

301 "You've got to adapt": Author interview with Eugenia Pruitt, October 11, 2016.

301 "We loved the Kayes": Author interview with Lisa Crockett, October 28, 2016.

302 "They were extremely friendly": The Kayes described their years on Tangier over two days of interviews at their home in Wilmington, Delaware, on September 20–21, 2016.

302 That fall, the Kayes: Tangier Town Council minutes of November 18, 2003. The Kayes told me they recalled starting this project further along in their stay—it wasn't unveiled in the rec center until more than six years later. But these minutes, and those of December 2, 2003, indicate that this is when it began.

304 Islanders had long assumed: Nina Pruitt's quote is from our July 13, 2016, interview.

305 So it was no small matter: The Kayes shared their open letter with me, which was dated April 18, 2009.

305 As the Kayes saw it: The council voted against the ferry at its meeting of July 22, 2009. From the minutes: "Neil Kaye sent an email to the Town asking if the Town had a place for a 65 foot boat could dock [*sic*] if the state gave the grant money for a year round ferry. The mayor and council voted that they didn't want a year round ferry."

306 Neil banged out: Anna Pruitt-Parks shared the August 21, 2010, email with me.

307 In hindsight: The council minutes I cite are from the meeting of August 23, 2010.

307 A week after their Facebook: The Kayes shared their August 28, 2010, email with me.

308 When I asked Ooker: Ooker made these comments while we crabbed on July 30, 2016.

308 The Kayes, who counted: The Kayes shared their email, which was dated September 17, 2010.

308 Nichols replied a few days: The Kayes shared Nichols's email, which was dated September 26, 2010.

CHAPTER TWENTY

311 The election even: Pastor Flood delivered this sermon on September 4, 2016.

312 In the Situation Room: This scene is from September 13, 2016.

314 One Saturday I sit: My visit with Beth Thomas occurred on October 8, 2016.

315 In the nineteenth century: The collapse of oystering is described in Schulte, "History of the Virginia Oyster Fishery," and Mariner, *God's Island*.

315 Then, in 1949: Schulte, "History of the Virginia Oyster Fishery," and Najjar et al., "Potential Climate-Change Impacts on the Chesapeake Bay."

315 Ten years later: Ibid.

316 When summer came: *Report of the Task Force on the Virginia Blue Crab Winter Dredge Fishery to the Governor and the General Assembly of Virginia* (Richmond: Commonwealth of Virginia, 2000), 1, and Swift, "The Tangierman's Lament."

316 For the catch to flatline: Ibid.

316 So in the 1990s: See Justin Blum, "Starting Next Year, Va. Crabbers Must Turn the Little Ones Loose," *Washington Post*, June 23, 1993; and Swift, "The Tangierman's Lament."

316 All of these fixes: Francis X. Clines, "Virginia's Desperate Step To

Protect the Blue Crab," *New York Times*, July 30, 2000; and Swift, "The Tangierman's Lament."

317 Of all the bay's crabbers: *Report of the Task Force on the Virginia Blue Crab Winter Dredge Fishery.*

317 So even with the new: Tom Pelton and Bill Goldsborough, *Bad Water and the Decline of Blue Crabs in the Chesapeake Bay* (Annapolis, Md.: Chesapeake Bay Foundation, 2008), available online at http://www.cbf.org/document-library/cbf-reports/CBF_BadWaters Report6d49.pdf (retrieved November 15, 2017).

317 By 2007, the fishery: Ibid. See also David A. Fahrenthold, "Despite Rescue Effort, Bay Crabs at an Ebb," *Washington Post*, November 17, 2007.

317 Virginia officials recognized: Scott Harper, "New State Ban on Dredging of Crabs Upheld by Judge," *Virginian-Pilot*, November 25, 2008; and Michael W. Fincham, "The Blue Crab Conundrum," *Chesapeake Quarterly*, July 2012, http://www.chesapeakequarterly.net /v11n2/main1/.

318 This did not have the drastic: *Report of the Task Force on the Virginia Blue Crab Winter Dredge Fishery*; and Scott Harper, "New state ban on dredging of crabs upheld by judge," *Virginian-Pilot*, November 25, 2008.

318 But the fall's cooling: The quotes, as noted, are from the New Testament service of September 25, 2016.

319 But by the time: Flood's comments are from the Swain service of October 5, 2016.

320 Still, this is: The airplane was abandoned beside the runway on July 15, 2016.

321 But except for damage: New Testament Church evening service, October 9, 2016.

321 The first Sunday: Swain Memorial morning service, November 6, 2016.

321 I pedal onward: The election results I cite are from the Accomack County registrar.

322 That Friday: The *Courtney Thomas* conversation took place on November 11, 2016.

322 I have borne: Situation Room session of October 11, 2016.

CHAPTER TWENTY-ONE

327 Duane Crockett has spent: Sermon of February 5, 2017.

329 In the face of this: Ryan B. Carnegie and Eugene M. Burreson, *Status of the Major Oyster Diseases in Virginia 2006-2008* (Gloucester Point, Va.: VIMS, December 2009).

329 Which is what prompts me: My oystering trip on the *Alona Rahab* took place on November 17, 2016.

335 One dark, frigid morning: I was aboard the *Claudine Sue* on December 14, 2016.

336 I think about that: I made the crossing on December 15, 2016.

339 And so winter: The snowstorm hit the weekend of January 7–8, 2017.

339 Still, the watermen: Carlene spoke at the New Testament service on January 15, 2017.

CHAPTER TWENTY-TWO

341 Many an islander: I interviewed numerous islanders about the weather conditions of April 24, among them Jason Charnock, Lonnie Moore, Allen Ray Crockett, John Wesley Charnock, Tracy Moore, Dean Dise, Freddie Wheatley, Allen Parks, and Mitchell Shores. The wind speeds cited here and elsewhere in this chapter are from Lonnie and Jason.

341 So it was: Jason Charnock described pot placement in our interview of October 6, 2017.

342 Mindful that the wind: Ibid.

342 Better luck waited: Ibid.

342 And so it was: Ibid. Jason told me that they headed for home with thirty-two bushels of sooks and four of jimmies. The weight I cite is based on a forty-pound bushel, a figure supplied to me by Lonnie Moore. Jason Charnock estimated the time at which he noticed the boat felt soft in our October 6, 2017, interview.

343 And the waves were high: The odd nature of the seas came up in multiple interviews—with Lonnie Moore on October 5, 2017; Tracy Moore on October 7; and John Wesley Charnock and Andy Parks on October 9.

343 Leon McMann has told me: Leon made this observation in the Situation Room on December 12, 2016. The boat's equipment is from my Jason Charnock interview of October 6, 2017.

343 The second was an automatic: Lonnie Moore explained bailers in our October 5, 2017, interview.

344 In the cabin, Jason: This and the following paragraphs were informed by my Jason Charnock interview of October 6, 2017.

344 Yes, everything's fine: Ibid., and Lonnie Moore interview of October 5, 2017.

344 Except that it didn't: Jason Charnock interview of October 6, 2017.

345 Jason shouted for help: Ibid., and interview with Billy Brown on October 8, 2017.

345 He scanned the cabin: This paragraph and all those to the end of the section are based on my October 6, 2017, interview with Jason Charnock.

346 Billy Brown's call: October 8, 2017, interview with Billy Brown. Sandra Parks told me she believed the call came from Dean Dise, but both Billy and Dean refute that account.

346 That one crabber: Andy Parks telephone interview of October 31, 2017.

346 Andy grabbed the phone: Ibid., and Kelly Wheatley telephone interview of October 31, 2017.

347 Down in the boat stalls: Telephone interviews with Dean Dise on October 31, 2017, and Freddie Wheatley on November 1, 2017.

347 It missed Lonnie Moore: Lonnie Moore interview of October 5, 2017; Loni Renee Charnock interview of October 6, 2017.

347 Then she heard: Ibid.

348 Lonnie had just reached: Lonnie Moore interview of October 5, 2017.

348 The Coast Guard: Liz Holland, "Coast Guard suspends search for missing Tangier waterman," at http://www.delmarvanow.com/story /news/local/virginia/2017/04/25/search-continues-missing-tangier -waterman/100878370/ (retrieved November 22, 2017).

348 Freddie Wheatley and Dean: Dise interview of October 31, 2017; Wheatley interview of November 1, 2017.

349 Lonnie Moore, meanwhile: Lonnie Moore interview of October 5, 2017.

349 The *Henrietta C.*'s bow: This entire section is based on my October 6, 2017, interview with Jason Charnock, and on the statement he provided the Coast Guard on April 27, 2017.

350 An hour after Jason's call: October 5, 2017, interview with Lonnie Moore.

351 The Coast Guard chopper: Ibid., and interview with Carol Moore on October 5, 2017.

352 From Jason Charnock's statement: I was in Annette Charnock's home on April 27, 2017, when Jason provided this statement to a visiting Coast Guard investigator. Annette lent me a transcript of his statement that afternoon.

352 Two nights later: I was present for the Swain Memorial service of April 26, 2017.

354 The next day I stop in: I visited Annette Charnock's home on April 27, 2017.

355 A thousand feet up: I made the search flight with Ooker and Kenny Carpenter on April 27, 2017. I sat in the back seat, left side; Ooker rode shotgun.

356 The following Sunday morning: I attended the class meeting at Swain Memorial on April 30, 2017.

357 An hour later: I hitched a ride on Mark Haynie's boat for the dragging expedition of April 30, 2017.

358 Ten days after the sinking: Interview with Tommy Eskridge of October 7, 2017. Tommy recovered the body with the help of Mark Crockett on May 4, 2017. Ed came home on the mailboat on May 17, 2017.

358 Later that afternoon: Situation Room session of May 17, 2017.

359 "A dry wood worm": Interview with Jerry Frank and Inez Pruitt of October 7, 2017.

359 He'd loved that boat: Annette Charnock interview of October 7, 2017; telephone interview with Sharon Marshall of Smith Island Drydock, November 2, 2017. The company's records, Sharon told me, showed that the *Henrietta C.* left the boatyard on November 28, 2016.

359 But he had not gotten around: Interview with Jerry Frank and Inez Pruitt of October 7, 2017; interview with Annette Charnock of October 7, 2017.

359 That Monday: Jason Charnock interview of October 6, 2017.

CHAPTER TWENTY-THREE

362 We finish the pots: My visit to Watts Island occurred on or about July 7, 1994.

363 Watts was about: My reference to the 1977 cable installation was informed by Mariner, *God's Island*, and Don Harrison, "Underwater Cable to Supply Electricity to 2 Bay Islands," *Virginian-Pilot*, April 29, 1977. Several islanders, among them Mark "Mooney" Crockett, told me about the gradual exposure of the junction box.

364 When I return: The weird crab appeared on the *Washington Post*'s website on June 1, 2017. The story, as picked up by other papers, is available at http://www.duluthnewstribune.com/news/4276610-crab-unlike-any -youve-ever-seen-has-been-pulled-chesapeake-bay (retrieved November 13, 2017). Incidentally, I saw the crab when I visited Ooker's shanty on June 4, 2017. It died a few hours later. Ooker subsequently froze it.

364 But aside from that: Carlene offered her prayer on June 4, 2017.

365 Leon's woes are compounded: Leon's mishap on the water and the subsequent Situation Room session described here, including the CNN visit, occurred on June 5, 2017.

367 The segment airs: The CNN report is preserved at Jennifer Gray, "Rising Seas May Wash Away This US Town," CNN, June 9, 2017. Video, 5:00. http://www.cnn.com/videos/us/2017/06/09/virginia-island -sea-washing-away-gray-lead-pkg.cnn (retrieved November 13, 2017).

368 The islanders I speak with: I interviewed Barb Baechtel by telephone on August 25, 2017.

368 We soon learn: The president's call came on the afternoon of June 12, 2017. My reconstruction of the conversation is based on Ooker's account of it to me on June 27, 2017, and numerous news reports—among them Carol Vaughn, "Trump Tells Tangier Island Mayor Not to Worry About Sea-Level Rise," *Daily Times* (Salisbury, Md.), June 13, 2017, https://www.usatoday.com/story/news/politics/2017/06/14/trump -tells-tangier-island-mayor-not-worry-sea-level-rise/394688001/; and Travis M. Andrews, "Trump Calls Mayor of Shrinking Chesapeake Island and Tells Him Not to Worry About It," *Washington Post,* June 14, 2017, https://www.washingtonpost.com/news/morning-mix /wp/2017/06/14/trump-calls-mayor-of-shrinking-chesapeake-island -and-tells-him-not-to-worry-about-it/?utm_term=.9f73153e673e.

369 Foreign reporters descend: The monologue on Tangier aired on Friday, June 16, 2017. It's available at *The Late Show with Stephen Colbert*, "Trump Says 'Not to Worry' About Rising Sea Levels," YouTube, June 17, 2017. Video, 2:11. https://youtu.be/o5AxKZF_xP8 (retrieved December 14, 2017).

370 This does not go down: The CNN "town hall" aired on August 1, 2017. The segment featuring Ooker is preserved at "Mayor of Disappearing Island Faces Al Gore and Shuts Down Global Warming Claim," YouTube, August 2, 2017. Video, 2:25. https://www .youtube.com/watch?v=VUfARQaVWsI (retrieved November 13, 2017). A longer excerpt can be found at https://www.youtube.com/watch ?v=sjvLagal2Wk (retrieved January 5, 2018); Ooker's segment begins at 1:35.

371 Instead, he takes: https://www.youtube.com/watch?v=sjvLagal2Wk. Al Gore's joke begins at 4:30.

372 For the rest: The report is available at Channel 4 News, "America's Climate Change Refugees Putting Their Faith in Donald Trump," YouTube, August 11, 2017. Video, 7:14. https://www.youtube.com /watch?v=-Lk0jYeVENo (retrieved November 13, 2017). It's worth watching for its depiction of the Situation Room. Lonnie's comment comes at about 5:40.

375 Late afternoon, early autumn: My visit to Uppards with Carol Moore took place on October 5, 2017.

INDEX

———